Lecture Notes in Comp ... 74

Edited by G. Goos, J. Hartmanis a...

Springer
Berlin
Heidelberg
New York
Barcelona
Budapest
Hong Kong
London
Milan
Paris
Santa Clara
Singapore
Tokyo

Gilles Dowek Jan Heering
Karl Meinke Bernhard Möller (Eds.)

Higher-Order Algebra, Logic, and Term Rewriting

Second International Workshop, HOA '95
Paderborn, Germany, September 21-22, 1995
Selected Papers

Springer

Series Editors

Gerhard Goos, Karlsruhe University, Germany

Juris Hartmanis, Cornell University, NY, USA

Jan van Leeuwen, Utrecht University, The Netherlands

Volume Editors

Gilles Dowek
INRIA-Rocquencourt
P.O. Box 105, F-78153 Le Chesnay, France

Jan Heering
Stichting Mathematisch Centrum, CWI
P.O. Box 94079, 1090 GB Amsterdam, The Netherlands

Karl Meinke
Department of Computer Science, University College of Swansea
Singleton Park, Swansea SA2 8PP, United Kingdom

Bernhard Möller
Institut für Mathematik, Universität Augsburg
Universitätsstr. 8, D-86135 Augsburg, Germany

Cataloging-in-Publication data applied for

Die Deutsche Bibliothek - CIP-Einheitsaufnahme

Higher order algebra, logic, and term rewriting : second international workshop ; selected papers / HOA '95, Paderborn, Germany, September 21 - 22, 1995. Gilles Dowek ... (ed.). - Berlin ; Heidelberg ; New York ; Barcelona ; Budapest ; Hong Kong ; London ; Milan ; Paris ; Santa Clara ; Singapore ; Tokyo : Springer, 1996
(Lecture notes in computer science ; Vol. 1074)
ISBN 3-540-61254-8
NE: Dowek, Gilles [Hrsg.]; HOA <2, 1995, Paderborn>; GT

CR Subject Classification (1991): F.3-4

ISBN 3-540-61254-8 Springer-Verlag Berlin Heidelberg New York

Printed in Germany

Typesetting: Camera-ready by author
SPIN 10512910 06/3142 - 5 4 3 2 1 0 Printed on acid-free paper

Preface

The Second International Workshop on Higher-Order Algebra, Logic, and Term Rewriting (HOA '95) was held in Paderborn (Germany) on 21st–22nd September 1995. Of the 24 papers submitted, 20 were selected for presentation at the workshop and inclusion in the participant's proceedings. On the basis of 39 referee's reports, 14 of these were selected for publication in the present volume. They appear here in revised final form, ordered alphabetically according to the name of the first author. Preceding them is an invited contribution by Jan Willem Klop. We would like to thank the referees for making this selection possible. The programme committee consisted of Gilles Dowek, Mike Gordon, Jan Heering, Karl Meinke, Bernhard Möller, and Tobias Nipkow.

The workshop was hosted by Universität-GH Paderborn. Hans Kleine Büning, Theo Lettmann, Stephan Bauerfeind, Ulf Dunker, Zhao Ping and Tanja Prior gave invaluable help in organizing it. We are grateful to the European Association for Computer Science Logic (EACSL), which sponsored the workshop and gave financial support. Finally, thanks to financial support provided by Digita International we were able to produce a striking poster with a reproduction of a painting by Patrick Heron.

February 1996

Gilles Dowek Jan Heering Karl Meinke Bernhard Möller

Table of Contents

Term Graph Rewriting
Jan Willem Klop (invited speaker) 1

Approximation and Normalization Results for Typeable Term Rewriting Systems
Steffen van Bakel and Maribel Fernández 17

Modular Properties of Algebraic Type Systems
Gilles Barthe and Herman Geuvers 37

Collapsing Partial Combinatory Algebras
Inge Bethke and Jan Willem Klop 57

A Complete Proof System for Nested Term Graphs
Stefan Blom 74

R^n- and G^n-Logics
Claus Hintermeier, Hélène Kirchner, Peter D. Mosses 90

The Variable Containment Problem
Stefan Kahrs 109

Higher-Order Equational Logic for Specification, Simulation and Testing
Karl Meinke 124

The Correctness of a Higher-Order Lazy Functional Language Implementation: An Exercise in Mechanical Theorem Proving
Sava Mintchev and David Lester 144

Assertions and Recursions
Bernhard Möller 163

Development Closed Critical Pairs
Vincent van Oostrom 185

Two *Different* Strong Normalization Proofs?
Jaco van de Pol 201

Third-Order Matching in the Polymorphic Lambda Calculus
Jan Springintveld 221

Higher-Order Algebra with Transfinite Types
L.J. Steggles 238

Abstraction of Hardware Construction
Li-Guo Wang and Michael Mendler 264

Term Graph Rewriting

Jan Willem Klop*

CWI, P.O. Box 94079, 1090 GB Amsterdam, The Netherlands
and
Vrije Universiteit, Department of Mathematics and Computer Science,
de Boelelaan 1081a, 1081 HV Amsterdam, The Netherlands

Abstract. We discuss some aspects of term graph rewriting based on systems of recursion equations. This is done for first-order signatures as well as lambda calculus. Also relations with infinitary rewriting are discussed.

0. Introduction. In this paper we will discuss in an informal way some aspects of térm graph rewriting. We will indicate why this subject falls in the scope of higher-order rewriting, and thus in the scope of the present workshop. Our discussion will be loosely structured by a numbered sequence of keywords and phrases, thereby following a talk given at the workshop.

In the theory of rewriting, term graph rewriting is a relatively new development, prompted by the actual practice in functional language implementations where subterm sharing is a matter of routine. There are several approaches to a theoretical foundation of graph rewriting, of which an important one is that based on category theory and single or double push-outs. We will not discuss this approach here, but instead refer to [SPvE93] where several references to the categorical treatment can be found.

1. The Equational Approach to Term Graph Rewriting. We will advocate the equational approach, that starts from term graphs as systems of recursion equations. An example of such a system is

$$\langle \alpha \mid \alpha = F(\beta,\gamma),\ \beta = G(\alpha),\ \gamma = H(\alpha,\alpha)\rangle$$

where α is the 'root' variable, G a unary function symbol, and F and H are binary function symbols from a first-order signature. We use the terminology 'term graph' since the graphs that we are concerned with look locally the same as terms (first-order terms or, later on, lambda terms). This means that in pictures like Fig. 1, the outgoing arcs of a node are in fact ordered from left to right; this is only suggested in the pictures but not made explicit. The nodes in a term graph are locations that in figures are literally filled with operator symbols; nodes will in general have names $\alpha, \beta, \gamma, \ldots$, but we will also admit unnamed nodes. (Thus, also ordinary terms are covered in our treatment; and term graphs may have parts that are 'term-like'.) As Fig. 1 shows, our term graphs may

* This work was partially supported by ESPRIT Working Group 6345 Semagraph, and by ESPRIT BRA 6454 CONFER.

contain cycles. Term graphs as just introduced are already studied in [CKV74] under the name 'systems of fixed-point equations'; they are a simple form of recursive program schemes. We will use the phrases 'recursion system', 'term graph', 'system of recursion equations' interchangeably. It is understood that the recursion variables (or node names) are subject to renaming (α-conversion) just as in λ-calculus. So $\langle \alpha \mid \alpha = F(\alpha)\rangle$ is identical to $\langle \beta \mid \beta = F(\beta)\rangle$.

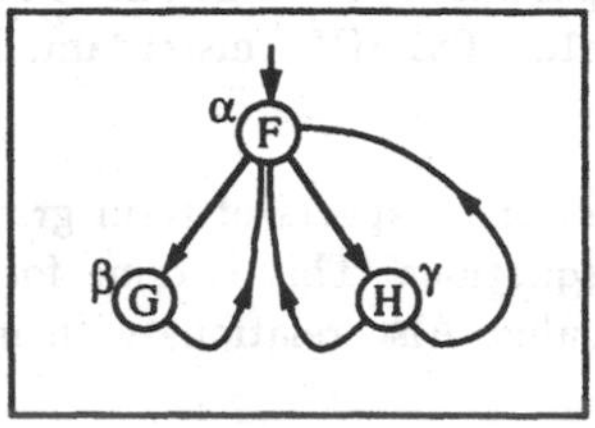

Fig. 1.

2. Basic Transformations: Substitution, Copying, Garbage Collection. Having introduced recursion systems as in (1), certain operations on these systems suggest themselves naturally. The first is substitution, denoted by $\rightarrow_{\mathsf{s}}$; it consists of replacing an occurrence of a recursion variable by its right-hand side. E.g.,

$$\langle \alpha \mid \alpha = F(\alpha)\rangle \rightarrow_{\mathsf{s}} \langle \alpha \mid \alpha = F(F(\alpha))\rangle.$$

Here the last term graph has an unnamed node, namely the one containing the second occurrence of F. Another basic operation is that of copying, denoted by $\rightarrow_{\mathsf{c}}$. It consists of creating copies $\alpha', \alpha'', \ldots$ of a node α and using these new names accordingly. E.g.,

$$\langle \alpha \mid \alpha = F(\alpha)\rangle \rightarrow_{\mathsf{c}} \langle \alpha \mid \alpha = F(\alpha'), \alpha' = F(\alpha''), \alpha'' = F(\alpha')\rangle.$$

More precisely, a copying step is the inverse of a step in which some recursion variables are identified, or collapsed together; identifying $\alpha, \alpha', \alpha''$ yields again $\langle \alpha \mid \alpha = F(\alpha)\rangle$. A third basic operation is naming, denoted by $\rightarrow_{\mathsf{n}}$, consisting of giving a name (a recursion variable) to a previously unnamed subexpression. Thus

$$\langle \alpha \mid \alpha = F(F(\alpha))\rangle \rightarrow_{\mathsf{n}} \langle \alpha \mid \alpha = F(\beta), \beta = F(\alpha)\rangle.$$

A transformation that is even more 'a priori' than the operations $\rightarrow_{\mathsf{c}}$, $\rightarrow_{\mathsf{s}}$, $\rightarrow_{\mathsf{n}}$ is that of garbage collection, notation $\rightarrow_{\mathsf{gc}}$, consisting of removing parts that are inaccessible from the root. E.g.,

$$\langle \alpha \mid \alpha = F(\beta), \gamma = F(\alpha)\rangle \rightarrow_{\mathsf{gc}} \langle \alpha \mid \alpha = F(\beta)\rangle.$$

(The last system has a free variable β. In the corresponding picture there will be an empty node with name β. Cf. Fig. 3.) There are some simple relations between $\rightarrow_{\mathsf{c}}$, $\rightarrow_{\mathsf{s}}$, $\rightarrow_{\mathsf{n}}$. E.g. : $\rightarrow_{\mathsf{c}} \cup \rightarrow_{\mathsf{s}} = \rightarrow_{\mathsf{c}} \circ \rightarrow_{\mathsf{n}}^{-1}$.

A further remark on naming: nodes may be unnamed — but only when they have in-degree (number of incoming arrows) 1. If the in-degree is 2 or more, then that node must have a name — otherwise the system could not be written down. To see this, note that, intuitively, there is an isomorphism between the term graph as picture and the term graph as expression in the linear format that we introduced. In particular, the picture has the same number of occurrences of function symbols as the corresponding expression. Now consider the term graph $\langle \alpha \mid \alpha = F(\beta, \beta),\ \beta = C \rangle$, and especially the picture corresponding to it. The node named β has in-degree 2. Suppose that this name β is removed. Then there is no way of writing down a term graph expression for this graph picture while maintaining the isomorphism between expression and picture that we strive for. This discussion touches on something that is vital for term graph rewriting: namely, how many times parts of a graph are allowed to be used. A formal calculus dealing with this issue of 'unique resources' is developed in [BS93].

3. Black Hole. Term graphs as introduced thus far have the form

$$\langle \alpha \mid \alpha_1 = t_1, \ldots, \alpha_n = t_n \rangle,$$

where the α_i are pairwise different recursion variables and the t_i are terms over the first-order signature at hand; the t_i should start with an operator symbol, i.e., they are not recursion variables. So equations $\alpha = \beta$ in the right-hand side of the $\langle \mid \rangle$-construct are not allowed. However, such equations may arise after rewriting, which is as yet not introduced. For instance, consider CL (Combinatory Logic) with its collapsing rewrite rule $Ix \rightarrow x$. Applied to $\langle \alpha \mid \alpha = I(\alpha) \rangle$ we may rewrite this to $\langle \alpha \mid \alpha = \alpha \rangle$. What to do with the circular equation $\alpha = \alpha$? We adopt a special constant 'black hole' for this situation, and replace $\alpha = \alpha$ by $\alpha = \bullet$. Likewise, $\langle \alpha \mid \alpha = \beta,\ \beta = \alpha \rangle$ will be replaced by $\langle \alpha \mid \alpha = \bullet \rangle$.

4. Horizontal and Vertical Sharing. Term graphs can be viewed as arising from two sources: one is the need for subterm sharing, the other is the need for recursion. We wish to make a precise distinction between these notions, and to that end we use the tentative terminology 'horizontal sharing' versus 'vertical sharing'. Vertical sharing is the kind of sharing exhibited typically by μ-expressions, to be discussed below. More precisely, we say that a term graph has only vertical sharing if there are no two different acyclic paths starting from the root to the same node. So $\langle \alpha \mid \alpha = F(F(\alpha)) \rangle$ has only vertical sharing; $\langle \alpha \mid \alpha = F(\beta, \gamma), \beta = C, \gamma = C \rangle$ has horizontal sharing. Of course, in general a term graph will have both kinds of sharing.

5. Bisimulation. In the theory of communicating processes, such as Milner's CCS, there is the key notion of bisimulation to compare processes. Also in the present setting of term graphs this is a very useful notion. Its definition is similar but simpler than that for processes: we say that term graphs g, h are bisimilar, notation $g \leftrightarrows h$, if there is a relation between the node set of g and the node set of h such that roots are related, and whenever nodes α and β are related, then their contents must coincide and the successors $\alpha_1, \ldots, \alpha_n$ of α must be related, respectively, to the successors $\beta_1, \ldots, \beta_n$ of β. Now, it is not hard to see that term graphs g, h are bisimilar iff they unwind to the same, possibly infinite, tree. An interesting special case of bisimilarity is encountered when the relation involved is actually a function from the first node set to the second one. In such a case we have a functional bisimulation, and we write $g \rightrightarrows h$ to denote that there is a functional bisimulation from g to h. This ties up nicely with the basic transformations discussed in (2): we have $g \rightrightarrows h$ iff $h \to_{\mathsf{c}} g$. A further interesting fact about $\rightrightarrows$ is that it is a partial order, and in fact a complete partial order, in fact even a complete lattice. This entails in turn that $\to_{\mathsf{c}}$ is a confluent operation. The complete lattices obtained in this way can be quite complicated: the one obtained by starting from $\langle \alpha \mid \alpha = F(\alpha) \rangle$, then taking all images under $\to_{\mathsf{c}}$, is in fact isomorphic to the lattice $(\mathbb{N}^+ \cup \{\infty\}, |)$ where $\mathbb{N}^+$ is the set of positive natural numbers, and the partial order $|$ is defined by: $n|m$ if n divides m; $n|\infty$; $\infty|\infty$. In other terminology ([Sme93], [Bar95]), functional bisimulations are also called rooted homomorphisms. In graph theory ([SS93]) the phrase 'graph covering' is used.

6. Non-confluence: the Even-Odd Phenomenon. As we have seen, via the notion of (functional) bisimilarity, the copying operation is confluent. It would be pleasant when this confluence result extended to the operation of substitution. However, $\to_{\mathsf{s}}$ and $\to_{\mathsf{cs}}$, the union of $\to_{\mathsf{c}}$ and $\to_{\mathsf{s}}$, are not confluent, as a simple counterexample demonstrates. Consider $\langle \alpha \mid \alpha = F(\beta),\ \beta = F(\alpha) \rangle$. Two substitution steps are possible. One leads to $\langle \alpha \mid \alpha = F(F(\alpha)) \rangle$ after garbage collection, the other to $\langle \alpha \mid \alpha = F(\beta), \beta = F(F(\beta)) \rangle$. These two systems are irreversibly separated now; no amount of further c- or s- steps can bring them together again. This is easy to see: the first system has an even number of F's, the second an odd number, and this remains invariant under further cs-steps. However, if we permit ourselves the use of naming $\to_{\mathsf{n}}$, then the resulting operation $\to_{\mathsf{csn}}$ (the union of $\to_{\mathsf{c}}, \to_{\mathsf{s}}$ and $\to_{\mathsf{n}}$) is confluent. Simple as this confluence counterexample is, it has a bearing on further developments in this theory of term graph rewriting, namely as an obstacle in the search for confluent systems for term graph rewriting (cf. item (9)).

7. Acyclic Substitution. Not surprisingly, the failure of confluence for $\to_{\mathsf{cs}}$ is caused by the presence of cycles. Now suppose we restrict substitution as follows. We say for recursion variables α, β in a recursion system that $\alpha > \beta$ if α depends on β but not vice versa, in the obvious notion of dependency. The quasi-order of the dependency relation is pictured in Fig. 2. Now suppose $\alpha > \beta$. Then the following is an acyclic substitution step:

$$\langle \gamma \mid \ldots, \alpha = t(\beta), \ldots, \beta = s, \ldots \rangle \to_{\mathsf{as}} \langle \gamma \mid \ldots, \alpha = t(s), \ldots, \beta = s, \ldots \rangle.$$

Here, in $t(\beta)$ just one occurrence of β is displayed and replaced by s. So in the figure the only as-steps are from δ in α, from β in α, from γ in β. Now we have that $\rightarrow_{\mathbf{cas}}$, the union of $\rightarrow_{\mathbf{c}}$ and $\rightarrow_{\mathbf{as}}$, is confluent. The proof is not trivial, because the rewrite relation $\rightarrow_{\mathbf{as}}$ is not a terminating one.

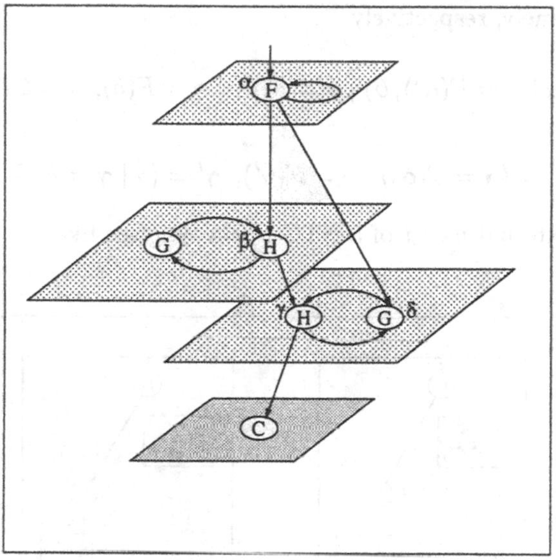

Fig. 2.

8. Nesting. An even more expressive format for graph rewriting arises when we admit nesting of recursion systems. Furthermore in the part to the left of '|', we admit a general expression t, not just a variable. So, our syntactic constructs now are of the form $\langle t \mid E\rangle$ where t is again an expression and E, the 'environment', is a set of recursion equations : $E \equiv \alpha_1 = t_1, \cdots, \alpha_n = t_n$ (with the α_i pairwise different). Graphically, we display a construct $\langle t \mid E\rangle$ as a box, filled with 'water and air' as in Fig. 3.

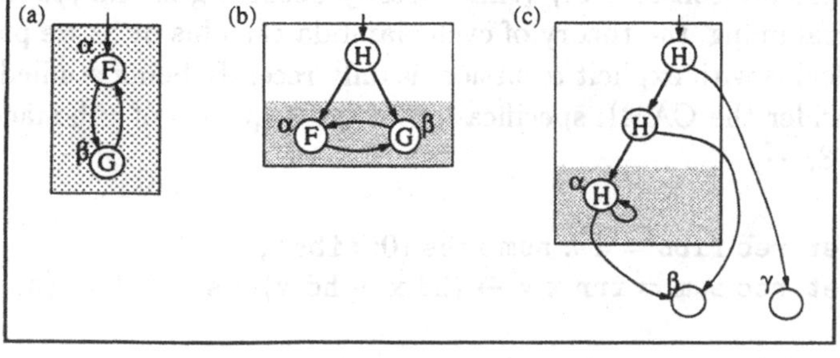

Fig. 3.

In $\langle t \mid E\rangle$, t, the external part is 'above the water', E, the environment, is still hidden under the water, waiting to be made explicit or extracted. The $\langle\ \mid\ \rangle$-construct can be used in a flexible way to define useful forms of term graph rewriting (see item (17) below). Figure 4 displays some further nested term graphs, namely, respectively,

$$\langle H(\langle\alpha \mid \alpha = F(\alpha)\rangle, \beta) \mid \beta = F(\langle\gamma \mid \gamma = F(\delta),\ \delta = G(\gamma)\rangle)\rangle,$$

$$\langle H(\alpha', \beta) \mid \alpha' = \langle\alpha = F(\alpha)\rangle,\ \beta = F(\gamma'),\ \gamma' = \langle\gamma \mid \gamma = F(\delta),\ \delta = G(\gamma)\rangle\rangle.$$

Note that the external part t of a NTG is always 'tree-like'.

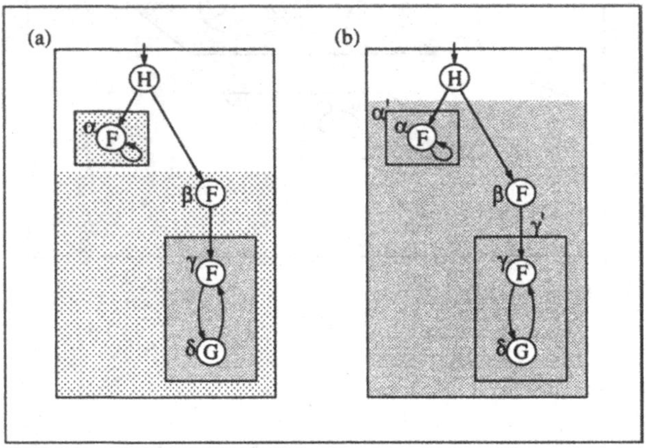

Fig. 4.

9. Lambda Term Graphs. Not only in the realm of first-order terms cycles are important, also for lambda calculus they constitute a useful new level of rewriting. (The classical reference to lambda term graphs is [Wad71], but there only acyclic graphs are considered.) While already occurring in the practice of functional programming, the theory of cyclic lambda calculus or as we prefer to say, lambda calculus with explicit recursion, is only recently being studied. As an example, consider the CAML specification of the sequence of Fibonacci numbers $1, 1, 2, 3, 5, 8, \cdots$.

```
# let rec fibs = 1 :: sum fibs (0::fibs);;
# let rec sum = fun x y → (hd x + hd y) :: sum (tl x) (tl y) ;;
```

Graphically, this is a cyclic lambda graph as in Fig. 5. (The heavy arrows point to the roots of the two β-redexes that are present in this graph.)

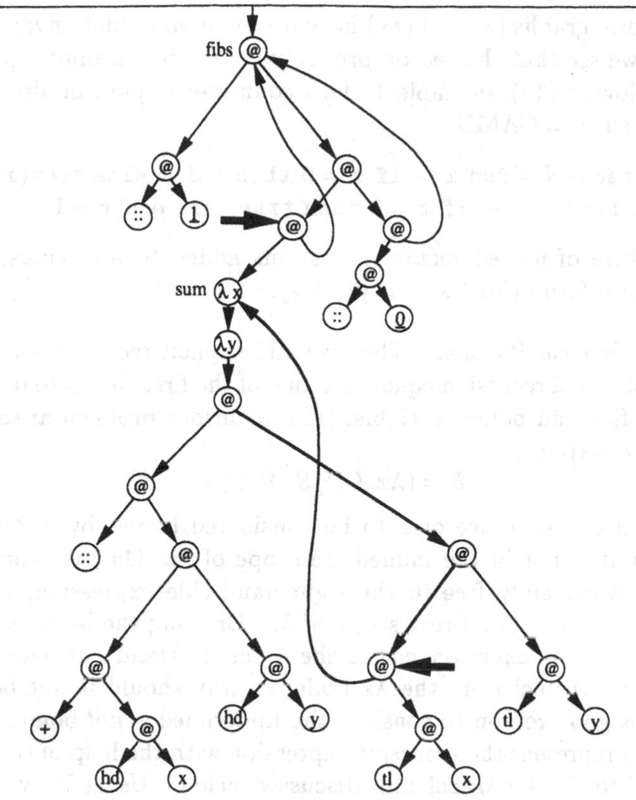

Cyclic lambda graph for computing the sequence of Fibonacci numbers

Fig. 5.

Confluence for cyclic lambda graphs is problematic, as witnessed by the following counterexample to confluence ([AK94], [AK96b]):

$$
\begin{aligned}
& \langle \alpha \mid \alpha = \lambda x.\delta(Sx),\ \delta = \lambda y.\alpha(Sy)\rangle \\
\rightarrow_{\mathbf{s}}\ & \langle \alpha \mid \alpha = \lambda x.(\lambda y.\alpha(Sy))(Sx),\ \delta = \lambda y.\alpha(Sy)\rangle \\
\rightarrow_{\beta}\ & \langle \alpha \mid \alpha = \lambda x.\alpha(S(Sx)),\ \delta = \lambda y.\alpha(Sy)\rangle \\
\rightarrow_{\mathbf{gc}}\ & \langle \alpha \mid \alpha = \lambda x.\alpha(S(Sx))\rangle \qquad (*)
\end{aligned}
$$

$$
\begin{aligned}
& \langle \alpha \mid \alpha = \lambda x.\delta(Sx),\ \delta = \lambda y.\alpha(Sy)\rangle \\
\rightarrow_{\mathbf{s}}\ & \langle \alpha \mid \alpha = \lambda x.\delta(Sx),\ \delta = \lambda y.\lambda x.\delta(Sx))(Sy)\rangle \\
\rightarrow_{\beta}\ & \langle \alpha \mid \alpha = \lambda x.\delta(Sx),\ \delta = \lambda y.\delta(S(Sy))\rangle \qquad (**)
\end{aligned}
$$

Here $\rightarrow_{\mathbf{s}}$ is substitution as in (2), and $\rightarrow_{\beta}$ is 'ordinary' β-reduction.

Now the term graphs (*) and (**) have no common reduct anymore. Analyzing why not, we see that the reason precisely is the 'discontinuity' phenomenon mentioned below in (15), in Table 1. This counterexample can also be phrased in a language such as CAML:

$$\begin{array}{l}\#\ \texttt{let rec}\ \mathit{odd} = \texttt{fun}\ x \rightarrow \texttt{if}\ x = 0\ \texttt{then false else}\ \mathit{even}(x-1)\\ \texttt{and}\ \mathit{even} = \texttt{fun}\ x \rightarrow \texttt{if}\ x = 0\ \texttt{then true else}\ \mathit{odd}(x-1)\ ;;\end{array}$$

With the feature of nested recursion systems added to λ-calculus, the β-rule takes a very nice form: $(\lambda\alpha.t)s \rightarrow \langle t \mid \alpha = s\rangle$.

10. The Free Variable Problem. When we add explicit recursion to lambda calculus, i.e., systems of recursion equations, one of the first things to do is to define the notion of free and bound variable. Here a curious problem arises. Consider the recursion equation

$$\delta = (\lambda x.F(\delta, Sx))(S\underline{x}).$$

Is the underlined occurrence of x to be considered bound by λx? On the one hand no since it is not in the immediate scope of λx. On the other hand, the underlined x is certainly free in the right-hand side expression, which is the same as δ. And δ is in the direct scope of λx. Drawing the lambda term graph corresponding to this equation makes the dilemma stand out even better: the underlined x is 'visible' from the λx-node; so why should it not be bound by it. Yet there is good reason to consider the underlined x *not* bound by λx. The argument is to represent the recursive expression with the help of the fixed point combinator Y (or in the $\lambda\mu$-calculus discussed below). Using Y, we would have

$$Y(\lambda\delta.(\lambda x.F(\delta, Sx))(S\underline{x}))$$

and in this representation it is clear that $\underline{x}$ is free, and not bound by λx. A similar consideration already applies for the simpler example

$$\langle\alpha \mid \alpha = \lambda x.\beta,\ \beta = Sx\rangle.$$

It is tempting to consider the x in Sx bound by λx; however, defining α in terms of Y again we obtain (using a standard definition for pairing $(,)$ and unpairing $()_1$, $()_2$):

$$\alpha = (Y\lambda\xi.(\lambda x.(\xi)_2, Sx))_1$$

containing a free x. In rejecting the option that x is bound by λx, we appeal to the intuition that containing a free variable occurrence should be invariant under a straightforward representation (using Y) as we considered. The conclusion that emerges in this way is that a λx cannot 'see through' a recursion variable or node name; node names are not transparant. A consequence is what we may call the scope cut-off phenomenon as in Fig. 6: when the path of a λx to an occurrence of x intersects the access path from the root to that λx, the λx does not bind the x. The scope of λx is cut off by its own access path. This is a consequence, because the node, where the access path to λx intersects with the path from λx to x, must have a name since it has two incoming arrows, according to the remark at the end of (2).

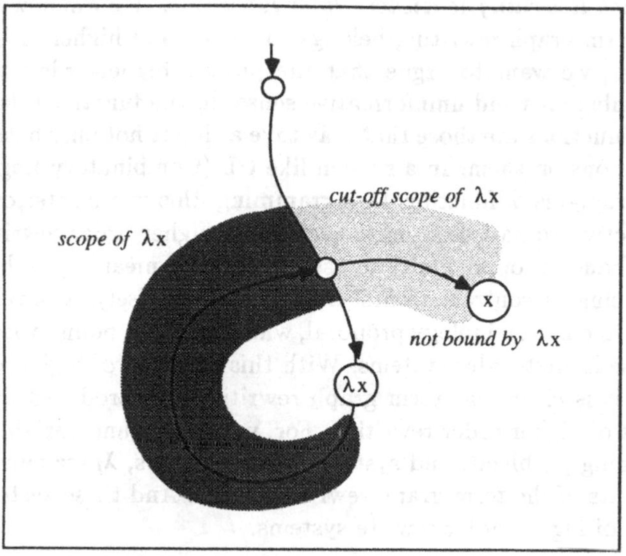

Fig. 6.

11. The μ-Calculus. This is a paradigm among higher-order rewrite systems. Its expressions are formed from some first-order signature, and if t is a μ-expression then $\mu\alpha.t$ is one as well. The μ-rule is

$$\mu\alpha.t(\alpha) \rightarrow t(\mu\alpha.t(\alpha))$$

meaning that all free occurrences of α in $t(\alpha)$ are replaced by the left-hand side. Note that μ-terms are a simple form of nested term graphs, via the translation $\mu\alpha.t \mapsto \langle\alpha \mid \alpha = t\rangle$.

12. Proof Theory. For the μ-calculus, when we adopt as semantics of μ-terms their possibly infinite tree unwinding, there is an elegant complete proof system (see [AK96a]), originally proposed by T. Nipkow. Nested term graphs are a generalization of μ-terms; μ-terms only exhibit 'vertical sharing', NTGs both horizontal and vertical. For NTGs a complete proof system has been given by [Blo96].

13. The λμ-Calculus. A rewrite system that is very interesting from the point of view of term graph rewriting is $\lambda\mu$-calculus, introduced by [Ros92], and [AK96a],[AK96b]. It consists of the combination of the μ-calculus as in (11) with λ-calculus (in its pure form, not with nested recursion systems added). The two rewrite rules are the μ-rule as in (11), and the β-rule. Now $\lambda\mu$ is in fact an orthogonal CRS (Combinatory Reduction System) or HRS (Higher-Order Rewrite System); and therefore it is confluent. An interesting question is how adequate or expressive $\lambda\mu$ is for term graph rewriting — can all of term graph rewriting be expressed in $\lambda\mu$? Anyway, $\lambda\mu$ with its nice properties like confluence, stands as a 'model' for the design of rewrite systems for term graph rewriting.

14. Term Graph Rewriting as Higher-Order Rewriting. We will now indicate why in our view term graph rewriting belongs to the area of higher-order rewriting. First, however, we want to argue that the phrase 'higher-order' is often used in a rather ambiguous and uninformative sense. In the functional language area higher-order functions are those that may take as input not only basic data types but also functions on them. In a system like CL (Combinatory Logic), which is one of the paradigms of functional programming, this is a matter of fact — but CL is a perfectly ordinary first-order system. In higher-order rewriting we have on the other hand an opportunity to assign a precise meaning to 'higher-order'. Namely, as being in contrast to first-order; more precisely, a rewrite system is higher-order, in our terminology proposal, when there are bound variables, which is not the case in first-order systems. With this meaning of higher-order rewriting in mind, it is clear that term graph rewriting as introduced and discussed above, is part of higher-order rewriting. For, we have bound variables with their typical renaming problems, and systems like μ-calculus, $\lambda\mu$-calculus are important inhabitants of the term graph rewriting scene. And these systems certainly are examples of higher-order rewrite systems.

15. Infinitary rewriting. The semantics of the cyclic term graphs that are our subject are the possibly infinite term trees or lambda trees that originate by infinite unwinding. Furthermore, a single rewrite step in a cyclic term graph will in general give rise to infinitely many rewrite steps in the corresponding infinite trees. Hence it makes sense to develop a theory of infinitary (or transfinite) rewriting; and this has been done, for the first-order case, in [KKSdV96]. Apart from the interest that infinitary rewriting has in itself, it also serves as a semantic adequacy criterion for term graph rewriting; often, dealing with cycles is conceptually more difficult than dealing with infinite terms and infinite sequences of rewrite steps. (For a general study of the adequacy of term graph rewriting for term rewriting we refer to [KKSdV94].) The crucial manœuvre to get infinite rewriting off the ground, is the formulation of the right notion of converging rewrite sequences. Namely, we have rewrite sequences which may take more than ω steps, where ω is the ordinal just after the natural numbers. So we need to know what is the *limit* of a rewrite sequence at limit ordinals λ. It turns out that the right notion of convergence towards a limit term is the one where not only an increasing part of the term is 'crystallized out', but also one in which the depth of the rewrite activity tends to infinity at every limit ordinal, $\omega, \omega.2, \omega.3, \cdots, \omega^2, \cdots$. Figure 7 pictures this situation.

Infinitary rewriting poses many new confluence problems. For orthogonal systems, some satisfactory results are stated in [KKSdV95]. Here we mention a curious phenomenon that is again somehow related to the non-confluence 'even-odd' phenomenon in item (6), and that plays a role in non-confluence of cyclic lambda graphs (see item (9)). Consider finite sequences of natural numbers, subject to the following rewrite rule: we may add to any number in the sequence one of its neighbours. A more formal rendering of this rewrite procedure would consist of adopting the infinitely many rules $\underline{n}(\underline{m}(x)) \rightarrow (\underline{n+m})(x)$ as in Table 1.

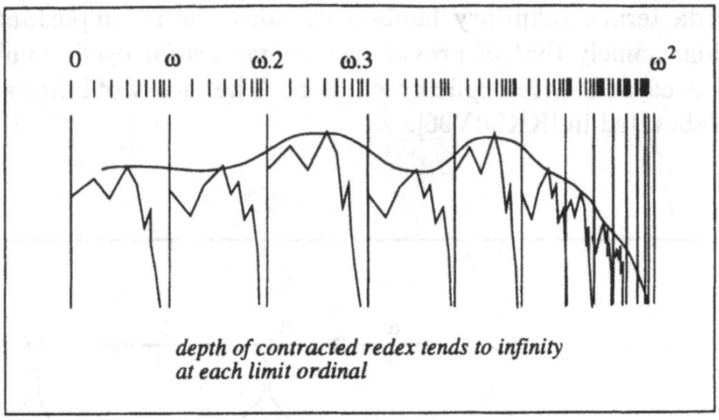

Fig. 7.

Table 1.

$\underline{n}(\underline{m}(x)) \to (\underline{n+m})(x)$
$1\ \underline{1}\ 1\ \underline{1}\ 1\ \underline{1}\ 1\ \underline{1}\ 1\ \underline{1}\ 1 \cdots \to$ $2\ \underline{1}\ 1\ \underline{1}\ 1\ \underline{1}\ 1\ \underline{1}\ 1 \cdots \to$ $2\quad 2\ \underline{1}\ 1\ \underline{1}\ 1\ \underline{1}\ 1 \cdots \to_\omega$ $2\quad 2\quad 2\quad 2\quad 2\quad \cdots$
$1\ \underline{1}\ 1\ \underline{1}\ 1\ \underline{1}\ 1\ \underline{1}\ 1\ \underline{1}\ 1 \cdots \to$ $1\quad 2\ \underline{1}\ 1\ \underline{1}\ 1\ \underline{1}\ 1\ \underline{1}\ 1 \cdots \to$ $1\quad 2\quad 2\ \underline{1}\ 1\ \underline{1}\ 1\ \underline{1}\ 1 \cdots \to_\omega$ $1\quad 2\quad 2\quad 2\quad 2\quad 2 \cdots$

So, e.g.,

$$2\,3\,5\,1 \to 5\,5\,1 \to 5\,6 \to 11\ .$$

Clearly this rewrite procedure is confluent, yielding as normal form the sum of all the numbers in the sequence. Now consider the infinite sequence $1111\cdots$ or 1^ω. In infinitely many steps, it can be rewritten to $2^\omega = 2222\cdots$. But then, we could also rewrite to $12222\cdots = 12^\omega$. These two different sequences however cannot be made to converge by further rewriting, not even in infinitely many steps – because a 'reduct' of 2^ω will consist entirely of even numbers, while any reduct of 12^ω contains at least one odd number.

16. Infinitary λ-Calculus. Infinitary rewriting is a point of view that can be applied to first-order rewriting, but also to the higher-order rewrite system of lambda calculus. Figure 8 displays a rewrite sequence of length $\omega + \omega$ involving infinite lambda terms. Infinitary lambda calculus has an important theoretical application: namely that of providing a semantics for cyclic lambda graph rewriting (or λ-calculus with explicit recursion). The theory of infinitary lambda calculus is elaborated in [KKSdV96].

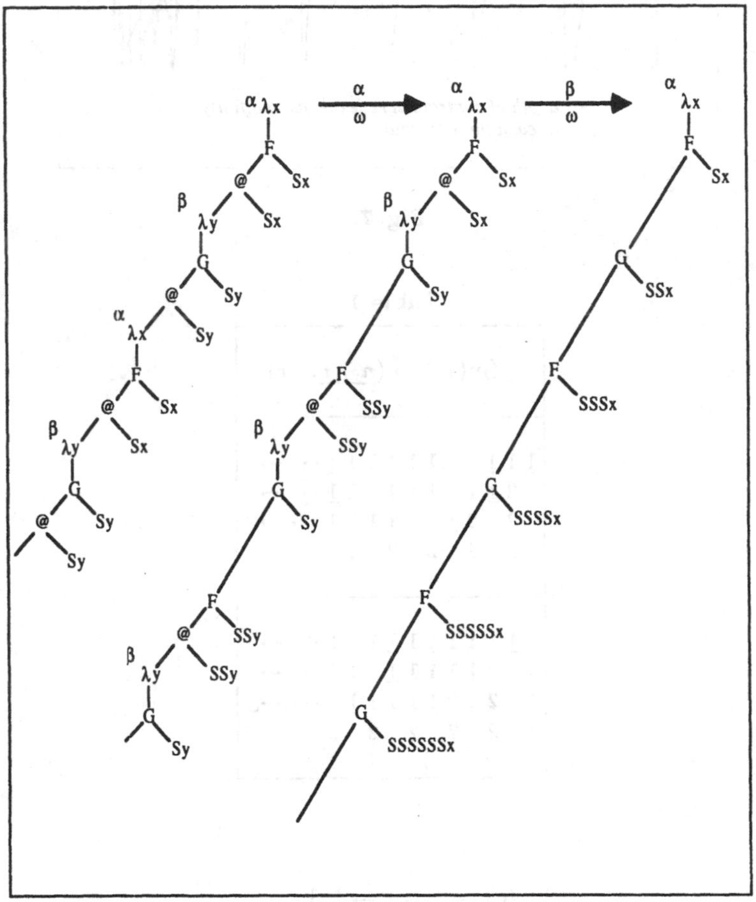

Fig. 8.

The infinite reductions displayed in Fig. 8 refer to another counterexample for confluence of β-reduction on cyclic lambda graphs [AK96a], [AK96b]. The point here is that the first two lambda trees are regular, but the third one is not; hence it cannot be the unwinding of a lambda term graph. It is for such comparisons that the infinitary lambda calculus is important, yielding conclusions about 'finite but cyclic' lambda calculus (lambda calculus with explicit recursion).

17. Term Graph Rewriting. Much of the interest of first-order or lambda term graphs as discussed thus far is that they are very suitable to be combined with actual rewriting. We already considered several rewrite rules, the β- and μ-rule, and also the rules corresponding to the basic transformations in (2), c,s,n,gc. However the latter transformations can be considered as being not proper rewrite rules; they concern the 'hardware' of the term graphs. But now let us consider the addition of a first-order TRS, e.g. CL. Then we have a smooth combination of the term graph rules and the first-order rules.

Below we give an example of a rewrite sequence of term graphs in Combinatory Logic, both as pictures in Fig. 9 and as expressions. (The curved lines in the term graphs denote the separation between external and internal part, as '|' in $\langle \mid \rangle$ does .)

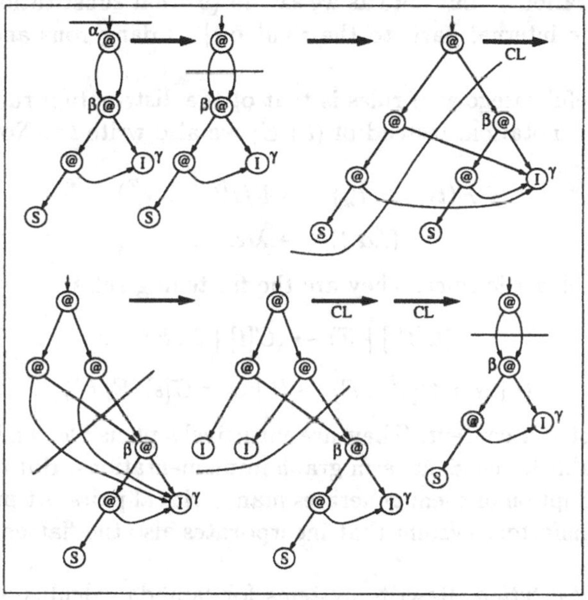

Fig. 9.

$$\begin{array}{ll}
\langle \alpha \mid \alpha = \beta\beta,\ \beta = \mathsf{S}\gamma\gamma,\ \gamma = \mathsf{I}\rangle & \rightarrow \\
\langle \beta\beta \mid \beta = \mathsf{S}\gamma\gamma,\ \gamma = \mathsf{I}\rangle & \rightarrow \\
\langle \mathsf{S}\gamma\gamma\beta \mid \beta = \mathsf{S}\gamma\gamma,\ \gamma = \mathsf{I}\rangle & \twoheadrightarrow_{\mathsf{CL}} \\
\langle \gamma\beta(\gamma\beta) \mid \beta = \mathsf{S}\gamma\gamma,\ \gamma = \mathsf{I}\rangle & \rightarrow \\
\langle \mathsf{I}\beta(\mathsf{I}\beta) \mid \beta = \mathsf{S}\gamma\gamma,\ \gamma = \mathsf{I}\rangle & \twoheadrightarrow_{\mathsf{CL}} \\
\langle \beta\beta \mid \beta = \mathsf{S}\gamma\gamma,\ \gamma = \mathsf{I}\rangle &
\end{array}$$

There are various possibilities to devise rewrite systems as combinations of a first-order TRS and rules for term graph manipulation. It is a matter of current experimentation to find elegant and expressive combinations, both for the first-order case and for the case of lambda calculus.

18. Flattening rules. In devising good term graph rewrite systems there are some more rules of importance. Some of them are obvious and easy to deal with, some of them are highly problematic. First the easy ones. What is indispensable is the rule for 'external substitution'; it was already used in the example of the CL-term graphs in (17):

$$\langle C[\alpha] \mid \alpha = s, E\rangle \to \langle C[s] \mid \alpha = s, E\rangle.$$

This means that in a recursion construct $\langle t \mid E\rangle$ we can substitute for occurrences of the recursion variables in the external part, one occurrence at the time. This is a harmless operation — but note as we saw in (6) that substitution for variables occurring in the internal part, to the right of $|$, is dangerous and may lead to non-confluence.

Another useful category of rules is that of the distribution rules. Let us first introduce a new notation: instead of $\langle t \mid E\rangle$ we also write t^E. Now we have the distribution rules:

$$F(t_1, \ldots, t_n)^E \to F(t_1^E, \ldots, t_n^E)$$
$$(\lambda\alpha.t)^E \to \lambda\alpha.t^E.$$

Finally the problematic rules. They are the flattening rules:

$$\langle C[t^E] \mid F\rangle \to \langle C[t] \mid E, F\rangle$$

$$\langle t \mid \alpha = C[s^E], F\rangle \to \langle t \mid \alpha = C[s], E, F\rangle.$$

Here $C[\;]$ denotes a context. They are intuitively plausible, and as argued in [AK96a], desirable for efficient term graph implementations. But they are highly problematic; adoption of them generates many critical pairs. At present we have not found a satisfactory system that incorporates also the flattening rules.

19. Explicit Substitution. Rewrite systems for lambda calculus with nested recursion systems, and the β-rule as mentioned in (9), together with the distribution rules mentioned in (18), are reminiscent of some of the rewrite systems for lambda calculus with explicit substitution ([ACCL91]), also called $\lambda\sigma$-calculi. This was first put forward by K.H. Rose (see [Ros92]), who coined the phrase 'explicit cyclic substitutions'. In [AK96b] the connection is further discussed.

20. `let` *and* `letrec`. Clearly, the above has much to do with the fundamental programming constructs `let` and `letrec`. E.g., $\langle \lambda\alpha.\alpha\gamma \mid \gamma = F_0\rangle$ stands for

`let` γ = F_0 `in` $\lambda\alpha.\alpha\gamma$ `end`,

whereas

$$\langle \alpha \mid \alpha = \lambda\gamma.F(\delta, \gamma),\; \delta = \lambda\eta.G(\alpha, \eta)\rangle$$

stands for

`letrec` α = $\lambda\gamma.F(\delta, \gamma)$, δ = $\lambda\eta.G(\alpha, \eta)$ `in` α `end` .

So we may say that much of the explorations discussed above are in fact attempts towards better understanding of these constructs, and towards devising suitable rewrite systems involving them.

Acknowledgements. The work on the equational approach to term graph rewriting was done in cooperation with Zena Ariola. The work on infinitary rewriting was done in cooperation with Richard Kennaway, Ronan Sleep, Fer-Jan de Vries, in the framework of ESPRIT BRA/WG Semagraph. Inspiring discussions with Kristoffer Rose on explicit recursion, with Femke van Raamsdonk and Vincent van Oostrom on higher-order rewriting, and with Stefan Blom and Erik Barendsen on nested term graphs are gratefully acknowledged. Finally, without the stimulation and the patience of Jan Heering and the invaluable help of Inge Bethke this paper would not have been written.

References

[ACCL91] M. Abadi, L. Cardelli, P.-L. Curien and J.-J. Lévy. Explicit substitutions. *Journal of Functional Programming*, 4(1): 375–416, 1991.

[AK94] Z.M. Ariola and J.W. Klop. Cyclic lambda graph rewriting. In *Proc. Ninth Symposium on Logic in Computer Science (LICS '94), Paris, France*, pages 416–425, 1994.

[AK96a] Z.M. Ariola and J.W. Klop. Equational Term Graph Rewriting. *To appear in Fundamenta Informaticae*, 1996.

[AK96b] Z.M. Ariola and J.W. Klop. Lambda calculus with explicit recursion. To appear as Report CS-R96xx, CWI Amsterdam, 1996.

[Bar95] E. Barendsen. *Types and Computations in Lambda Calculi and Graph Rewrite Systems.* PhD thesis, University of Nijmegen, 1995.

[Blo96] S.C.C. Blom. A Complete Proof System for Nested Term Graphs. In this proceedings.

[BS93] E. Barendsen and J.E.W. Smetsers. Conventional and uniqueness typing in graph rewrite systems. Technical Report CSI-R9328, Computing Science Institute, University of Nijmegen, 1993.

[CKV74] B. Courcelle, G. Kahn and J. Vuillemin. Algorithmes d'équivalence et de réduction à des expressions minimales dans une classe d'équations récursives simples. In J. Loeckx, editor, *Proc. 2nd Colloquium on Automata, Languages and Programming (ICALP '74), University of Saarbrucken*, Springer-Verlag LNCS 14, pages 200–213, 1974.

[KKSdV94] J.R. Kennaway, J.W. Klop, M.R. Sleep and F.J. de Vries. The adequacy of term graph rewriting for simulating term rewriting. *Transactions on Programming Languages and Systems*, 16(3): 493–523, 1994.

[KKSdV95] J.R. Kennaway, J.W. Klop, M.R. Sleep and F.J. de Vries. Transfinite reductions in orthogonal term rewriting systems. *Information and Computation*, 119(1):18–38, 1995.

[KKSdV96] J.R. Kennaway, J.W. Klop, M.R. Sleep and F.J. de Vries. Infinitary lambda calculus. *To appear in Theoretical Computer Science*, 1996.

[Ros92] K.H. Rose. Explicit cyclic substitutions. In M. Rusinowitch and J.L. Rémy, editors, *Proc. 3rd International Workshop on Conditional Term Rewriting Systems (CTRS '92), Pont-à-Mousson, France*, Springer-Verlag LNCS 656, pages 36–50, 1992.

[Sme93] J.E.W. Smetsers. Graph Rewriting and Functional Languages. PhD thesis, University of Nijmegen, 1993.

[SPvE93] M.R. Sleep, M.J. Plasmeijer and M.C.D.J. van Eekelen, editors. *Term Graph Rewriting: Theory and Practice.* John Wiley & Sons, 1993.

[SS93] T. Schmidt and T. Ströhlein. *Relations and Graphs.* EATCS Monographs on Theoretical Computer Science, Springer-Verlag, 1993.

[Wad71] C. Wadsworth. *Semantics and Pragmatics of the Lambda-Calculus.* PhD thesis, University of Oxford, 1971.

Approximation and Normalization Results for Typeable Term Rewriting Systems

Steffen van Bakel and Maribel Fernández

Department of Computing,
Imperial College,
180 Queens Gate, London SW7 2BZ,
svb@doc.ic.ac.uk

DMI - LIENS (CNRS URA 1327)
École Normale Supérieure
45 Rue d'Ulm, 75005 Paris, France
maribel@dmi.ens.fr

Abstract. We consider an intersection type assignment system for term rewriting systems extended with application, and define a notion of (finite) approximation on terms. We then prove that for typeable rewrite systems satisfying a general scheme for recursive definitions, every typeable term has an approximant of the same type. This approximation result, and the proof technique developed to obtain it, allow us to deduce in a direct way a head-normalization, a normalization, and a strong normalization theorem, for different classes of typeable terms.

1 Introduction

Term rewriting systems (TRS) can be seen as a model of computation, as programming or specification languages, or as formulae manipulating systems that can be used in various applications such as program optimization or automated theorem proving. *Confluence* and *termination* (also called *strong normalization*) are the most important properties of these systems: confluence ensures determinacy, whereas termination ensures that all rewrite sequences are finite. For some applications, in particular when TRS are used as a programming language, weaker properties than termination suffice. For instance, during computation *head-normalization* is a useful property: a system is head-normalizing if every term can be reduced to a term which will never reduce at the root.

In this paper we focus on normalization properties of TRS. We consider a class of systems, called *Curryfied Term Rewriting Systems* (@TRS), that are first-order TRS with a binary function symbol *Ap* that allows for *partial application* of function symbols. This function symbol can be used to define Curryfied versions of other function symbols. An advantage of having *Ap* in the language is that it makes it easy to code higher-order languages, like Combinator Systems and λ-calculus (LC), as @TRS (i.e. first-order rewrite systems).

It is well-known that in the study of normalization of reduction systems, the notion of *types* plays an important role. The type system used in this paper, the *intersection type discipline* [10] (see also [8, 9, 2]) is an extension of Curry's type assignment system for LC [11]. The extension consists of that terms and term-variables are (essentially) allowed to have more than one type. Intersection types are constructed by having, in addition to the type constructor '$\rightarrow$', the type constructor '$\cap$' and the type constant 'ω'. Using intersection types a characterization of typeable λ-terms can be given:

- the set of terms typeable without using ω is the set of strongly normalizable terms,
- the set of terms typeable with type σ from a basis B, such that ω does not occur in B and σ, is the set of normalizable terms, and
- the set of terms typeable with type $\sigma \neq \omega$ is the set of terms having a head-normal form.

(see, for instance, [19, 2]). Perhaps less well-known is the fact that the notion of *approximant* can be useful in the study of the relation between typeability and normalization. Intuitively, an approximant of a term is a finite description of its (possibly infinite) normal forms (see, e.g., [23] for a definition of approximants for LC). The aim of this paper is to investigate the relation between normalization and typeability through the use of approximants within the framework of @TRS.

In [6] and [7] we presented an intersection type assignment system for @TRS and showed that, provided the rewrite rules are typeable and satisfy a *general scheme of recursion* (inspired by the scheme of Jouannaud and Okada [16]):

- if ω is not used, typeable terms are strongly normalizable,
- non-Curryfied terms (i.e. terms without Curryfied functions and *Ap*) typeable with type σ from a basis B, such that ω does not occur in B and σ, are normalizable, and
- terms typeable with type $\sigma \neq \omega$ have a head-normal form.

Each of the above properties was proved separately. In this paper we will show that these properties can in fact be derived from one result. To that aim, we first present the formal construction needed to show that any typeable term in a typeable @TRS has an approximant of the same type (the *approximation theorem*), and then show that all of the above properties can be derived from this result in a straightforward manner. This then confirms our initial claim that the notion of approximant is a useful tool in the study of the relations between typeability and normalization in the rewriting framework.

In order to define approximants of terms in @TRS, we introduce a special symbol $\perp$ (*bottom*) into our systems, and follow the ideas of Thatte [22], which in turn are based on the definition of Ω-normal forms of Huet and Lévy [15]. As shown by Thatte, it is possible to define a fully abstract model of a term rewriting language by interpreting terms by sets of approximants. Our approximation theorem shows the connections between the intersection type system and the semantics of the language: any typeable term has a "meaning" in the model (i.e. an approximant different from $\perp$). In [1] (see also [17]), Ariola et al. discuss several notions of "undefined" (or "meaningless") terms. It turns out that our meaningless terms (which, as a consequence of the approximation theorem are untypeable) do not have a head-normal form. In the terminology of [1], the set of terms without head-normal form is the set U_{rs} (terms without a root stable form). Hence, our notion of undefined terms can be seen as an extension to typeable systems of the latter one.

In order to prove the approximation theorem, we define a rewrite relation on type derivations (called ***derivation reduction***), and prove that under some restrictions on the use of recursion in the rewrite rules (the general scheme) this relation is strongly normalizing. We use the technique of Computability Predicates for this proof (see, e.g., [14]). We then show that an approximant of t can be obtained simply by replacing untypeable subterms by $\bot$ in the normal form of the derivation for t. The strong normalization of the standard rewrite relation on typeable terms when ω is not used, is an immediate consequence of the strong normalization theorem for derivation reduction, whereas the head-normalization and the normalization theorems are direct consequences of the approximation theorem. These results hold in particular for Combinator Systems, as a particular case of @TRS that satisfies the required conditions.

This paper is organized as follows: in Section 2 we recall the definition of @TRS, and in Section 3 the intersection type assignment system for @TRS. Section 4 deals with derivation reduction and strong normalization of type derivations. In Section 5 we define approximants, prove the approximation theorem, and then the head-normalization, normalization and strong normalization theorems. Section 6 contains the conclusions and directions for future work.

2 Curryfied Term Rewriting Systems

Roughly speaking, Curryfied Term Rewriting Systems (@TRS) are first-order TRS with a binary function symbol *Ap* that allows for partial application of function symbols. @TRS extend the function-constructor systems used in most functional programming languages, in that not only constructor symbols can be used in the operand space of the left-hand side of rewrite rules, but all function symbols. In the following we assume familiarity with the basic notions and notations of term rewriting systems; we refer the reader to the surveys [18, 12] for a detailed account and examples.

We consider a ***signature*** Σ consisting of a denumerable set $\mathcal{X}$ of ***variables***, a finite set $\mathcal{F}$ of ***function symbols*** (each equipped with an ***arity***, which is a natural number), and a binary operator *Ap* for ***application***. The set $T(\mathcal{F},\mathcal{X})$ of ***terms*** is defined inductively as usual, but from $\mathcal{X}$ and $\mathcal{F}\cup\{Ap\}$. $Var(t)$ denotes the set of variables that appear in t.

We will call 'term-substitution' the operation that replaces term-variables by terms (we reserve the word 'substitution' for the operation that replaces type-variables by types). To denote a term-substitution, we will use capital characters like 'R', instead of Greek characters like 'σ', which will be used to denote types. The application of the term-substitution R to the term t will be denoted by t^{R}. We will use the notation $\{x_1 \mapsto t_1, \ldots, x_n \mapsto t_n\}$ to denote a term-substitution.

Definition 1. 1. Given a signature Σ with a set $\mathcal{X}$ of variables and a set $\mathcal{F}$ of function symbols, a ***rewrite rule*** in Σ is a pair (l,r) of terms in $T(\mathcal{F},\mathcal{X})$. Often a rewrite rule will get a name, e.g. $\mathbf{r}$, and we write $\mathbf{r} : l \to r$. As usual, two conditions are imposed:
   (a) l is not a variable.
   (b) The variables occurring in r are contained in l.

If $\mathbf{r} : F(t_1, \ldots, t_n) \to r$ and, for $1 \leq i \leq n$, either t_i is not a variable, or t_i is variable and there is a $1 \leq i \neq j \leq n$ such that $t_i = t_j$, then t_i is called a *pattern* of $\mathbf{r}$.

2. A *Curryfied Term Rewriting System* (@TRS) is a pair $(\Sigma, \mathbf{R})$ of a signature $\Sigma = (\mathcal{F}, \mathcal{X})$ and a set $\mathbf{R}$ of rewrite rules in Σ, such that, for every $F \in \mathcal{F}$ of arity $n > 0$, there exist n additional function symbols $F_{n-1}, \ldots, F_1, F_0$ in $\mathcal{F}$, the *Curryfied-versions of* F, and $\mathbf{R}$ contains the n rewrite rules:

$$\begin{array}{lll} Ap\,(F_{n-1}\,(x_1,\ldots,x_{n-1}),x_n) & \to & F\,(x_1,\ldots,x_n) \\ \vdots & & \\ Ap\,(F_1\,(x_1),x_2) & \to & F_2\,(x_1,x_2) \\ Ap\,(F_0,x_1) & \to & F_1\,(x_1) \end{array}$$

If F_i is a Curryfied version of a function symbol F, then its Curryfied versions coincide with the corresponding Curryfied versions of F: $F_{i-1}, \ldots, F_0$. Moreover, we assume that for any rule $\mathbf{r} : l \to r$ in $\mathbf{R}$, if Ap occurs in l, then $\mathbf{r}$ is of the shape:

$$Ap\,(F_{i-1}\,(x_1,\ldots,x_{i-1}),x_i) \;\to\; F_i\,(x_1,\ldots,x_i)$$

for some Curryfied version F_{i-1}, and that Curryfied versions do not appear in the root of any left-hand side of a rule in $\mathbf{R}$.

3. A rewrite rule $\mathbf{r} : l \to r$ determines a set of *reductions* $l^{\mathrm{R}} \to r^{\mathrm{R}}$ for all term-substitutions R. The term l^{R} is called a *redex*; it may be replaced by its *contractum* r^{R} inside a context C[ ]; this gives rise to *rewrite steps*: $\mathrm{C}[\,l^{\mathrm{R}}] \to_{\mathbf{r}} \mathrm{C}[\,r^{\mathrm{R}}]$. We write $t \to_{\mathbf{R}} t'$, if there is a $\mathbf{r} \in \mathbf{R}$ such that $t \to_{\mathbf{r}} t'$. Concatenating rewrite steps we have (possibly infinite) *rewrite sequences* $t_0 \to t_2 \to \cdots$. If $t_0 \to \cdots \to t_n$ ($n \geq 0$) we also write $t_0 \to^* t_n$, and t_n is a *reduct* of t_0.

Terms that do *not* contain Curryfied versions of function symbols or Ap will be called *non-Curryfied terms.*

Because of the extra rules for $F_{n-1}, \ldots, F_1, F_0$, etc., the rewrite systems are called *Curry-closed.* When presenting a rewrite system sometimes we will omit the rules that define the Curryfied versions.

Example 2. Curryfied Combinatory Logic (CCL) is the @TRS with function symbols $\mathcal{F} = \{S, S_2, S_1, S_0, K, K_1, K_0, I, I_0\}$, and rewrite rules

$$\begin{array}{lll} S\,(x,y,z) & \to & Ap\,(Ap\,(x,z), Ap\,(y,z)) \\ K\,(x,y) & \to & x \\ I\,(x) & \to & x \end{array}$$

Because CCL is Curry-closed, it inherits combinatory completeness from Combinatory Logic (every lambda term can be translated into a term in CCL).

Definition 3. 1. A term is in *normal form* if it is irreducible.

2. A term t is in *head-normal form* if for all t' such that $t \to^* t'$:
   (a) t' is not itself a redex, and
   (b) if $t' = Ap\,(v,u)$ then v is in head-normal form.

The notions *(head-)normalizable* and *strongly normalizing* are defined as usual.

Our definition of head-normal form is an extension to systems with *Ap* of the notion of root stable form defined by Ariola et al. [1]i (see also [17]). Note that the head of a term of the form $Ap\,Ap\,p(v,u)$ is in v, since we can think of *Ap* as an invisible symbol.

Example 4. Take the @TRS $F(G,H) \to A$, $B(C) \to G$, then $F(B(C),H)$ and $Ap(F(B(C),H))$ are not redexes, but are neither head-normal forms since $F(B(C),H)$ reduces to $F(G,H)$ which is a redex.

In a rewrite rule, the leftmost, outermost symbol in the left-hand side that is not an *Ap*, is called *the defined symbol* of that rule; the set $\mathcal{F}$ of function symbols can be divided into ***defined symbols***, and other symbols, the ***constructors***.

We will assume that rules are ***not*** mutually recursive.

Definition 5. A symbol F ***depends on*** G if G occurs in the right-hand side of a rule that defines F. A @TRS whose dependency-graph is an ordered a-cyclic graph, is called ***hierarchical***.

The rewrite rules of a hierarchical @TRS can be regrouped in such a way that they are ***incremental*** definitions of the defined symbols $F^1, \ldots, F^k$, so that the rules defining F^i only depend on $F^1, \ldots, F^{i-1}$.

Example 6. Since *Ap* is never a defined symbol, $D(x) \to Ap(x,x)$ is ***not*** considered a recursive system. Notice that, for example, the term $D(D_0)$ has no normal form (this term plays the role of $(\lambda x.xx)(\lambda x.xx)$ in LC). This means that, in the formalism of this paper, there exist non-recursive @TRS that are not normalizing.

3 Type assignment in @TRS

In this section we recall a variant of the intersection type assignment system for @TRS presented in [3]. This notion of type assignment is partial in the sense of [20]: not only will we define how terms and rewrite rules can be typed, but we will also assume that every function symbol already ***has*** a type, provided by an ***environment*** (i.e. a mapping from function symbols to types).

3.1 Bases, Types, and Operations

We consider a set of ***sorts*** (names of domains), the constant types of our system. When sorts are not taken into account, the ***strict intersection types*** defined below are the representatives for equivalence classes of the types considered in the system of [8]. We will define the type ω as an intersection of zero types: if $n = 0$, then $\sigma_1 \cap \cdots \cap \sigma_n = \omega$.

Definition 7. 1. $\mathcal{T}_s$, the set of ***strict types***, and $\mathcal{T}_S$, the set of ***strict intersection types***, are defined through mutual induction by:

(a) i. All type-variables $\varphi_0, \varphi_1, \ldots \in \mathcal{T}_s$.
ii. All sorts $s_1, s_2, \ldots \in \mathcal{T}_s$.

iii. If $\tau \in \mathcal{T}_s$ and $\sigma \in \mathcal{T}_S$, then $\sigma \rightarrow \tau \in \mathcal{T}_s$.
(b) If $\sigma_1, \ldots, \sigma_n \in \mathcal{T}_s$ $(n \geq 0)$, then $\sigma_1 \cap \cdots \cap \sigma_n \in \mathcal{T}_S$.

2. On $\mathcal{T}_S$, the relation $\leq_S$ is defined by:
   (a) $\forall 1 \leq i \leq n$ $(n \geq 1)$ $[\sigma_1 \cap \cdots \cap \sigma_n \leq_S \sigma_i]$.
   (b) $\forall 1 \leq i \leq n$ $(n \geq 0)$ $[\sigma \leq_S \sigma_i] \Rightarrow \sigma \leq_S \sigma_1 \cap \cdots \cap \sigma_n$.
   (c) $\sigma \leq_S \tau \leq_S \rho \Rightarrow \sigma \leq_S \rho$.
3. We define the relation $\leq$ on $\mathcal{T}_S$ like the relation $\leq_S$, but we add an extra alternative:
   (d) $\rho \leq \sigma \;\&\; \tau \leq \mu \Rightarrow \sigma \rightarrow \tau \leq \rho \rightarrow \mu$.
4. On $\mathcal{T}_S$, the relation $\sim$ is defined by: $\sigma \sim \tau \Leftrightarrow \sigma \leq \tau \leq \sigma$.

$\mathcal{T}_S$ will be considered modulo $\sim$, then $\leq$ becomes a partial order.

Notice that ω does not occur in an intersection subtype. Moreover, intersection types (so also ω) occur in strict types only as subtypes at the left-hand side of an arrow type. In general, according to the previous definition, if $\sigma_1 \cap \cdots \cap \sigma_n$ is used to denote a type, then all $\sigma_1, \ldots, \sigma_n$ are strict. Notice also that $\mathcal{T}_s$ is a proper subset of $\mathcal{T}_S$.

A ***statement*** is an expression of the form $t{:}\sigma$, where $t \in T(\mathcal{F}, \mathcal{X})$ and $\sigma \in \mathcal{T}_S$. t is the ***subject*** and σ the ***predicate*** of $t{:}\sigma$. A ***basis*** is a set of statements with only distinct variables as subjects. If $\sigma_1 \cap \cdots \cap \sigma_n$ is a predicate in a basis, then $n \geq 1$. The relations $\leq$ and $\sim$ are extended to bases by: $B \leq B' \Leftrightarrow \forall x{:}\sigma' \in B'$ $\exists x{:}\sigma \in B$ $[\sigma \leq \sigma']$, and $B \sim B' \Leftrightarrow B \leq B' \leq B$.

If $B_1, \ldots B_n$ are bases, then $\Pi\{B_1, \ldots B_n\}$ is the basis defined as follows: $x{:}\sigma_1 \cap \cdots \cap \sigma_m \in \Pi\{B_1, \ldots B_n\}$ if and only if $\{x{:}\sigma_1, \ldots, x{:}\sigma_m\}$ is the set of all statements whose subject is x that occur in $B_1 \cup \ldots \cup B_n$. If $n = 0$, then $\Pi\{B_1, \ldots B_n\} = \emptyset$. Often $B \cup \{x{:}\sigma\}$ (or $B, x{:}\sigma$) will be written for the basis $\Pi\{B, \{x{:}\sigma\}\}$, when x does not occur in B.

We will use three different operations on types (that extend to pairs of $\langle$***basis***, ***type***$\rangle$), namely ***substitution***, ***expansion***, and ***lifting***. These were first defined in [4] to show that the strict type assignment system of [2] has the principal type property. Substitution is the operation that instantiates a type (i.e. that replaces type variables by types). The operation of expansion replaces types by the intersection of a number of copies of that type. The operation of lifting replaces basis and type by a smaller basis and a larger type, in the sense of $\leq$. See [3] for formal definitions.

Definition 8. A *chain Ch* is a sequence $[O_1, \ldots, O_n]$, where each O_i is an operation of expansion, substitution, or lifting, and

$$[O_1, \ldots, O_n](\langle B, \sigma \rangle) = O_n(\cdots(O_1(\langle B, \sigma \rangle))\cdots).$$

3.2 Intersection type assignment in @TRS

Definition 9. Let $(\Sigma, \mathbf{R})$ be a @TRS. A mapping $\mathcal{E} : \mathcal{F} \cup \{Ap\} \rightarrow \mathcal{T}_s$ is called an *environment* if $\mathcal{E}(Ap) = (\varphi_1 \rightarrow \varphi_2) \rightarrow \varphi_1 \rightarrow \varphi_2$, and for every $F \in \mathcal{F}$ with arity n, $\mathcal{E}(F) = \mathcal{E}(F_{n-1}) = \cdots = \mathcal{E}(F_0)$.
Since $\mathcal{E}$ maps all $F \in \mathcal{F}$ to types in $\mathcal{T}_s$, no function symbol is mapped to ω.

The notion of type assignment on @TRS will be defined in two stages. In the next definition we consider type assignment on terms, in Def. 12 we will define type assignment on term rewrite rules.

Definition 10. 1. ***Type assignment*** and ***derivations*** are defined by the following natural deduction system (where all types displayed are in $\mathcal{T}_s$, except for σ in rule $(\leq)$, and $\sigma_1, \ldots, \sigma_n$ in rule $(\rightarrow\mathrm{E})$):

$$(\leq): \frac{x{:}\sigma \in B \quad \sigma \leq \tau}{B \vdash_{\mathcal{E}} x{:}\tau} \qquad (\cap\mathrm{I}): \frac{B \vdash_{\mathcal{E}} t{:}\sigma_1 \quad \ldots \quad B \vdash_{\mathcal{E}} t{:}\sigma_n}{B \vdash_{\mathcal{E}} t{:}\sigma_1 \cap \cdots \cap \sigma_n} \ (n \geq 0)$$

$$(\rightarrow\mathrm{E}): \frac{B \vdash_{\mathcal{E}} t_1{:}\sigma_1 \quad \ldots \quad B \vdash_{\mathcal{E}} t_n{:}\sigma_n}{B \vdash_{\mathcal{E}} F(t_1, \ldots, t_n){:}\sigma} \ (\exists\, Ch \ [Ch\,(\mathcal{E}(F)) = \sigma_1 \rightarrow \cdots \rightarrow \sigma_n \rightarrow \sigma])$$

2. We write $B \vdash_{\mathcal{E}} t{:}\sigma$ if and only if there is a derivation that has $B \vdash_{\mathcal{E}} t{:}\sigma$ as conclusion. A derivation D for $B \vdash_{\mathcal{E}} t{:}\sigma$ will be denoted by $\mathrm{D} :: B \vdash_{\mathcal{E}} t{:}\sigma$ (or simply D when the conclusion $B \vdash_{\mathcal{E}} t{:}\sigma$ is clear from the context).

Notice that, by $(\cap\mathrm{I})$, for every B and t, $B \vdash_{\mathcal{E}} t{:}\omega$. However, we will limit the expression *typeable terms* to those terms that have a type different from ω.

Note that the types that can be assigned to occurrences of function symbols or *Ap* are obtained from the type provided by the environment by using a chain of operations. The use of an environment in rule $(\rightarrow\mathrm{E})$ introduces a notion of *polymorphism* into our system.

To guarantee the subject reduction property, as shown in [3], it is sufficient to define type assignment on rewrite rules using a notion of principal pairs.

Definition 11. $\langle P, \pi \rangle$ is called ***a principal pair for t with respect to*** $\mathcal{E}$, if $P \vdash_{\mathcal{E}} t{:}\pi$ and, if $B \vdash_{\mathcal{E}} t{:}\sigma$, there is a chain Ch such that $Ch\,(\langle P, \pi \rangle) = \langle B, \sigma \rangle$.

Definition 12. Let $(\Sigma, \mathbf{R})$ be a @TRS, and $\mathcal{E}$ an environment. We say that $l \rightarrow r \in \mathbf{R}$ with defined symbol F ***is typeable with respect to*** $\mathcal{E}$, if there are basis P, type $\pi \in \mathcal{T}_s$, and an assignment of types to l and r such that:

1. $\langle P, \pi \rangle$ is a principal pair for l with respect to $\mathcal{E}$, and $P \vdash_{\mathcal{E}} r{:}\pi$.
2. In $P \vdash_{\mathcal{E}} l{:}\pi$ and $P \vdash_{\mathcal{E}} r{:}\pi$, the type actually used for each occurrence of F (or Curryfied versions of F) is $\mathcal{E}\,(F)$.

We say that $(\Sigma, \mathbf{R})$ ***is typeable with respect to*** $\mathcal{E}$ if every $\mathbf{r} \in \mathbf{R}$ is.

Theorem 13. Subject Reduction. ([3]) *Let* $(\Sigma, \mathbf{R})$ *be a typeable* @TRS *with respect to an environment* $\mathcal{E}$. *If* $B \vdash_{\mathcal{E}} t{:}\sigma$ *and* $t \rightarrow^*_{\mathbf{R}} t'$, *then* $B \vdash_{\mathcal{E}} t'{:}\sigma$.

The type assignment system we have presented is undecidable, but its restriction to intersection types of Rank 2 is decidable, as shown in [5]. The properties we will show in the following sections hold also in the decidable Rank 2 system.

4 Strong normalization of derivation reduction

In this section, we will define the notion of reduction on derivations (here called *derivation reduction*) as a generalization of cut-elimination: reductions on a derivation $D :: B \vdash'_{\mathcal{E}} t{:}\sigma$ will follow standard reduction, by contracting redexes that have a type different from ω in D, and replacing the derivation for the redex by the derivation for the right-hand side of the rewrite rule. Derivation reduction has an interesting property: it is strongly normalizing when the rewrite rules are typeable and satisfy certain syntactical conditions (a general scheme of recursion). This property will be used in the following section to prove the approximation theorem (for the full system), as well as to deduce the properties of head-normalization, normalization, and strong normalization of terms according to the set of assignable types.

To prove strong normalization of derivation reduction, however, we will restrict the definition of type assignment: for technical reasons, the derivation rule ($\leq$) of Def. 10 is too powerful, and we will limit its use to that of a rule ($\cap$E), thus defining a notion $\vdash'_{\mathcal{E}}$.

Definition 14. 1. *Strict type assignment* and *strict derivations* are defined by the following natural deduction system (where all types displayed are in $\mathcal{T}_s$, except for $\sigma_1,\ldots,\sigma_n$ in rule ($\to$E)):

$$(\cap E): \frac{x{:}\sigma_1\cap\cdots\cap\sigma_n \in B}{B \vdash'_{\mathcal{E}} x{:}\sigma_i}\ (1 \leq i \leq n) \quad (\cap I): \frac{B \vdash'_{\mathcal{E}} t{:}\sigma_1 \quad \ldots \quad B \vdash'_{\mathcal{E}} t{:}\sigma_n}{B \vdash'_{\mathcal{E}} t{:}\sigma_1\cap\cdots\cap\sigma_n}\ (n \geq 0)$$

$$(\to E): \frac{B \vdash'_{\mathcal{E}} t_1{:}\sigma_1 \ldots \quad B \vdash'_{\mathcal{E}} t_n{:}\sigma_n}{B \vdash'_{\mathcal{E}} F(t_1,\ldots,t_n){:}\sigma}\ (\exists\, Ch\ [Ch\,(\mathcal{E}(F)) = \sigma_1\to\cdots\to\sigma_n\to\sigma])$$

2. We write $B \vdash'_{\mathcal{E}} t{:}\sigma$ if and only if there is a derivation that has $B \vdash'_{\mathcal{E}} t{:}\sigma$ as conclusion. A derivation D for $B \vdash'_{\mathcal{E}} t{:}\sigma$ will be denoted by $D :: B \vdash'_{\mathcal{E}} t{:}\sigma$, or simply D if the conclusion $B \vdash'_{\mathcal{E}} t{:}\sigma$ is clear from the context.

The relation between the two notions of type assigment is formulated by:

Lemma 15. 1. If $B \vdash_{\mathcal{E}} t{:}\sigma$, then there is a B' such that $B \leq B'$, and $B' \vdash'_{\mathcal{E}} t{:}\sigma$.
2. If $B \vdash'_{\mathcal{E}} t{:}\sigma$, and B' is such that $B' \leq B$, then $B' \vdash_{\mathcal{E}} t{:}\sigma$.

Rewriting preserves types also in the strict type assignment system:

Theorem 16. Strict Subject Reduction. *Let $(\Sigma, \mathbf{R})$ be a typeable* GTRS *with respect to an environment $\mathcal{E}$. If $B \vdash'_{\mathcal{E}} t{:}\sigma$, and $t \to^*_{\mathbf{R}} t'$, then $B \vdash'_{\mathcal{E}} t'{:}\sigma$.*

Definition 17. We will use a short-hand notation for strict derivations.

1. We write $\langle\cap E\rangle :: B \vdash'_{\mathcal{E}} x{:}\sigma$ to denote the derivation of $B \vdash'_{\mathcal{E}} x{:}\sigma$ that consists of nothing but an application of rule ($\cap$E).
2. We write $D = \langle D_1,\ldots,D_n,\cap I\rangle$, if there are t, $\sigma_1,\ldots,\sigma_n$ such that $D_i :: B \vdash'_{\mathcal{E}} t{:}\sigma_i$, for every $1 \leq i \leq n$, and D is obtained from $D_1, \ldots, D_n$ by applying rule ($\cap$I).
3. We write $D = \langle D_1,\ldots,D_n,\to E\rangle$, if there are $F \in \mathcal{F} \cup \{Ap\}$, $t_1,\ldots,t_n$, and $\sigma_1,\ldots,\sigma_n$ such that $D_i :: B \vdash'_{\mathcal{E}} t_i{:}\sigma_i$, for every $1 \leq i \leq n$, and D is obtained from $D_1, \ldots, D_n$ by applying rule ($\to$E).

We can prove the following properties, that are needed further on:

Lemma 18. 1. If $D :: B \vdash'_{\mathcal{E}} x{:}\tau$, then there is $x{:}\sigma \in B$ such that $\sigma \leq_S \tau$.

2. $B \vdash'_{\mathcal{E}} F_n(t_1, \ldots, t_n){:}\sigma$ & $\sigma \in \mathcal{T}_s \Rightarrow \exists \alpha \in \mathcal{T}_S, \beta \in \mathcal{T}_s \ [\sigma = \alpha \to \beta]$.

In order to define derivation reduction, we need a notion of substitution on derivations (for lack of space, we do not give a formal definition):

Definition 19. Let $B' = \{x_1{:}\sigma_1, \ldots, x_n{:}\sigma_n\}$, $D :: B' \vdash'_{\mathcal{E}} t{:}\tau$, and let R, B be such that, for every $1 \le i \le n$, there is a $D_i :: B \vdash'_{\mathcal{E}} x_i{}^{R}{:}\sigma_i$. Each leaf of the form $x_i{:}\sigma_i \in B'$ that appears in D is the premise for a $(\cap E)$-rule. More precisely, for each of those leaves there is a subderivation

$$D_{i,j} :: \quad \frac{x_i{:}\sigma_i \in B'}{B' \vdash'_{\mathcal{E}} x_i{:}\rho_j^i} \ (\cap E)$$

in D, where, by Lem. 18-1, $\sigma_i \le_S \rho_j^i$. Now, there are two possibilities: either $\sigma_i = \rho_j^i$, or $\sigma_i = \rho_1^i \cap \cdots \cap \rho_{m_i}^i$ and hence the last step in the derivation D_i has to be $(\cap I)$, i.e. D_i is of the shape:

$$\frac{\overline{D_1^i} \quad B \vdash'_{\mathcal{E}} x_i{}^{R}{:}\rho_1^i \qquad \cdots \qquad \overline{D_{m_i}^i} \quad B \vdash'_{\mathcal{E}} x_i{}^{R}{:}\rho_{m_i}^i}{B \vdash'_{\mathcal{E}} x_i{}^{R}{:}\sigma_i}(\cap I)$$

The derivation $D[D_i/x_i{:}\sigma_i]$ is defined as the derivation obtained from D by replacing all occurrences of $D_{i,j}$ such that $\sigma_i = \rho_j^i$ by D_i, and the others by the corresponding D_j^i, and making in t the corresponding replacement of x_i by $x_i{}^{R}$.

We now give the definition of reduction on derivations.

Definition 20. The *derivation reduction* relation, denoted by $D :: B \vdash'_{\mathcal{E}} t{:}\sigma \ \to_D D' :: B \vdash'_{\mathcal{E}} t'{:}\sigma$, is defined as follows: Suppose there is a rewrite rule $l \to r$ where $Var(l) = \{x_1, \ldots, x_n\}$, and a subterm of t at position p (denoted $t|_p$) such that: $t|_p = l^{R} = F(t_1, \ldots, t_m)$ where $R = \{x_1 \mapsto u_1, \ldots, x_n \mapsto u_n\}$. Assume moreover that for $t|_p$, D contains at least one subderivation D_0 of the form:

$$\frac{\overline{D_1} \quad B \vdash'_{\mathcal{E}} u_1{:}\sigma_1 \qquad \cdots \qquad \overline{D_n} \quad B \vdash'_{\mathcal{E}} u_n{:}\sigma_n}{\overline{D_0} \quad B \vdash'_{\mathcal{E}} F(t_1, \ldots, t_m){:}\tau}$$

such that $\tau \ne \omega$, and the root of D_0 is the first occurrence of the statement $t|_p{:}\tau$ in a path from the root of D to a leaf. Then by the Strict Subject Reduction Theorem (Thm. 16), there exists $D'_0 :: \{x_1{:}\sigma_1, \ldots, x_n{:}\sigma_n\} \vdash'_{\mathcal{E}} r{:}\tau$. Let D' be obtained from D by replacing each subderivation D_0 of $t|_p$ satisfying the previous conditions by the corresponding

$$D'_0[D_1/x_1{:}\sigma_1, \ldots, D_n/x_n{:}\sigma_n] :: B \vdash'_{\mathcal{E}} r^{R}{:}\tau,$$

and propagating the replacement of $t|_p$ by r^{R} along all the derivation tree. Let t' be obtained from t by replacing $t|_p$ with r^{R}. Then we write $D :: B \vdash'_{\mathcal{E}} t{:}\sigma \ \to_D D' :: B \vdash'_{\mathcal{E}} t'{:}\sigma$.

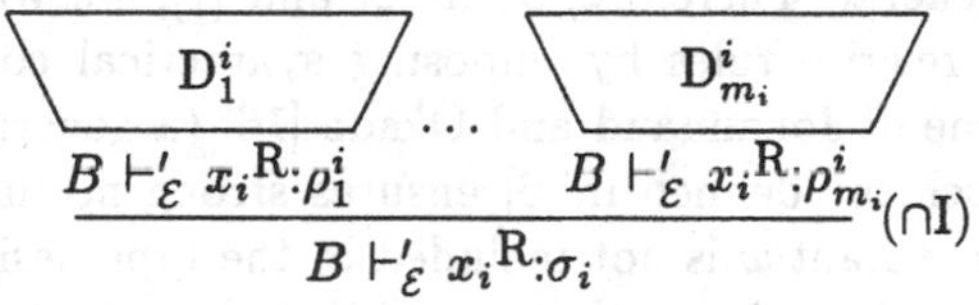

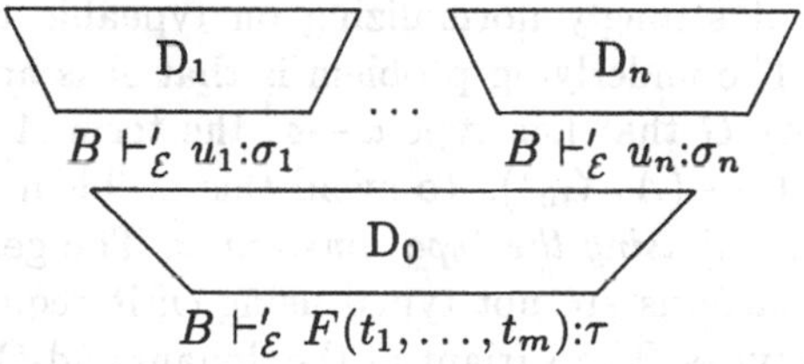

The reflexive and transitive closure of $\rightarrow_{\mathcal{D}}$ is denoted by $\rightarrow^*_{\mathcal{D}}$, and we write $SN(\mathrm{D})$ to indicate that D is strongly normalizable with respect to $\rightarrow_{\mathcal{D}}$.

Lemma 21. *1. If* $\mathrm{D} :: B \vdash'_{\mathcal{E}} t{:}\sigma \rightarrow_{\mathcal{D}} \mathrm{D}' :: B \vdash'_{\mathcal{E}} t'{:}\sigma$, *then* $t \rightarrow t'$.

2. Let $\mathrm{D} :: B \vdash'_{\mathcal{E}} Ap(t,u){:}\sigma = \langle D_1, D_2, \rightarrow\mathrm{E}\rangle$, *where* $\mathrm{D}_1 :: B \vdash'_{\mathcal{E}} t{:}\tau{\rightarrow}\sigma$ *and* $\mathrm{D}_2 :: B \vdash'_{\mathcal{E}} u{:}\tau$, *then:* $SN(\mathrm{D}) \Rightarrow SN(\mathrm{D}_1)$ & $SN(\mathrm{D}_2)$.

3. Let $\mathrm{D} :: B \vdash'_{\mathcal{E}} t{:}\sigma_1{\cap}\sigma_2 = \langle D_1, D_2, \cap\mathrm{I}\rangle$, *where* $\mathrm{D}_1 :: B \vdash'_{\mathcal{E}} t{:}\sigma_1$ *and* $\mathrm{D}_2 :: B \vdash'_{\mathcal{E}} t{:}\sigma_2$. *Then* $\mathrm{D} \rightarrow_{\mathcal{D}} \mathrm{D}' :: B \vdash'_{\mathcal{E}} t'{:}\sigma_1{\cap}\sigma_2$ *if and only if* $\mathrm{D}_1 \rightarrow_{\mathcal{D}} \mathrm{D}'_1 :: B \vdash'_{\mathcal{E}} t'{:}\sigma_1$, *or* $\mathrm{D}_2 \rightarrow_{\mathcal{D}} \mathrm{D}'_2 :: B \vdash'_{\mathcal{E}} t'{:}\sigma_2$.

4. If $\mathrm{D} :: B \vdash'_{\mathcal{E}} t{:}\sigma_1{\cap}\sigma_2 = \langle D_1, D_2, \cap\mathrm{I}\rangle$, *where* $\mathrm{D}_1 :: B \vdash'_{\mathcal{E}} t{:}\sigma_1$ *and* $\mathrm{D}_2 :: B \vdash'_{\mathcal{E}} t{:}\sigma_2$, *then* $SN(\mathrm{D}_1)$ *and* $SN(\mathrm{D}_2)$, *if and only if* $SN(\mathrm{D})$. ■

Note that in part 3 we have an 'or' because the redex might appear only in one of the derivations with a type different from ω.

It is easy to see that in the presence of recursive rules, $\rightarrow_{\mathcal{D}}$ is not strongly normalizing in general. Therefore, as in [6] and [7], we will control the use of recursion in the rewrite rules by imposing syntactical conditions inspired by the general scheme of Jouannaud and Okada [16] (a generalization of primitive recursion). The scheme defined in [6] ensures strong normalization of typeable terms when the constant ω is not included in the type assignment system. The version of the recursive scheme that we will use here takes also the presence of the type constant ω into account.

In a type system with ω, there are two kinds of typeable recursion: the one explicitly present in the syntax, as well as the one obtained by the so-called *fixed-point combinators*. Take, for example, the rewrite system

$$F(C(x)) \rightarrow F(x), \quad A(x,y) \rightarrow Ap(y, Ap(Ap(x,x),y)),$$

that satisfies the general recursive scheme of [6], and is typeable with respect to

$$\mathcal{E}(F) = \omega{\rightarrow}\sigma,\ \mathcal{E}(C) = \omega{\rightarrow}\sigma,\ \mathcal{E}(A) = ((\alpha{\rightarrow}\mu{\rightarrow}\beta){\cap}\alpha){\rightarrow}((\beta{\rightarrow}\rho){\cap}\mu){\rightarrow}\rho.$$

Then we can derive $\vdash'_{\mathcal{E}} F(A(A_0, C_0)){:}\sigma$, but notice that we also have the reduction $F(A(A_0,C_0)) \rightarrow^*_{\mathbf{R}} F(C(A(A_0,C_0))) \rightarrow_{\mathbf{R}} F(A(A_0,C_0))$. Hence this rewrite system is not strongly normalizing on typeable terms (and not even head-normalizing). The underlying problem is that A is acting as a fixed-point combinator: for every G that has type $\omega{\rightarrow}\sigma$, the term $A(A_0,G_0)$ has type σ, and $A(A_0,G_0) \rightarrow^*_{\mathbf{R}} G(A(A_0,G_0))$. To avoid this problem we will demand that *patterns are never typed using the type constant* ω. The general scheme defined in [7] ensures that patterns are not typed using ω: it requires constructor patterns, with sorts as types. The variant of the Jouannaud-Okada scheme used in this paper generalizes the one in [7] (still requiring that patterns are such that they cannot be assigned the type ω, but taking into account type derivations). It is defined as follows:

Definition 22. *Safety scheme.* Let Σ be a signature with a set of function symbols $\mathcal{F}_n = \mathcal{C} \cup \{F^1, \ldots, F^n\}$, where $F^1, \ldots, F^n$ will be the defined symbols that are not Curryfied-versions, and $\mathcal{C}$ the set of constructors and Curryfied versions of symbols. Assume that $F^1, \ldots, F^n$ are defined incrementally, by rules that satisfy the *general scheme*:

$$F^i(\bar{C}[\bar{x}], \bar{y}) \to C'[F^i(\overline{C_1}[\bar{x}], \bar{y}), \ldots, F^i(\overline{C_m}[\bar{x}], \bar{y}), \bar{x}, \bar{y}],$$

where $\bar{x}$, $\bar{y}$ are sequences of variables, and $\bar{x} \subseteq \bar{y}$. Also, $\bar{C}[\,]$, $C'[\,]$, $\overline{C_1}[\,]$, and $\overline{C_m}[\,]$ are sequences of contexts in $T(\mathcal{F}_{i-1}, \mathcal{X})$, and, for $1 \leq j \leq m$, $\bar{C}[\bar{x}] \rhd_{mul} \bar{C}_j[\bar{x}]$ where $\lhd$ is the strict subterm ordering ($\rhd$ denotes strict superterm), and *mul* denotes multiset extension.

If, moreover, patterns cannot be typed with ω (i.e. for any typing of $F^i(\bar{C}[\bar{x}], \bar{y})$, no variable typed with ω occurs twice in $F^i(\bar{C}[\bar{x}], \bar{y})$, and no non-variable subterm of $\bar{C}[\bar{x}]$ can be typed with ω), and in any typing the derivations for the arguments $\bar{C}_j[\bar{x}]$ of the recursive calls in the right-hand side are sub-derivations of those of $\bar{C}[\bar{x}]$, then the rewrite system is called *safe*.

Note that the rewrite system of the example above is not safe: the pattern $C(x)$ in the first rule can be typed with ω in a derivation of $F(C(x)) : \sigma$. However, the system containing only the second rule is safe.

Example 23. Combinator Systems are safe by definition, since all left-hand sides of rules have the form $C(x_1, \ldots, x_n)$, where $x_1, \ldots, x_n$ are different variables, and right-hand sides contain only variables and *Ap*. In particular, CCL (see Ex. 2) is typeable in the environment $\mathcal{E}_{CL}$:

$$\begin{aligned}
\mathcal{E}_{CL}(S) &= (\varphi_1 \to \varphi_2 \to \varphi_3) \to (\varphi_4 \to \varphi_2) \to \varphi_1 \cap \varphi_4 \to \varphi_3,\\
\mathcal{E}_{CL}(K) &= \varphi_5 \to \omega \to \varphi_5,\\
\mathcal{E}_{CL}(I) &= \varphi_6 \to \varphi_6.
\end{aligned}$$

We could also use the standard environment with arrow types only, but more types are typeable using $\mathcal{E}_{CL}$.

More examples of safe systems can be found in [7]. The definition of safe system given in [7] is a particular case of this one: if patterns are constructor terms, and constructors have ground types (with sorts only), then the conditions of the previous definition are asatisfied. Hence, all the systems that are safe in the sense of [7] are also safe according to the previous definition.

The rest of this section will be devoted to the proof of strong normalization of $\to_{\mathcal{D}}$ for all typeable rewrite systems that are safe. We will use the well-known method of Computability Predicates [21] (see also [14]), adapted to the rewriting framework. The proof will have two parts; in the first one we define the predicate *Comp* on derivations, and prove that if *Comp* (D) holds, then *SN*(D). In the second part *Comp* is shown to hold for each derivation (this second part differs from the usual proofs for LC in that the structure of the rewrite rules can vary; here is where the general scheme plays an important role).

Definition 24. 1. Let B be a basis, $t \in T(\mathcal{F}, \mathcal{X})$, σ a type, and $D :: B \vdash'_{\mathcal{E}} t{:}\sigma$ a derivation. We define the Computability Predicate *Comp* $(D :: B \vdash'_{\mathcal{E}} t{:}\sigma)$ recursively on σ by:

(a) If $\sigma = \varphi$, or $\sigma = s$, then *Comp* (D) $\Leftrightarrow$ *SN*(D).
(b) If $\sigma = \alpha \to \beta$, then *Comp* (D) $\Leftrightarrow \forall D' :: B' \vdash'_{\mathcal{E}} u{:}\alpha$ [*Comp* (D') $\Rightarrow$ [*Comp* $(\langle D, D', \to E\rangle :: \Pi\{B, B'\} \vdash'_{\mathcal{E}} Ap(t,u){:}\beta)$]].
(c) If $\sigma = \sigma_1 \cap \cdots \cap \sigma_n$, then $D = \langle D_1, \ldots, D_n, \cap I\rangle$, and *Comp* (D) $\Leftrightarrow$ $\forall 1 \leq i \leq n$ [*Comp* $(D_i :: B \vdash'_{\mathcal{E}} t{:}\sigma_i)$].

2. We say that a term-substitution R is *computable in a basis B with respect to a basis B'* if for every $x{:}\sigma \in B$, there is $\mathrm{D} :: B' \vdash'_{\mathcal{E}} x^{\mathrm{R}}{:}\sigma$, such that *Comp* (D).

Note that a derivation $\mathrm{D} :: B \vdash'_{\mathcal{E}} t{:}\omega$ is trivially computable, by Def. 24-1c.

The class of *neutral* terms plays a crucial role in strong normalization proofs. A notion of neutrality can be defined for derivations as well:

Definition 25. 1. A term is *neutral* if it is not of the form $F_i\,(t_1, \ldots, t_i)$, where F_i is a Curryfied version of a function symbol F.
2. A derivation $\mathrm{D} :: B \vdash'_{\mathcal{E}} t{:}\sigma$ is *neutral* if t is a neutral term.

We will prove that *Comp* satisfies the standard properties of computability predicates. First we need a lemma:

Lemma 26. Let $\mathrm{D} :: B \vdash'_{\mathcal{E}} t{:}\sigma$ *be a derivation, and* u *be a subterm of* t *that is only typed with* ω *in* D. *Let* D′ *be a derivation obtained from* D *by replacing the subderivations of the form* $B' \vdash'_{\mathcal{E}} u{:}\omega$ *by* $B' \vdash'_{\mathcal{E}} u'{:}\omega$, *for* u' *an arbitrary term (and propagating the replacements of* u *by* u' *along the derivation). Then Comp* (D) $\Rightarrow$ *Comp* (D′).

Property 27. *C1*: If *Comp* (D), then *SN*(D).
C2: If *Comp* (D) and $\mathrm{D} \to_{\mathcal{D}} \mathrm{D}'$, then *Comp* (D′).
C3: Let $\mathrm{D} :: B \vdash'_{\mathcal{E}} t{:}\sigma$ be a neutral derivation. Then for all D′ such that $\mathrm{D} \to_{\mathcal{D}} \mathrm{D}'$, *Comp* (D′) holds, then *Comp* (D). ■

In order to prove the Strong-Normalization Theorem we shall prove a stronger property, for which we will need the following ordering.

Definition 28. Let $\trianglerighteq$ stand for the well-founded encompassment ordering, i.e. $u \trianglerighteq v$ if $u \neq v$ modulo renaming of variables, and $u|_p = v^{\mathrm{R}}$ for some position $p \in u$ and term-substitution R. The subderivation relation (as well as the subterm ordering) will be denoted by $\triangleleft$, i.e. we write $\mathrm{D}' :: B' \vdash'_{\mathcal{E}} t'{:}\sigma' \triangleleft \mathrm{D} :: B \vdash'_{\mathcal{E}} t{:}\sigma$ if D′ is a derivation for a strict subterm t' of t, contained in D. Let $(\Sigma, \mathbf{R})$ be a @TRS. We define the ordering $\gg$ on triples – consisting of a pair of natural numbers, a term, and a multiset of derivations – as the object

$$((>_{\mathbb{N}}, >_{\mathbb{N}})_{lex},\ \trianglerighteq\ , (\to_{\mathcal{D}} \cup \triangleright)_{mul})_{lex}$$

where $>_{\mathbb{N}}$ denotes the standard ordering on natural numbers, and *lex*, *mul* denote respectively the *lexicographic* and *multiset* extension of an ordering.

We now come to the main theorem of this section, in which we show that for any derivation $\mathrm{D} :: B \vdash'_{\mathcal{E}} t{:}\sigma$ and computable term-substitution R in $B = \{x_1{:}\sigma_1, \ldots, x_n{:}\sigma_n\}$, also the derivation $\mathrm{D}' :: B' \vdash'_{\mathcal{E}} t^{\mathrm{R}}{:}\sigma$ obtained by substituting the computable derivations $\mathrm{D}_i :: B' \vdash'_{\mathcal{E}} {x_i}^{\mathrm{R}}{:}\sigma_i$ in D is computable. The strong normalization result then follows, using Prop. *C1*, for any derivation for t, taking for R the identity.

Property 29. Let R a computable term-substitution in $B = \{x_1{:}\sigma_1, \ldots, x_n{:}\sigma_n\}$, i.e. for $1 \leq i \leq n$, *Comp* $(\mathrm{D}_i :: B' \vdash'_{\mathcal{E}} {x_i}^{\mathrm{R}}{:}\sigma_i)$, and $\mathrm{D} :: B \vdash'_{\mathcal{E}} t{:}\sigma$. Let D′ be the derivation

$$D' = D[D_1/x_1{:}\sigma_1, \ldots, D_n/x_n{:}\sigma_n] :: B' \vdash'_{\mathcal{E}} t^{R}{:}\sigma.$$

Then *Comp* (D').

Proof: We will interpret D' by the triple $\mathcal{I}(D' :: B' \vdash'_{\mathcal{E}} t^{R}{:}\sigma) = \langle (i,j), t, \{R\} \rangle$ where i is the maximal super-index of the function symbols belonging to t, j is the minimum of the differences $arity(F^i)$ - $arity(F^i_k)$ such that F^i_k occurs in t, and $\{R\}$ is the multiset of derivations $\{D_{ij} :: B' \vdash'_{\mathcal{E}} x_i{}^{R}{:}\rho_{ij}\}$ that are substituted for the $x_i{:}\sigma_i$ in the leaves of D to obtain D'. These triples are compared in the ordering $\gg$.

When R is computable, then by Def. 24-1c and Def. 19, the derivations in $\{R\}$ are all computable, and hence by Prop. *C1* they are strongly normalizable, so $\rightarrow_{\mathcal{D}}$ is well-founded on $\{R\}$. Also, it is easy to see that the union of $\rhd$ with the terminating relation $\rightarrow_{\mathcal{D}}$ is well-founded. Hence, when restricted to computable term-substitutions, $\gg$ is a well-founded ordering. With the help of this ordering, we can prove the property by noetherian induction.

We will assume that $\sigma \neq \omega$, since otherwise the property is trivial. If $\sigma = \sigma_1 \cap \cdots \cap \sigma_n$, then, by Def. 24-1c, we have to prove *Comp* $(D'_i :: B' \vdash'_{\mathcal{E}} t^{R}{:}\sigma_i)$ for very $1 \leq i \leq n$. So, without loss of generality we can consider $\sigma \in \mathcal{T}_s$.

We distinguish the cases:

1. $D :: B \vdash'_{\mathcal{E}} t{:}\sigma$ is a neutral derivation.
   If t is a variable then, by Lem. 18-1, there is a τ such that $x{:}\tau \in B$, and $\tau \leq_S \sigma$. *Comp* $(D'' :: B' \vdash'_{\mathcal{E}} t^{R}{:}\tau)$ holds by assumption, and then, by Def. 24-1c, *Comp* $(D' :: B' \vdash'_{\mathcal{E}} t^{R}{:}\sigma)$.
   If t is not a variable, then also t^{R} is neutral, and we can use Prop. *C3*: If $D' :: B' \vdash'_{\mathcal{E}} t^{R}{:}\sigma$ is irreducible, then *Comp* (D') holds by *C3*. Otherwise, let $D' :: B' \vdash'_{\mathcal{E}} t^{R}{:}\sigma \rightarrow_{\mathcal{D}} D'' :: B' \vdash'_{\mathcal{E}} w{:}\sigma$ at position p in t^{R}. In the following we will prove either *Comp* $(D' :: B' \vdash'_{\mathcal{E}} t^{R}{:}\sigma)$ itself, or prove *Comp* $(D'' :: B' \vdash'_{\mathcal{E}} w{:}\sigma)$ and apply *C3*.

   (a) $p = qp'$, $t|_q = x_i \in \mathcal{X}$, so the rewriting takes place in a subterm of t^{R} that is introduced by the term-substitution. Let $\tau_1, \ldots, \tau_n$ be the types of $t|_q$ in D. Let z be a new term-variable.
   Take $R' = R \cup \{z \mapsto w|_q\}$, and note that D' has a subderivation D'_j such that $D'_j :: B' \vdash'_{\mathcal{E}} t^{R}|_q{:}\tau_j \rightarrow_{\mathcal{D}} D''_j :: B' \vdash'_{\mathcal{E}} w|_q{:}\tau_j$ at position p'. Since $t|_q = x_i$, and R is assumed to be computable, *Comp* $(D'_j :: B' \vdash'_{\mathcal{E}} t^{R}|_q{:}\tau_j)$ holds for each subderivation of D that has root $t|_q{:}\tau_j$. So *Comp* $(D''_j :: B' \vdash'_{\mathcal{E}} w|_q{:}\tau_j)$ holds by Prop. *C2*, hence R' is computable in $B \cup \{z{:}\tau_1 \cap \cdots \cap \tau_n\}$.
   Now, if the variable x_i $(= t|_q)$ has exactly *one* occurrence in t, then $t = t[z]_q$ modulo renaming of term-variables, and otherwise $t \rhd t[z]_q$. In the first case (since $\{R\}$ contains a derivation that is rewritten to get $\{R'\}$) we have $\mathcal{I}(D' :: B' \vdash'_{\mathcal{E}} t^{R}{:}\sigma) \gg_3 \mathcal{I}(D'' :: B' \vdash'_{\mathcal{E}} t[z]_q{}^{R'}{:}\sigma)$, and $\mathcal{I}(D' :: B' \vdash'_{\mathcal{E}} t^{R}{:}\sigma) \gg_2 \mathcal{I}(D'' :: B' \vdash'_{\mathcal{E}} t[z]_q{}^{R'}{:}\sigma)$ in the second case. Both cases yield, by induction, *Comp* $(D'' :: B' \vdash'_{\mathcal{E}} t[z]_q{}^{R'}{:}\sigma)$ and note that $t[z]_q{}^{R'} \equiv w$.

   (b) Now assume that p is a non-variable position in t. We analyze separately the cases:

   i. p is not the root position. Then $t \rhd t|_p$. Let $\tau_1, \ldots, \tau_n$ be the types assigned to $t|_p$ in the derivation $\mathrm{D} :: B \vdash'_{\mathcal{E}} t{:}\sigma$. Since $\mathcal{I}(\mathrm{D}' :: B' \vdash'_{\mathcal{E}} t^{\mathrm{R}}{:}\sigma) \gg_2 \mathcal{I}(\mathrm{D}''_j :: B' \vdash'_{\mathcal{E}} t|_p{}^{\mathrm{R}}{:}\tau_j)$ then *Comp* $(\mathrm{D}''_j :: B \vdash'_{\mathcal{E}} t|_p{}^{\mathrm{R}}{:}\tau_j)$ holds by induction. Let z be a new variable, and $\mathrm{R}' = \mathrm{R} \cup \{z \mapsto t^{\mathrm{R}}|_p\}$, then R' is computable in $B \cup \{z{:}\tau_1 \cap \cdots \cap \tau_n\}$, and $B \cup \{z{:}\tau_1 \cap \cdots \cap \tau_n\} \vdash'_{\mathcal{E}} t[z]_p{:}\sigma$. Now $t \rhd t[z]_p$, hence $\mathcal{I}(\mathrm{D}' :: B' \vdash'_{\mathcal{E}} t^{\mathrm{R}}{:}\sigma) \gg_2 \mathcal{I}(\mathrm{D}' :: B' \vdash'_{\mathcal{E}} t[z]_p{}^{\mathrm{R}'}{:}\sigma)$, hence *Comp* $(\mathrm{D}' :: B' \vdash'_{\mathcal{E}} t^{\mathrm{R}}{:}\sigma)$ by induction.

ii. p is the root position. Then the possible cases for t are:

A. $t \equiv F(t_1, \ldots, t_n)$, where at least one of the t_i is not a variable, and F is either a defined symbol of arity n or $F \equiv Ap$ and $n = 2$. Take $\mathrm{R}' = \mathrm{R} \cup \{z_1 \mapsto t_1{}^{\mathrm{R}}, \ldots, z_n \mapsto t_n{}^{\mathrm{R}}\}$. Then if $\mathrm{D}_i :: B \vdash'_{\mathcal{E}} t_i{:}\sigma_i$ is a subderivation of D, *Comp* $(\mathrm{D}'_i :: B' \vdash'_{\mathcal{E}} t_i{}^{\mathrm{R}}{:}\sigma_i)$ holds by induction, since $t \rhd t_i$. Hence, R' is computable in $B \cup \{z_1{:}\sigma_1, \ldots, z_n{:}\sigma_n\}$, and $B \cup \{z_1{:}\sigma_1, \ldots, z_n{:}\sigma_n\} \vdash'_{\mathcal{E}} F(z_1, \ldots, z_n){:}\sigma$. But, since $t \rhd F(z_1, \ldots, z_n)$, $\mathcal{I}(\mathrm{D}' :: B' \vdash'_{\mathcal{E}} t^{\mathrm{R}}{:}\sigma) \gg_2 \mathcal{I}(\mathrm{D}' :: B' \vdash'_{\mathcal{E}} F(z_1, \ldots, z_n)^{\mathrm{R}'}{:}\sigma)$. Hence *Comp* $(\mathrm{D}' :: B' \vdash'_{\mathcal{E}} t^{\mathrm{R}}{:}\sigma)$.

B. $t \equiv F^k(z_1, \ldots, z_n)$ where $z_1, \ldots, z_n$ are different term-variables. (If $z_i = z_j$ for some $i \neq j$, we can reason as in part 1a.) Then t^{R} must be an instance of the left-hand side of a rule defining F^k, that is, there is a rule

$$F^k(\bar{\mathrm{C}}[\bar{x}], \bar{y}) \to \mathrm{C}'[F^k(\overline{\mathrm{C}_1}[\bar{x}], \bar{y}), \ldots, F^k(\overline{\mathrm{C}_m}[\bar{x}], \bar{y}), \bar{x}, \bar{y}],$$

such that $t^{\mathrm{R}} = F^k(z_1, \ldots, z_n)^{\mathrm{R}} = F^k(\bar{\mathrm{C}}[\overline{M}], \overline{N}) \to_{\mathbf{R}} \mathrm{C}'[F^k(\overline{\mathrm{C}_1}[\overline{M}], \overline{N}), \ldots, F^k(\overline{\mathrm{C}_m}[\overline{M}], \overline{N}), \overline{M}, \overline{N}] = w$, where $\bar{\mathrm{C}}[\overline{M}], \overline{N}$ are all terms in the image of R hence they have computable derivations by assumption.

Now, we will deduce *Comp* $(\mathrm{D}'' :: B' \vdash'_{\mathcal{E}} w{:}\sigma)$ in three steps:

Step I: Let R' be the term-substitution that maps the left-hand side of the rewrite rule into t^{R}, so $\overline{x^{\mathrm{R}'}} = \overline{M}$, $\overline{y^{\mathrm{R}'}} = \overline{N}$. Since $\bar{x} \subseteq \bar{y}$ and R is computable, by Def. 24-1c, also R' is computable. For every $1 \leq j \leq m$, F^k does not occur in $\overline{\mathrm{C}_j}$ (by definition of the general scheme), hence $\mathcal{I}(\mathrm{D}' :: B' \vdash'_{\mathcal{E}} F^k(z_1, \ldots, z_n)^{\mathrm{R}}{:}\sigma) \gg_1 \mathcal{I}(\mathrm{D}_j :: B' \vdash'_{\mathcal{E}} \mathrm{C}_j[\bar{x}]^{\mathrm{R}'}{:}\sigma_j)$, so also the derivations $\mathrm{D}_j :: B' \vdash'_{\mathcal{E}} \mathrm{C}_j[\bar{x}]^{\mathrm{R}'}{:}\sigma_j$ are computable.

Step II: Let, for $1 \leq j \leq m$, R_j be the computable term-substitution such that $t^{\mathrm{R}_j} = F^k(\overline{\mathrm{C}_j}[\bar{x}], \bar{y})^{\mathrm{R}'}$. Since $\bar{\mathrm{C}}[\bar{x}] \rhd_{mul} \overline{\mathrm{C}_j}[\bar{x}]$, and by Def. 22, the derivation for $\overline{\mathrm{C}_j}^{\mathrm{R}'}[\bar{x}]$ is a subderivation of the one for $\bar{\mathrm{C}}^{\mathrm{R}'}[\bar{x}]$, also $\overline{\mathrm{D}_l} :: B' \vdash'_{\mathcal{E}} \mathrm{C}[\bar{x}]^{\mathrm{R}'}{:}\tau \rhd_{mul} \overline{\mathrm{D}_{r_j}} :: B' \vdash'_{\mathcal{E}} \mathrm{C}_j[\bar{x}]^{\mathrm{R}'}{:}\tau$, hence

$$\mathcal{I}(\mathrm{D}' :: B' \vdash'_{\mathcal{E}} F^k(z_1, \ldots, z_n)^{\mathrm{R}}{:}\sigma) \gg_3 \mathcal{I}(\mathrm{D}'_j :: B' \vdash'_{\mathcal{E}} F^k(z_1, \ldots, z_n)^{\mathrm{R}_j}{:}\sigma_j),$$

and therefore the derivation $\mathrm{D}'_j :: B' \vdash'_{\mathcal{E}} F^k(z_1, \ldots, z_n)^{\mathrm{R}_j}{:}\sigma_j$ is computable.

Step III: Let v be the term obtained by replacing, in the right-hand side of the rule, the terms $F^k(\overline{\mathrm{C}_1}[\overline{M}], \overline{N}), \ldots, F^k(\overline{\mathrm{C}_m}[\overline{M}], \overline{N}), \overline{M}, \overline{N}$ by fresh variables. Let R'' be the term-substitution such that $C'[F^k(\overline{\mathrm{C}_1}[\overline{M}], \overline{N}), \ldots, F^k(\overline{\mathrm{C}_m}[\overline{M}], \overline{N}), \overline{M}, \overline{N}] = v^{\mathrm{R}''}$, then $t^{\mathrm{R}} \to_{\mathbf{R}} v^{\mathrm{R}''}$, and since $\sigma \neq \omega$, $\mathrm{D}' :: B' \vdash'_{\mathcal{E}} t^{\mathrm{R}}{:}\sigma \to_{\mathcal{D}} \mathrm{D}'' :: B' \vdash'_{\mathcal{E}} v^{\mathrm{R}''}{:}\sigma$. Notice that above we have shown that R'' is computable. When an F^j occurs in v, then by definition of the general scheme j ¡ k, and therefore $\mathrm{D}' \gg_1 \mathrm{D}''$, hence D'' is computable, and since $w = v^{\mathrm{R}''}$, we get *Comp* $(\mathrm{D}'' :: B' \vdash'_{\mathcal{E}} w{:}\sigma)$.

C. $t = Ap(z_1, z_2)$ where $z_1, z_2 \in \mathcal{X}$. By assumption, the derivations for $z_1{}^{\mathrm{R}}$ and $z_2{}^{\mathrm{R}}$ are computable, and since t is well-typed, z_1 must have an arrow type. Then, by Def. 24, $D' :: B' \vdash'_{\mathcal{E}} Ap(z_1{}^{R}, z_2{}^{R}){:}\sigma$ is computable. But $Ap(z_1{}^{R}, z_2{}^{R})$ is the same as $Ap(z_1, z_2)^{R}$.

2. $D :: B \vdash'_{\mathcal{E}} t{:}\sigma$ is not neutral. Let $t \equiv F_n(t_1, \ldots, t_n)$. There are two cases:

(a) Assume that at least one of the t_i is not a term-variable. Since $t \trianglerighteq t_i$ for $1 \leq i \leq n$, by induction, for each subderivation D_i of D with root $t_i : \tau_i$, we have *Comp* $(D'_i :: B' \vdash'_{\mathcal{E}} t_i{}^{R}{:}\tau_i)$. Hence, also the term-substitution $R' = \{z_1 \mapsto t_1{}^{R}, \ldots, z_n \mapsto t_n{}^{R}\}$ is computable. Since $t \trianglerighteq F_n(z_1, \ldots, z_n)$, we have $\mathcal{I}(D' :: B' \vdash'_{\mathcal{E}} t^{R}{:}\sigma) \gg_2 \mathcal{I}(D'' :: B' \vdash'_{\mathcal{E}} t^{R'}{:}\sigma)$, and D″ is computable by induction. Note that $t^{R'} = t^{R}$, and $D'' = D'$.

(b) All t_i are variables. Since $B \vdash'_{\mathcal{E}} t{:}\sigma$, by Lem. 18-2 $\sigma = \alpha{\to}\beta$. We have to prove *Comp* $(D' :: B' \vdash'_{\mathcal{E}} t^{R}{:}\alpha{\to}\beta)$, that is, if *Comp* $(D'' :: B'' \vdash'_{\mathcal{E}} u{:}\alpha)$, then *Comp* $(\langle D', D'', {\to}E\rangle :: \Pi\{B', B''\} \vdash'_{\mathcal{E}} Ap(t^{R}, u){:}\beta)$, for every D″. Take $D_0 = \langle D', D'', {\to}E\rangle :: \Pi\{B', B''\} \vdash'_{\mathcal{E}} Ap(t^{R}, u){:}\beta$. Since $Ap(t^{R}, u)$ is neutral, by Prop. *C3*, it is sufficient to prove that, if $D_0 \to_{\mathcal{D}} D''' :: \Pi\{B', B''\} \vdash'_{\mathcal{E}} v{:}\beta$, then *Comp* (D‴).

This will be proved by induction on the sum of the lengths of the rewrite sequences out of $D'' :: B'' \vdash'_{\mathcal{E}} u{:}\alpha$ and out of the substitution. Note that since both are computable, by Prop. *C1*, *SN*$(D'' :: B'' \vdash'_{\mathcal{E}} u{:}\alpha)$, and *SN*$(\{R\})$.

Base: If the type-derivations for u and $\{R\}$ are in normal form, the only reduction step out of D_0 could be:

$$\langle D', D'', {\to}E\rangle :: \Pi\{B', B''\} \vdash'_{\mathcal{E}} Ap(F_n(z_1, \ldots, z_n)^{R}, u){:}\beta \to_{\mathcal{D}}$$
$$D''' :: \Pi\{B', B''\} \vdash'_{\mathcal{E}} F_{n+1}(z_1{}^{R}, \ldots, z_n{}^{R}, u){:}\beta,$$

then $\mathcal{I}(D' :: B' \vdash'_{\mathcal{E}} t^{R}{:}\sigma) \gg_1 \mathcal{I}(D''')$ is computable.

Induction step: If the reduction out of D_0 takes place inside u or inside t^{R} (in the last case it must be inside $\{R\}$ since the rewrite system is safe) then D‴ is computable by induction. If $\langle D', D'', {\to}E\rangle \to_{\mathcal{D}} D''' :: \Pi\{B', B''\} \vdash'_{\mathcal{E}} F_{n+1}(z_1{}^{R}, \ldots, z_n{}^{R}, u){:}\beta$, we proceed as in the base case. ■

Theorem 30. Strong Normalization of Derivation Reduction. *If* $(\Sigma, \mathbf{R})$ *is typeable in* $\vdash'_{\mathcal{E}}$ *and safe, then for every* $D :: B \vdash'_{\mathcal{E}} t{:}\sigma$, *SN*(D).
Proof: From Prop. 29 and *C1*, taking R such that $x^{R} = x$. ■

Note that for the strong normalization property to hold, it is not enough to define a reduction relation that considers only typeable redexes in a term. It is crucial to preserve the derivations, as the following example shows:

Example 31. Consider the GTRS with rules:

$$E(x, y) \to Ap(y, Ap(Ap(x, x), y)),$$
$$C(x, y) \to y,$$

This rewrite system is typeable and safe with respect to

$$\mathcal{E}(E) = \omega{\to}(\omega{\to}(\alpha{\to}\alpha)) \to(\alpha{\to}\alpha),$$
$$\mathcal{E}(C) = \omega{\to}\alpha{\to}\alpha.$$

Notice that the term $E(E_0, C_0)$ is typeable by $\alpha{\to}\alpha$ in this environment. If instead of reducing derivations we would allow to reduce a redex whenever it has a type different from ω, then there is an infinite reduction sequence out of this term:

$E(E_0, C_0) \rightarrow_{\mathbf{R}} Ap(C_0, Ap(Ap(E_0, E_0), C_0)) \rightarrow_{\mathbf{R}} C_1(Ap(Ap(E_0, E_0), C_0)) \rightarrow^*_{\mathbf{R}} C_1(E(E_0, C_0)) \ldots$

However, any derivation for $\vdash'_{\mathcal{E}} E(E_0, C_0) : \alpha \rightarrow \alpha$ is strongly normalizable according to $\rightarrow_{\mathcal{D}}$. Take for instance the derivation that assigns ω to E_0 and $\omega \rightarrow \alpha \rightarrow \alpha$ to C_0 (we will annotate function symbols with types instead of writing the type-derivation):

$$\vdash'_{\mathcal{E}} E(E_0{:}\omega, C_0{:}\omega \rightarrow \alpha \rightarrow \alpha){:}\alpha \rightarrow \alpha \rightarrow_{\mathcal{D}}$$
$$\vdash'_{\mathcal{E}} Ap(C_0{:}\omega \rightarrow \alpha \rightarrow \alpha, Ap(Ap(E_0, E_0), C_0){:}\omega){:}\alpha \rightarrow \alpha \rightarrow_{\mathcal{D}}$$
$$\vdash'_{\mathcal{E}} C_1(Ap(Ap(E_0, E_0), C_0){:}\omega){:}\alpha \rightarrow \alpha$$

and the last derivation is in normal form, since the type assigned in this derivation to $Ap(Ap(E_0, E_0), C_0)$ is ω.

5 Approximation and normalization properties

In this section, the results of the previous section will be used to prove four theorems. The first is an approximation theorem; our definition of approximants is a combination of the one given by Wadsworth for the Lambda Calculus [23], and the approximants for Term Rewriting Systems defined by Thatte [22], based on the notion of Ω-normal forms of Huet and Lévy [15]. The last three theorems will formulate a relation between assignable types and head-normalization, strong normalization and normalization properties, respectively.

In order to define approximants of terms, we start by introducing a special symbol $\perp$ (*bottom*) into the language (so $\perp$ is not in $\mathcal{X}$, and neither in $\mathcal{F}$), that is intended to represent meaningless terms. (The definition of this new set of terms $T(\mathcal{F}, \mathcal{X}, \perp)$ is straightforward.) To define type assignment on $T(\mathcal{F}, \mathcal{X}, \perp)$, the type assignment rules given in Def. 10 need not be changed, it suffices that terms are allowed to be in $T(\mathcal{F}, \mathcal{X}, \perp)$. Since $\perp \notin \mathcal{F} \cup \{Ap\}$, this implies that $\perp$ can only be given the type ω, or appear in subterms that are given the type ω.

Terms in $T(\mathcal{F}, \mathcal{X}, \perp)$ can be ordered using the relation $\sqsubseteq$:

Definition 32. 1. $t \sqsubseteq u$ is inductively defined by:
   (a) For every $u \in T(\mathcal{F}, \mathcal{X}, \perp)$, $\perp \sqsubseteq u$.
   (b) For every $t \in T(\mathcal{F}, \mathcal{X}, \perp)$, $t \sqsubseteq t$.
   (c) $F(t_1, \ldots, t_n) \sqsubseteq F(u_1, \ldots, u_n)$, if and only if, for all $1 \leq i \leq n$, $t_i \sqsubseteq u_i$.
2. We write $t \uparrow u$ (and say that t and u *are compatible*) if there is a $v \in T(\mathcal{F}, \mathcal{X}, \perp)$ such that $t \sqsubseteq v$ and $u \sqsubseteq v$. We write $t \uparrow V$ if there is an $l \in V$ such that $t \uparrow l$.

By abuse of notation, we will use the symbol $\perp$ also for the term-substitution that replaces term-variables by $\perp$: $\perp = \{x \mapsto \perp \mid x \in \mathcal{X}\}$. In the following we consider a given GTRS $(\Sigma, \mathbf{R})$, and $Lhs^{\perp} = \{l^{\perp} \mid \exists r\ [l \rightarrow r \in \mathbf{R}]\}$.

We will now develop the notion of approximant of a term with respect to a given GTRS. A particular difference with the definition of approximant for lambda terms [23] is that our definition is 'static', whereas the other notion was defined as normal forms with respect to an extended notion of reduction. This approach would not be appropriate for our paper, because, to name just one problem, we would not be able to prove a subject reduction result for such a notion of reduction. Instead, we will recursively replace redexes by $\perp$. While doing this, it can be that a term is created that itself is not a redex, but looks like one, in the sense that is compatible to a left-hand side of a rewrite rule (where variables are replaced by $\perp$). Also such 'possible redexes' will be replaced by $\perp$.

Definition 33. $\mathcal{DA}(t)$, the *direct approximant of* t with respect to $(\Sigma, \mathbf{R})$ is defined by:

1. $t = x$. $\mathcal{DA}(x) = x$.
2. $t = F(t_1, \ldots, t_n)$; let, for $1 \leq i \leq n$, $a_i = \mathcal{DA}(t_i)$.
   $\mathcal{DA}(t) = \perp$, if $F(a_1, \ldots, a_n) \uparrow Lhs^{\perp}$; otherwise, $\mathcal{DA}(t) = F(a_1, \ldots, a_n)$.
3. $t = Ap(t_1, t_2)$; let $a_1 = \mathcal{DA}(t_1)$, and $a_2 = \mathcal{DA}(t_2)$.
   $\mathcal{DA}(t) = \perp$, if $a_1 = \perp$, or $a_1 = F_i(a_1, \ldots, a_i)$; otherwise, $\mathcal{DA}(t) = Ap(a_1, a_2)$.

Approximants of terms are obtained by taking direct approximants of their reducts (and making a downward closure).

Definition 34. 1. $\mathcal{DA}$, the set of *approximate normal forms* is defined as

$$\mathcal{DA} = \{a \in T(\mathcal{F}, \mathcal{X}, \perp) \mid \mathcal{DA}(a) = a\}.$$

2. $\mathcal{A}(t)$, the *set of approximants of* t, is defined by:

$$\mathcal{A}(t) = \{a \in \mathcal{DA} \mid \exists u\, [t \rightarrow^* u \;\&\; a \sqsubseteq \mathcal{DA}(u)]\}.$$

Intuitively, the terms whose only approximant is $\perp$ are undefined (i.e. meaningless). We will see below (Cor. 38) that typeable terms cannot be undefined. We will also see that this implies that typeable terms are head-normalizable. We introduce more notation now:

Definition 35. For any derivation $\mathrm{D} :: B \vdash_{\mathcal{E}} t{:}\sigma$, we denote by t_{D} the term obtained from t by replacing all its subterms that do not have a type different from ω in D, by $\perp$.

In other words, t_{D} is obtained by replacing any subterm u of t such that in D we only have $B' \vdash_{\mathcal{E}} u{:}\omega$, by $\perp$.

Lemma 36. *1. Let* $\mathrm{D} :: B \vdash_{\mathcal{E}} t{:}\sigma$. *Then there exists a derivation* $\mathrm{D}' :: B \vdash_{\mathcal{E}} t_{\mathrm{D}}{:}\sigma$.
2. If $\mathrm{D} :: B \vdash_{\mathcal{E}} t{:}\sigma$ *is irreducible with respect to* $\rightarrow_{\mathcal{D}}$ *then* $t_{\mathrm{D}} \in \mathcal{A}(t)$.
3. If $\mathcal{DA}(t) \neq \perp$, *then* t *is in head-normal form.*

With Thm. 30 and Lem. 36 we are able to prove:

Theorem 37. Approximation Theorem. *If* $(\Sigma, \mathbf{R})$ *is typeable in* $\vdash_{\mathcal{E}}$ *and safe, then for every* t *such that* $B \vdash_{\mathcal{E}} t{:}\sigma$ *there is an* $a \in \mathcal{A}(t)$ *such that* $B \vdash_{\mathcal{E}} a{:}\sigma$.
Proof: Let $B \vdash_{\mathcal{E}} t{:}\sigma$, then by Lem. 15-1 there is a B' such that $B \leq B'$ and $B' \vdash'_{\mathcal{E}} t{:}\sigma$. Let D be a derivation for this last result, then by Thm. 30, $SN(\mathrm{D})$. Let $\mathrm{D}' :: B' \vdash'_{\mathcal{E}} t'{:}\sigma$ be the normal form of D with respect to $\rightarrow_{\mathcal{D}}$ and let $a = t'_{\mathrm{D}'}$. Then, by Lem. 36-2, $a \in \mathcal{A}(t')$, and by Lem. 21-1, $t \rightarrow_{\mathbf{R}} t'$. Hence $a \in \mathcal{A}(t)$. By Lem. 36-1, $B' \vdash'_{\mathcal{E}} a{:}\sigma$, and by Lem. 15-2, $B \vdash_{\mathcal{E}} a{:}\sigma$. ∎

Corollary 38. Let $(\Sigma, \mathbf{R})$ *be typeable with respect to* $\mathcal{E}$ *and safe. If* $B \vdash_{\mathcal{E}} t{:}\sigma$, *and* $\sigma \neq \omega$, *then there exists* $a \in \mathcal{A}(t)$ *such that* $a \neq \perp$.
In other words, all typeable terms are meaningful.

Theorem 39. Head Normalization Theorem. *If $B \vdash_{\mathcal{E}} t{:}\sigma$, and $\sigma \neq \omega$, then t has a head-normal form.*
Proof: If $B \vdash_{\mathcal{E}} t{:}\sigma$, then by Cor. 38, there exists $a \in \mathcal{A}(t)$ such that $a \neq \bot$. Since $a \in \mathcal{A}(t)$, there is a v such that $t \rightarrow^* v$ and $a \sqsubseteq \mathcal{DA}(v)$. Hence $\mathcal{DA}(v) \neq \bot$. Then, by Lem. 36-3, v is in head-normal form, so, in particular, t has a head-normal form. ■

We have seen that typeable terms have an approximant different from $\bot$, and a head-normal form. Ariola et al. [1] define a notion of meaningless terms as terms without a root stable form; the set of undefined terms according to this notion is called U_{rs}. Untypeable terms are the meaningless terms in our system. Let U_t be the set of untypeable terms. By Thm. 39, $U_{rs} \subseteq U_t$.

Theorem 40. Strong Normalization Theorem. *Let $(\Sigma, \mathbf{R})$ be a safe rewrite system, typeable without using ω at all. If $B \vdash_{\mathcal{E}} t{:}\sigma$, and ω is not used to derive this result, then t is strongly normalizable.*
Proof: Let $\mathrm{D} :: B \vdash_{\mathcal{E}} t{:}\sigma$. By Thm. 30, D is strongly normalizable. Since ω is not used, and the system without ω has the Subject Reduction Property [6], by definition of $\rightarrow_{\mathcal{D}}$ and Lem. 21-1, we obtain strong normalization of t. ■

Notice that the converses of the two previous theorems do not hold because the environment is given (and fixed).

In the intersection system for LC, it is well-known that terms that are typeable without ω in base and type are normalizable. This is not true for @TRS, even if one considers safe recursive systems only. Then, as in [7], we will restrict the study of normalization properties of @TRS to non-Curryfied terms. Actually, to get a normalization result similar to that of LC we also need to impose the following condition on the @TRS:

Definition 41. A @TRS is *complete* if whenever a non-Curryfied term t that is assigned the ω-free type σ is reducible at a position p such that $t|_p$ can be assigned a type containing ω, there exists $q < p$ such that $t|_q$ can be assigned an ω-free type and $t|_q[x]_p$ is not in head-normal form.

Intuitively, in a complete @TRS a non-Curryfied term $F(t_1, \ldots, t_n)$ that has an ω-free type, and where there is a redex t_i that has a type containing ω, will be reducible either at the root (without taking t_i into account), or in some t_j with an ω-free type. This means that the rules defining F cannot have patterns that can be assigned types with ω, and also that constructors cannot accept arguments having a type which contains ω. Moreover, if a defined function accepts arguments having types with ω then its definition must be exhaustive.

Defined functions of safe systems satisfy the first condition. So, a safe system is complete whenever constructors have ground types without ω, and for all defined function F that accepts arguments with types that contain ω, the patterns of the rules defining F cover all possible cases.

Example 42. Combinator Systems are complete, since there is a rule for each combinator.

Theorem 43. Normalization Theorem. *Let t be a non-Curryfied term in a typeable, safe, and complete @TRS. If $B \vdash_{\mathcal{E}} t{:}\sigma$ and ω does not appear in B, σ, then t is normalizable.*

Proof: By Thm. 30, $SN(\mathrm{D} :: B \vdash_{\mathcal{E}} t{:}\sigma)$. Let D′ be a normal form of D, i.e. $\mathrm{D}' :: B \vdash_{\mathcal{E}} t'{:}\sigma$, and B,σ are ω-free. We prove that t' is in normal form by case analysis.

Since t' is a non-Curryfied term, two cases are possible:

1. t' is a variable, hence it is in normal form.
2. $t' = F(t_1, \ldots, t_n)$ In this case, t' cannot be a redex itself, because the derivation $B \vdash_{\mathcal{E}} t'{:}\sigma$ is in normal form and $\sigma \neq \omega$. Moreover, only subterms that are typed with ω can be reducible. But since the system is complete, the existence of redex of type ω implies the existence of redex of type different from ω, which leads to a contradiction. Then t' is in normal form. ∎

6 Conclusions and future work

Combinator Systems are @TRS that are trivially safe and complete (see Ex. 23, 42), hence all the results presented in this paper hold in particular for these systems. Dezani and Hindley presented a type assignment system for Combinator Systems that are combinatory complete [13]. Our system can be seen as an extension of this one, since we do not require the systems to be combinatory complete. The results we showed also apply to the type assignment system of Dezani and Hindley.

Approximants can be used to characterize equality in models of term rewriting languages, as shown by Thatte [22]: a fully abstract model can be built by interpreting terms as sets of approximants (more precisely, the interpretation of t is $\mathcal{A}(t)$). Our approximation theorem shows the connection between the intersection type assignment system and the semantics of the language: any typeable term has an interpretation different from $\perp$ in the model (i.e. it has a "meaning"). Approximants can be used to characterize equality in lambda models as well [23]. In the future, we will look at these properties in the setting of the combination of TRS and LC.

References

1. Z. Ariola, R. Kennaway, J.W. Klop, R. Sleep, and F-J. de Vries. Syntactic definitions of undefined: on defining the undefined. In *TACS '94*, *LNCS*, 789, pages 543–554, 1994.
2. S. van Bakel. Complete restrictions of the Intersection Type Discipline. *Theoretical Computer Science*, 102:135–163, 1992.
3. S. van Bakel. Partial Intersection Type Assignment in Applicative Term Rewriting Systems. In *TLCA '93*, *LNCS* 664, pages 29–44, 1993.
4. S. van Bakel. Principal type schemes for the Strict Type Assignment System. *Logic and Computation*, 3(6):643–670, 1993.

5. S. van Bakel. Rank 2 Intersection Type Assignment in Term Rewriting Systems. *Fundamenta Informaticae*, 1996. To appear.
6. S. van Bakel and M. Fernández. Strong Normalization of Typeable Rewrite Systems. In *HOA '93*, *LNCS* 816, pages 20–39, 1994.
7. S. van Bakel and M. Fernández. (Head-)Normalization of Typeable Rewrite Systems. *RTA '95*, *LNCS* 914, pages 279–293, 1995.
8. H. Barendregt, M. Coppo, and M. Dezani-Ciancaglini. A filter lambda model and the completeness of type assignment. *Journal of Symbolic Logic*, 48(4):931–940, 1983.
9. F. Cardone and M. Coppo. Two Extensions of Curry's Type Inference System. In *Logic and Computer Science*, pages 19–75, 1990.
10. M. Coppo and M. Dezani-Ciancaglini. An Extension of the Basic Functionality Theory for the λ-Calculus. *Notre Dame Journal of Formal Logic*, 21(4):685–693, 1980.
11. H.B. Curry and R. Feys. *Combinatory Logic*, volume 1, 1958.
12. N. Dershowitz and J.P. Jouannaud. Rewrite systems. In *Handbook of Theoretical Computer Science*, volume B, chapter 6, pages 245–320, 1990.
13. M. Dezani-Ciancaglini and J.R. Hindley. Intersection types for combinatory logic. *Theoretical Computer Science*, 100:303–324, 1992.
14. J.-Y. Girard, Y. Lafont, and P. Taylor. *Proofs and Types*. Cambridge Tracts in Theoretical Computer Science, 1989.
15. G. Huet and J.J. Lévy. Computations in Orthogonal Rewriting Systems. In *Computational Logic. Essays in Honour of Alan Robinson*, 1991.
16. J.P. Jouannaud and M. Okada. Executable higher-order algebraic specification languages. In *LiCS '91*, pages 350–361, 1991.
17. R. Kennaway, V. van Oostrom, and F.J. de Vries. Meaningless terms in rewriting. Submitted for publication. Obtainable as: http://wwwbroy.informatik.tu-muenchen.de/ oostrom, 1996.
18. J.W. Klop. Term Rewriting Systems. In *Handbook of Logic in Computer Science*, volume 2, chapter 1, pages 1–116, 1992.
19. D. Leivant. Typing and computational properties of lambda expressions. *Theoretical Computer Science*, 44:51–68, 1986.
20. F. Pfenning. Partial Polymorphic Type Inference and Higher-Order Unification. In *LISP and Functional Programming Languages '88*, pages 153–163, 1988.
21. W.W. Tait. Intensional interpretation of functionals of finite type I. *Journal of Symbolic Logic*, 32(2):198–223, 1967.
22. S.R. Thatte. Full Abstraction and Limiting Completeness in Equational Languages. *Theoretical Computer Science*, 65:85–119, 1989.
23. C.P. Wadsworth. The relation between computational and denotational properties for Scott's D_∞-models of the lambda-calculus. *SIAM J. Comput.*, 5:488–521, 1976.

Modular Properties of Algebraic Type Systems

Gilles Barthe [1]* gilles@cwi.nl
Herman Geuvers [2,3] herman@win.tue.nl

[1] CWI, Amsterdam, The Netherlands
[2] Faculty of Mathematics and Informatics, University of Nijmegen, The Netherlands
[3] Fac. of Math. and Informatics, Techn. Univ. of Eindhoven, The Netherlands

Abstract. We introduce the framework of algebraic type systems, a generalisation of pure type systems with higher order rewriting *à la* Jouannaud-Okada, and initiate a generic study of the modular properties of these systems. We give a general criterion for one system of this framework to be strongly normalising. As an application of our criterion, we recover all previous strong normalisation results for algebraic type systems.

1 Introduction

Algebraic-functional languages, introduced by Jouannaud and Okada in [19], are based on a very powerful paradigm combining type theory and higher-order rewriting systems. These languages embed in typed λ-calculi higher-order rewriting and hence allow the definition of abstract data types as it is done in equational languages such as OBJ. Examples of such languages which have been studied in the literature include the algebraic simply typed λ-calculus ([19]), algebraic type assignments systems ([2]) and the algebraic calculus of constructions ([3]). In this paper, we introduce a very general framework to study the combination of type theories with higher-order rewriting systems. The combination is based on pure type systems ([4]); the resulting framework of *algebraic type systems* covers in particular the systems of the algebraic λ-cube, a generalisation of Barendregt's cube studied in [3, 19]. A particular interest of algebraic type systems is to offer the possibility to initiate a generic study of the meta-theory of the combination between type theory and rewriting. First, basic meta-theoretic results, such as the substitution lemma or the generation lemma ([4, 16]) can be proved for arbitrary algebraic type systems. Second, one can address modularity results in a very abstract way, as it has been successfully done in term-rewriting (some striking examples can be found in [21, 26]). The main contribution of this paper is to give a general criterion for an algebraic type system to be strongly normalising. As an application of our criterion, we obtain a new proof of the modularity of strong normalisation for the algebraic cube ([2, 3, 10, 11, 19] for

* This work was performed while working at the University of Nijmegen (The Netherlands) and visiting the University of Manchester (United Kingdom).

subsystems). We also derive a strong normalisation result for algebraic higher-order logic (the algebraic extension of λHOL [16]) and the algebraic calculus of constructions with universes (with left-linear and confluent rewriting systems). In our view, the distinctive features of our approach are its generality (all the known results on modularity of termination for algebraic type systems can be obtained as a corollary of our result), its simplicity (the complexity of the proof is similar to the corresponding strong normalisation argument for pure type systems) and its flexibility (it is easy to adapt the proof to variants of pure type systems).

The paper is organised as follows: in the next section, we introduce algebraic type systems. In section 3, we give an alternative syntax in which variables come labelled with a potential type and show the 'equivalence' between the two formulations. Besides we formulate a general criterion for an algebraic type system to be strongly normalising. In section 4, we prove strong normalisation for those systems satisfying the criterion by a general model construction. Section 5 focuses on the applications of the result to existing systems. The last section contains some final remarks about the work as well as directions for future research. We assume the reader to be reasonably familiar with pure type systems and their basic meta-theory, as presented for example in [15], [4] or [16].

2 Combining higher-order rewriting systems and pure type systems

2.1 Higher-order rewriting systems

In this section, we introduce higher-order rewriting systems. The framework we consider is slightly less general than the one of [3, 12, 19] and has been chosen for clarity of presentation. For examples and applications of the general schema, the reader is refered to [12, 19].

Let Λ be a set. Elements of Λ are called base data[2]. The set of data is defined inductively as follows:

- every base datum is a datum;
- if $\sigma_1, \ldots, \sigma_n$ are data and τ is a base datum, then $(\sigma_1, \ldots, \sigma_n) \to \tau$ is a datum.

For convenience and without loss of generality, we can always assume the type of a function symbol to be of the form $(\sigma_1, \ldots, \sigma_m, \tau_1, \ldots, \tau_n) \to \tau_{n+1}$ where the σ_i's are data of arrow type and the τ_i's are base data. Such data are called *higher-order* data. The set of *first-order data* is the subset of higher-order data for which $m = 0$, i.e. a first-order datum is one of the form $(\tau_1, \ldots, \tau_n) \to \tau_{n+1}$ where the τ_i's are base data. The set of higher-order data is denoted by Λ^*. When there is no risk of confusion, we will simply talk about data.

Definition 1 *A* higher-order signature Σ *over* Λ *consists of an indexed family of (pairwise disjoint) sets* $(\mathcal{F}_w)_{w \in \Lambda^*}$.

[2] Usually elements of Λ are called sorts. We prefer to keep this name for the sorts of the pure type system.

Elements of the $\mathcal{F}_w$'s are called function symbols. A function symbol is first-order if it belongs to $\mathcal{F}_w$ for some first-order datum w and higher-order otherwise. For every datum σ, the set $T_{(\Sigma,\sigma)}$ of terms of datum σ is defined inductively. As usual, we start from a countably infinite set V_σ for each datum σ. The rules are:

- elements of V_σ are terms of datum σ;
- if $x \in V_{(\sigma_1,\ldots,\sigma_n)\to\tau}$ and t_i has datum σ_i for $i = 1,\ldots,n$, then $x\ t_1\ \ldots\ t_n$ has datum τ;
- if $f \in \mathcal{F}_{(\sigma_1,\ldots,\sigma_n)\to\tau}$ and t_i has datum σ_i for $i = 1,\ldots,n$, then $f\ t_1\ \ldots\ t_n$ has datum τ.

A term is *first-order* if all variables occurring in it are of base datum and all function symbols occurring in it are of first-order datum and is *higher-order* otherwise. In other words, first-order terms are of the form $f\ t_1\ \ldots\ t_n$ where f is a first-order function symbol and the t_i's are first-order terms. Higher-order terms are of the form $F\ X_1\ \ldots\ X_m\ t_1\ \ldots\ t_n$ where the X_i's are higher-order variables and the t_i's are terms of base datum. Note that all terms are fully applied in the sense that only variables can be of higher-order datum[3]. The set var of variables of a term, occurences and substitutions are defined as usual.

Definition 2 *A* rewrite rule *is a pair* (s,t) *(written* $s \to t$*) of terms of the same datum such that* $\mathsf{var}(t) \subseteq \mathsf{var}(s)$ *and* s *is not a variable. A rewrite rule is* first-order *if the terms are and* higher-order *otherwise.*

In [19], Jouannaud and Okada define a general schema for higher-order rewrite rules.

Definition 3 ([3, 19]) *A higher-order rewrite rule* $F\ X_1\ \ldots\ X_m\ t_1\ \ldots\ t_n \to v$ *satisfies* the general schema *if*

1. *F is a higher-order function symbol;*
2. *F does not occur in any of the t_i's;*
3. *the higher-order variables occuring in the t_i's belong to $\{X_1,\ldots,X_m\}$;*
4. *for every subterm of v of the form $F\ X'_1\ \ldots\ X'_m\ r_1\ \ldots\ r_n$, one has* $\mathbf{t} \rhd_{mul} \mathbf{r}$ *where* $\rhd_{mul}$ *is the multiset extension of the strict subterm ordering.*

Condition 3 is not essential but ensures that $F\ X_1\ \ldots\ X_m\ t_1\ \ldots, t_n$ is rewritable in the sense of [12]. Note that as a consequence of the definition, F does not occur in any subterm of v of the form $F\ X'_1\ \ldots\ X'_m\ r_1\ \ldots\ r_n$ except in head position. Higher-order rewrite rules are a mild generalisation of the rules of primitive recursion.

Definition 4 *A* higher-order rewriting system *is a set of rewrite rules such that:*

- *first-order rules are non-duplicating*[4]*;*

---

[3] Using fully applied terms is important if one wants to consider type systems with η-reduction, see [6].

[4] Recall that a rewrite rule $s \to t$ is *non-duplicating* if the number of occurences of each variable x in t is lesser or equal to the number of occurences of x in s.

- *higher-order rules satisfy the general schema;*
- *there are no mutually recursive definitions of higher-order function symbols.*

The last requirement is not essential but has been added to simplify proofs. In the sequel, we let $\rightarrow_R$ denote the algebraic reduction relation.

2.2 Algebraic type systems

In this paragraph, we extend the framework of pure type systems with higher-order rewriting *à la* Jouannaud-Okada. The resulting framework of algebraic type systems covers a large class of algebraic-functional languages and provides a suitable basis to study modular properties of these languages.

Definition 5 *An* algebraic type system *(or apts for short) is specified by a quintuple* $\lambda S = (\mathcal{R}, S, \mathsf{sortax}, \mathsf{rules}, \mathsf{datax})$ *where*

- $\mathcal{R}$ *is a finite list of higher-order rewriting systems* $\mathcal{R}_i = (\Lambda_i, \Sigma_i, R_i)$[5] *for* $i = 1, \ldots, n$*;*
- S *is a set of sorts;*
- $\mathsf{sortax} : S \rightharpoonup S$, $\mathsf{rules} : S \times S \rightharpoonup S$ *are partial functions;*
- $\mathsf{datax} : \{\Lambda_1, \ldots, \Lambda_n\} \rightharpoonup S$ *is a total function.*

Note that the definition implicitely requires the algebraic type system to be functional in the sense of [16] (such systems are called singly-sorted in [4]). This is not a real restriction as one can hardly imagine a non-functional pure type system of interest.

Definition 6 *Let* V *be an arbitrary infinite set of variables. The set of pseudo-terms* Pseudo *of an algebraic type system* $\lambda S = (\mathcal{R}, S, \mathsf{sortax}, \mathsf{rules}, \mathsf{datax})$ *is defined as follows:*

- *elements of* V*, sorts and data are pseudo-terms;*
- *if* A, B *are pseudo-terms and* $x \in V$*, then* $A\,B$*,* $\lambda x : A.B$ *and* $\Pi x : A.B$ *are pseudo-terms;*
- *if* f *is a function symbol of some signature* Σ_i *of datum* $(\tau_1, \ldots, \tau_n) \rightarrow \tau$ *and* $t_1, \ldots, t_n$ *are pseudo-terms, then* $f t_1 \cdots t_n$ *is a pseudo-term.*

There are two notions of reduction on pseudo-terms: algebraic reduction $\rightarrow_R$ inherited from the term-rewriting systems and β-reduction. The combined reduction is denoted by $\rightarrow_{mix}$. The rules for derivation for λS are:

---

[5] That is, Λ_i is a set of (base) data, Σ_i is a higher-order signature over Λ_i and $\mathcal{R}_i$ is a higher-order rewriting system over Σ_i.

Axiom	$\overline{\vdash c : s}$	if datax $\Lambda = s$ and $c \in \Lambda$ or sortax $c = s$
Function	$\dfrac{\Gamma \vdash t_i : \sigma_i \quad \text{for } i = 1, \ldots, n}{\Gamma \vdash f\, t_1 \ldots\, t_n : \tau}$	if f is a function symbol with arity $(\sigma_1, \ldots, \sigma_n) \to \tau$
Start	$\dfrac{\Gamma \vdash A : s}{\Gamma, x : A \vdash x : A}$	if $x \notin \Gamma$
Weakening	$\dfrac{\Gamma \vdash t : A \quad \Gamma \vdash B : s}{\Gamma, x : B \vdash t : A}$	if $x \notin \Gamma$
Product	$\dfrac{\Gamma \vdash A : s_1 \quad \Gamma, x : A \vdash B : s_2}{\Gamma \vdash \Pi x : A.B : s_3}$	if rules$(s_1, s_2) = s_3$
Application	$\dfrac{\Gamma \vdash t : \Pi x : A.B \quad \Gamma \vdash u : A}{\Gamma \vdash tu : B[u/x]}$	
Abstraction	$\dfrac{\Gamma, x : A \vdash t : B \quad \Gamma \vdash \Pi x : A.B : s}{\Gamma \vdash \lambda x : A.t : \Pi x : A.B}$	
Exp/Red	$\dfrac{\Gamma \vdash u : A \quad \Gamma \vdash B : s}{\Gamma \vdash u : B}$	if $A \to_{mix} B$ or $B \to_{mix} A$

In an algebraic type system, the reduction relation is not confluent on the set of pseudo-terms; as a result, the usual proofs of subject reduction and of other results relying on subject reduction, such as strengthening cannot be extended. This motivates the following definition (see Section 6 for a longer discussion on subject reduction).

Definition 7 *An algebraic type system* $\lambda S = (\mathcal{R}, S, \text{sortax}, \text{rules}, \text{datax})$ *has the* subject reduction property *if for all pseudo-terms* M, N, A *with* $M \to_\beta N$ *and pseudo-context* Γ,

$$\Gamma \vdash M : A \quad \Rightarrow \quad \Gamma \vdash N : A$$

As subject reduction for R-reduction holds in an arbitrary algebraic type system, it is easy to conclude that in an algebraic type system with the subject reduction property,

$$\Gamma \vdash M : A \quad \Rightarrow \quad \Gamma \vdash N : A$$

for every pseudo-context Γ and all pseudo-terms M, N, A with $M \to_{mix} N$.

Terminology For the sake of exposition, we conclude this paragraph by introducing some terminology.

Definition 8 *An algebraic type system* $\lambda S = (\mathcal{R}, S, \text{sortax}, \text{rules}, \text{datax})$ *is* $\mathcal{R}$-confluent (resp. $\mathcal{R}$-terminating, resp. $\mathcal{R}$-canonical, resp. $\mathcal{R}$-left-linear) *if all its rewriting systems are confluent (resp. terminating, resp. canonical, resp. left-linear).*

In order to name algebraic type systems, it is useful to consider their underlying pure type systems. In the sequel, we will sometimes refer to an algebraic type system $\lambda S = (\mathcal{R}, S, \text{sortax}, \text{rules}, \text{datax})$ as an *algebraic extension* of the pure type system $\lambda S' = (S, \text{sortax}, \text{rules})$.

3 A criterion for strong normalisation

In [25], Terlouw gives a general criterion for a type system to be strongly normalising. We adapt his criterion to algebraic type systems and give an equivalent criterion in terms of algebraic type systems with labelled variables. The advantage of the second characterisation is that it eliminates the need to reason on contexts.

3.1 Stratified algebraic type systems

Recall that an *environment* is a family $\Gamma = (x_i : A_i)_{i\in\mathbf{N}}$ where for every i, x_i is a variable and A_i is a pseudo-term such that for some sort s_{i+1}, $x_0 : A_0, \ldots, x_i : A_i \vdash A_{i+1} : s_{i+1}$.

Definition 9 *Let $\Gamma = (x_i : A_i)_{i\in\mathbf{N}}$ be an environment.*

- *A pseudo-term M is a* prototype *w.r.t Γ if there exists a natural i, a sort s and pseudo-terms $P_1, \ldots, P_n$ such that $x_0 : A_0, \ldots, x_i : A_i \vdash M\ P_1\ \ldots\ P_n : s$.*
- *The relation $\prec_\Gamma$ on pseudo-terms is defined as the smallest relation such that for all $M, N \in$ Pseudo, if MN is a prototype w.r.t. Γ, then $N \prec M$ and $MN \prec M$.*
- *An algebraic type system is* stratified *if the relation $\prec_\Gamma$ is well-founded for every environment Γ.*

The main result of the paper is the following criterion for strong normalisation.

Theorem 10 *Every stratified $\mathcal{R}$-terminating algebraic type system with the subject reduction property is strongly normalising.*

As a corollary, we recover the standard results on strong normalisation of algebraic type systems as well as some new results.

Corollary 11
- *$\mathcal{R}$-terminating extensions of systems of the λ-cube are strongly normalising ([3, 19]).*
- *$\mathcal{R}$-terminating extensions of higher-order logic are strongly normalising.*
- *$\mathcal{R}$-canonical and $\mathcal{R}$-left-linear extensions of the algebraic calculus of constructions with universes are strongly normalising.*

Note that for the first result, we use the fact that algebraic extensions of systems of the λ-cube have subject reduction ([3]). For the third result, note that left-linearity of $\mathcal{R}$ (i.e. variables may only occur once in a left hand side of a rewrite rule) is a real restriction. However, there are interesting examples of higher-order rewrite rules that are left-linear, e.g.

$$\text{Maplist}\ X\ \text{nil} \rightarrow_R \text{nil},$$
$$\text{Maplist}\ X\ (\text{cons}\ al) \rightarrow_R \text{cons}\ (Xa)(\text{Maplist}\ Xl).$$

The restriction to left-linearity is made, because if $\mathcal{R}$ is left-linear, then $\rightarrow_{mix}$ is confluent, hence we have the subject reduction property for the system and hence Theorem 10 applies.

3.2 Labelled variables

In this section, we introduce a technical variant of (algebraic) type systems in which variables are "typed". This is reminiscent of some presentations of simply typed λ-calculus in which each type τ comes equipped with a set of variables of type τ. In algebraic type systems, terms and types are defined simultaneously so the naive approach taken for simply typed λ-calculus cannot be used any longer. Our solution is to assign to every variable a pseudo-term, which will be its unique type if the variable is well-typed. In the sequel, we consider a fixed algebraic type system $\lambda S = (\mathcal{R}, S, \mathsf{sortax}, \mathsf{rules}, \mathsf{datax})$; as usual, its set of pseudo-terms is denoted by T.

Definition 12 *A* variable labelling *is a map* $\epsilon : V \to T$ *such that the set* $\{x \in V | \epsilon x = t\}$ *is infinite for every* $t \in T$.

Of course, such maps always exist if V is sufficiently large (the cardinal of V is determined by the cardinal of S). One nice aspect of variable labelling is that it eliminates the need to manipulate contexts. In the sequel, we assume we are given a fixed labelling ϵ. We can define a notion of derivation w.r.t. ϵ; the rules are

Axiom	$\dfrac{}{\vdash_\epsilon c : s}$	if $\mathsf{datax}\ \Lambda = s$ and $c \in \Lambda$ or $\mathsf{sortax}\ c = s$
Function	$\dfrac{\vdash_\epsilon t_i : \sigma_i \quad \text{for } i = 1, \ldots, n}{\vdash_\epsilon f\ t_1\ \ldots\ t_n : \tau}$	if f is a function symbol with arity $(\sigma_1, \ldots, \sigma_n) \to \tau$
Start	$\dfrac{\vdash_\epsilon A : s}{\vdash_\epsilon x : A}$	if $\epsilon x = A$ and x is fresh in A
Product	$\dfrac{\vdash_\epsilon A : s_1 \quad \vdash_\epsilon B : s_2}{\vdash_\epsilon \Pi x : A.B : s_3}$	if $\mathsf{rules}(s_1, s_2) = s_3$ and $\epsilon x = A$
Application	$\dfrac{\vdash_\epsilon t : \Pi x : A.B \quad \vdash_\epsilon u : A}{\vdash_\epsilon tu : B[u/x]}$	
Abstraction	$\dfrac{\vdash_\epsilon t : B \quad \vdash_\epsilon \Pi x : A.B : s}{\vdash_\epsilon \lambda x : A.t : \Pi x : A.B}$	
Conversion	$\dfrac{\vdash_\epsilon u : A \quad \vdash_\epsilon B : s}{\vdash_\epsilon u : B}$	if $A \to_{mix} B$ or $B \to_{mix} A$

It is not difficult to check that algebraic type systems with variable labelling are essentially equivalent to algebraic type systems for systems with subject reduction.

Proposition 13 *Assume the algebraic type system has subject reduction.*

- *If* $\vdash_\epsilon M : A$, *then* $\Gamma \vdash M : A$ *for some context* Γ.
- *If* $\Gamma \vdash M : A$, *then* $\vdash_\epsilon \rho M : \rho A$ *for some variable renaming* ρ.

Proof sketch: The proof of the second part is by first renaming the bound and free variables in Γ, M and A in such a way that, if $x : B$ occurs in Γ, M or

A, then $\epsilon x = B$. The statement we obtain, say $\Gamma' \vdash M' : A'$, is still derivable. Now one proves $\Gamma' \vdash M' : A' \Rightarrow \vdash_\epsilon M' : A'$ by induction on the derivation. For the proof of the first part, we first have to prove that, if subject reduction holds, then we have ***strengthening***, which is the following property

$$\text{If } \Gamma, x : A, \Delta \vdash M : B,\ x \notin \mathsf{FV}(\Delta, M, B), \text{ then } \Gamma, \Delta \vdash M : B.$$

Using the fact that the underlying pure type system is functional, we can use the 'standard' proof (e.g. in [16]) prove strengthening. Using strengthening, one also proves a ***permutation property***, which states the following.

$$\text{If } \Gamma, x : A, y : C, \Delta \vdash M : B,\ x \notin \mathsf{FV}(C), \text{ then } \Gamma, y : C, x : A, \Delta \vdash M : B.$$

Now, the following slight extension of the first result above can be proved by induction on the derivation (using strengthening and the permutation property).

If $\vdash_\epsilon M : A$, then $\Gamma \vdash M : A$

for all Γ such that Γ ***respects*** ϵ and $\mathsf{dom}(\Gamma) = \mathsf{FV}(M, A)$.

Here, $\Gamma(= x_1 : C_1, \ldots, x_n : C_n)$ ***respects*** ϵ means that $\epsilon x_i = C_i$ and $x_{i+1} \notin \mathsf{FV}(C_1, \ldots, C_i)$ for all i. Furthermore, $\mathsf{dom}(\Gamma)$ denotes the set $\{x_1, \ldots, x_n\}$.□

It follows that strong normalisation and subject reduction of the system with labelled variables (or ***labelled system*** for short) is equivalent to strong normalisation and subject reduction of the original system. Besides, one can reformulate the criterion for systems with labelled variables.

Definition 14 ***Let λS be an algebraic type system with a variable labelling ϵ. A*** **prototype** ***is a pseudo-term M for which there exist $N_1, \ldots, N_p \in$*** **Pseudo** ***and $s \in S$ such that***

$$\vdash_\epsilon M\ N_1\ \ldots\ N_p : s$$

The set of prototypes is denoted by Proto. As before, we consider the relation $\prec$ defined as the smallest relation such that

$$\forall M, N \in \mathsf{Pseudo}[(M\ N) \in \mathsf{Proto} \Rightarrow N \prec M\ \wedge\ (M\ N) \prec M]$$

Definition 15 ***λS is*** **stratified** ***if the relation $\prec$ is well-founded.***

Theorem 10 can now be rephrased as:

Theorem 16 ***Every $\mathcal{R}$-terminating stratified labelled type system with the subject reduction property is strongly normalising.***

Theorem 10 follows easily from Theorem 16.

4 The proof of the main theorem

In this section, we prove Theorem 16. The proof is divided in two parts: in the first part, we prove that algebraic reduction is strongly normalising on legal terms. In the second part, we give a model-construction for stratified algebraic type systems. Strong normalisation is derived easily from the model construction.

4.1 Strong normalisation of algebraic reduction

Strong normalisation of algebraic reduction on legal terms is established directly by advocating modularity results from [13] for example.

Proposition 17 *$\rightarrow_R$ is strongly normalising on legal terms.*

Proof: the technique is inspired from [5] and consists of viewing λ-calculus as an algebraic signature. In this way, we define for every $\mathcal{R}$-algebraic type system $\lambda S = (\mathcal{R}, S, \mathsf{sortax}, \mathsf{rules}, \mathsf{datax})$ an algebraic signature $\Sigma_{\lambda S}$ extending the signatures of the rewrite systems and upon which algebraic reduction is terminating. Then we show that all legal terms can be obtained from the terms of $\Sigma_{\lambda S}$ by an erasure map $\lceil . \rceil$ which reflects reduction. Strong normalisation of algebraic reduction on legal terms follows easily. In the sequel, we consider a finite sequence of terminating higher-order rewriting systems $\mathcal{R}_i = (\Lambda_i, \Sigma_i, R_i)$ for $i = 1, \ldots, n$. Let $\Lambda = \bigcup_{i=1,\ldots,n} \Lambda_i$ and let $\Sigma_{\lambda S} = (\bigcup_{i=1,\ldots,n} \Sigma_i) \cup \Sigma_0$ where Σ_0 is the signature with function symbols:

- $\overline{s}_\tau : \tau$ for $s \in S$ and τ a datum,
- $\overline{\sigma}_\tau : \tau$ for σ, τ data,
- $\overline{\Pi}_{x,\tau_1,\tau_2,\tau_3}$, $\overline{\lambda}_{x,\tau_1,\tau_2,\tau_3} : \tau_1 \times \tau_2 \rightarrow \tau_3$ for every variable x and τ_1, τ_2, τ_3 data,
- $\mathsf{Appl}_{\tau_1,\tau_2,\tau_3} : \tau_1 \times \tau_2 \rightarrow \tau_3$ for every τ_1, τ_2, τ_3 data.

The union R_0 of the R_i's is a higher-order rewriting system over $\Sigma_{\lambda S}$. By hypothesis, its first-order reduction relation is terminating, so R_0 is terminating (see [13]).

To conclude the proof of the Proposition, first note that by subject reduction for $\rightarrow_R$, we only need to prove that there is no infinite reduction through legal terms. To this end, we define a map from the terms of $\Sigma_{\lambda S}$ to pseudo-terms. For the sake of simplicity, we assume that the set of variables for every sort τ is $\{x^\tau \mid x \in V\}$. The map $\lceil . \rceil$ is defined as follows:

$$\begin{aligned}
\lceil x^\tau \rceil &= x \\
\lceil f(t_1, \ldots, t_n) \rceil &= f \lceil t_1 \rceil \cdots \lceil t_n \rceil \\
\lceil \overline{\Pi}_{x,\tau_1,\tau_2,\tau_3}(t_1, t_2) \rceil &= \Pi x : \lceil t_1 \rceil . \lceil t_2 \rceil \\
\lceil \overline{\lambda}_{x,\tau_1,\tau_2,\tau_3}(t_1, t_2) \rceil &= \lambda x : \lceil t_1 \rceil . \lceil t_2 \rceil \\
\lceil \mathsf{Appl}_{\tau_1,\tau_2,\tau_3}(t_1, t_2) \rceil &= \lceil t_1 \rceil \ \lceil t_2 \rceil
\end{aligned}$$

The map is surjective on the set of legal terms. Moreover, every infinite R-reduction sequence on legal terms can be lifted to an infinite R_0-reduction sequence on the terms of $\Sigma_{\lambda S}$.□

4.2 The model construction

In this section, we present a model construction for stratified algebraic type systems with the subject reduction property. The construction is based on saturated sets and is a generalisation of strong normalisation proofs for pure type systems, such as the polymorphic λ-calculus ([18, 24, 14]) or the calculus of constructions ([17, 25]). The model is heavily inspired by [25]. Before giving a proof of Theorem 16, we need some preliminaries on saturated sets.

Saturated sets Traditionally, saturated sets are defined as sets of β-strongly normalisable untyped λ-terms. Here we consider a slightly different notion of saturated sets, more adapted to our framework: we define saturated sets as sets of pseudo-terms rather than as sets of λ-terms. This is not really important but makes the proof slightly more elegant. Moreover, we consider typed saturated sets as in [20, 25] rather than untyped saturated sets. This means that the notion of saturated sets is defined relative to a set of pseudo-terms. This is not important for pure type systems but turns out to be crucial for algebraic type systems (otherwise, we cannot use the results of the principal case).

Recall that a pseudo-term M is ***strongly normalising*** if all reduction sequences starting from M are finite. The set of strongly normalising terms is denoted by SN. Saturated sets will be defined as subsets of SN with certain closure properties.

Definition 18 *A* base term *is a term of the form $x\ P_1\ \ldots\ P_n$ where $x \in V$ and $P_1, \ldots, P_n \in \mathsf{SN}$.*

The set of base terms is denoted by Base. Note that all base terms are strongly normalising.

Definition 19 Key-reduction $\rightarrow_k$ *is the smallest relation on pseudo-terms such that for all pseudo-terms $M, N, O, P_1, \ldots, P_n$*

$$(\lambda x : M.N)\ O\ P_1\ \ldots\ P_n \rightarrow_k N[O/x]\ P_1\ \ldots\ P_n$$

Note that a term has at most one key-redex. The term obtained from M by contracting its key redex is denoted by $\mathbf{kred}(M)$.

Definition 20 *Let $U \subseteq \mathsf{Pseudo}$. A set X of pseudoterms is* saturated *in U if*

(i) $X \subseteq \mathsf{SN} \cap U$;
(ii) $\mathsf{Base} \cap U \subseteq X$;
(iii) If $\mathbf{kred}(M) \in X$ and $M \in \mathsf{SN} \cap U$, then $M \in X$.

*The collection of all saturated sets in U is denoted by $SAT(U)$. For $M \in \mathsf{Pseudo}$, we use $SAT(M)$ to denote the set of saturated sets in $\{N \in \mathsf{Pseudo} \mid \vdash_\epsilon N : M\}$. If $X \in \mathsf{SAT}(M)$, we say X is a M-*saturated set.

We list some closure properties of saturated sets.

Fact 21 *Let $U, U' \subseteq \mathsf{Pseudo}$.*

- $\mathsf{SN}(U) = \mathsf{SN} \cap U$ *is a saturated set in U.*
- *The set of saturated sets in U is closed under arbitrary non-empty intersections.*
- *If X is saturated in U and Y is saturated in U', then $X \rightarrow Y$ defined by*

$$X \rightarrow Y = \{M \in W | \forall N \in X. M\ N \in Y\}$$

is saturated in W provided that $\mathsf{Base} \cap W \subset X \rightarrow Y$ (i.e. for every $w \in \mathsf{Base} \cap W$ and $x \in X$, $wx \in Y$).

- *If X is saturated in U and Y_x is saturated in U'_x for $x \in X$, then $\Pi x \in X.Y_x$ defined by*

$$\Pi x \in X.Y_x = \{M \in W | \forall N \in X.M\ N \in Y_N\}$$

is saturated in W provided $\mathsf{Base} \cap W \subset \Pi x \in X.Y_x$ (i.e. for every $w \in \mathsf{Base} \cap W$ and $x \in X$, $wx \in Y_x$).

If $M \in \mathsf{Pseudo}$, then $\mathsf{SN}(M)$ is the saturated set of strongly normalising terms of type M.

The principal case The key fact in the model construction for algebraic type systems is that the sets of strongly normalising terms of base datum enjoy suitable closure properties.

Proposition 22 *Let f be a function symbol of datum $(\sigma_1, \ldots, \sigma_n) \to \tau$. Then for all pseudo-terms $t_1, \ldots, t_n$,*

$$t_i \in \mathsf{SN}(\sigma_i) \quad \text{for } i = 1, \ldots, n \quad \Rightarrow \quad f t_1 \cdots t_n \in \mathsf{SN}(\tau)$$

The proof is an adaptation of [19, 2]. This key fact ensures that the model construction for algebraic type systems can be carried out in exactly the same way as for pure type systems.

Intuition behind the proof The idea of the proof is to give a model construction in which types are interpreted as (saturated) sets and legal terms as pseudo-terms such that the following soundness condition is satisfied:

$$\vdash_\epsilon M : A \quad \Rightarrow \quad (\![M]\!) \in \langle\!\langle A \rangle\!\rangle$$

where $\langle\!\langle A \rangle\!\rangle$ is the saturated set interpretation of A and $(\![M]\!)$ is the pseudo-term interpretation of M. For simple systems, such as the (algebraic) simply typed λ-calculus $\lambda_\to$, the definition of $\langle\!\langle A \rangle\!\rangle$ can be given inductively on the structure of A and the soundness condition can be proved by induction on the derivation. For the polymorphic λ-calculus $\lambda 2$, one is forced to parameterise interpretations by valuations. One then has to prove that if a valuation ρ satisfies certain properties, then

$$\vdash_\epsilon M : A \quad \Rightarrow \quad (\![M]\!)_\rho \in \langle\!\langle A \rangle\!\rangle_\rho$$

In a system with dependent types such as λP or $\lambda P\omega$, terms might occur in types so one cannot any longer define $\langle\!\langle A \rangle\!\rangle$ by induction on A. The standard solution is to define $\langle\!\langle A \rangle\!\rangle$ as a partial interpretation and show that it is well-defined on legal types. This requires the introduction of a new interpretation $a(M)$ which assigns to a term its possible values. (In this context valuations are of the form (ρ, ζ) where ρ assigns to every variable (in some domain) a pseudo-term and ζ assigns to every variable (in some domain) a saturated set.) The idea is that $a(M)$ should be defined for every type M and be a set of saturated sets such that under suitable conditions

$$\vdash_\epsilon M : s \quad \Rightarrow \quad \langle\!\langle M \rangle\!\rangle_{\rho,\zeta} \in a(M)$$

Here we see that dependent types introduce a new difficulty: we have indexed families of types, i.e. terms of type $B \to *$[6]. These terms, which we have defined earlier as prototypes, will also need to be intepreted. To interpret them as families of types, we use induction on their structure: if M is of type $B \to C \to *$, we want to define $a(M)$ as the set of families of maps $\{(f_b)_{b:B} \mid f_b : a(b) \to a(M\ b)\}$. This requires $a(b)$ and $a(M\ b)$ to be already defined. This requirement matches exactly the definition of $\prec$: the assumption that $\prec$ is well-founded enables us to define the interpretation $a(M)$ by $\prec$-induction. The other two interpretations will be defined as usual by induction on the structure of the terms.

Convention From now on, we drop the subscript in $\vdash_\epsilon$.

The construction The set Data of data is defined as the union of the set of base data of the rewriting systems. The set Type of types is defined by

$$\mathsf{Type} = \{M \in \mathsf{Pseudo} \mid\ \vdash M : s \text{ for some } s \in S\}$$

Definition 23 *The map* a : Pseudo $\to$ *Set is defined by case distinction as follows.*

- *if* $M \in \mathsf{Type} \setminus \mathsf{Data}$, $a(M) = \mathsf{SAT}(M)$;
- *if* $M \in \mathsf{Proto}$, $a(M) = \{(f_B)_{B\in\mathsf{cone}(M)} | f_B : a(B) \to a(M\ B)\}$;
- *if* $M \in \mathsf{Data}$, $a(M) = \{\mathsf{SN}(M)\}$;
- *otherwise*, $a(M) = \{\{\emptyset\}\}$;

where $\mathsf{cone}(M) = \{B \in \mathsf{Pseudo} | (M\ B) \in \mathsf{Proto}\}$. *Define* $\mathbf{A} = \bigcup_{M\in\mathsf{Pseudo}} a(M)$.

Definition 24 *A* valuation *is a pair* (ρ, ζ) *such that* $\rho : V \to$ Pseudo *and* $\zeta : V \to \mathbf{A}$.

The extension $(\![.]\!)_\rho$: Pseudo $\to$ Pseudo of ρ is defined as the unique capture-avoiding substitution extending ρ.

Definition 25 *The map* ζ *is extended to terms by defining a map* $\langle\!\langle . \rangle\!\rangle_{\rho\zeta}$: Pseudo $\to \mathbf{A}$ *as follows.*

$$\begin{array}{rll}
\langle\!\langle x \rangle\!\rangle_{\rho\zeta} &= \zeta(x) & \text{if } x \in V,\ \rho(x) \in \mathsf{Proto} \\
\langle\!\langle \Pi x : A.B \rangle\!\rangle_{\rho\zeta} &= \{P \in \mathsf{Pseudo} | \forall (N,Q) \in \mathcal{E}_{\rho\zeta}(A). & \\
& \qquad PN \in \langle\!\langle B \rangle\!\rangle_{\rho(x:=N),\zeta(x:=Q)}\} & \text{if } (\![\Pi x : A.B]\!)_\rho \in \mathsf{Type} \\
\langle\!\langle M\ N \rangle\!\rangle_{\rho\zeta} &= (\langle\!\langle M \rangle\!\rangle_{\rho\zeta})_{(\![N]\!)_\rho}\ \langle\!\langle N \rangle\!\rangle_{\rho\zeta} & \text{if } (\![MN]\!)_\rho \in \mathsf{Proto} \\
\langle\!\langle \lambda x : A.b \rangle\!\rangle_{\rho\zeta} &= (\boldsymbol{\lambda} c \in a(B).\langle\!\langle b \rangle\!\rangle_{\rho(x:=B),\zeta(x:=c)})_{B\in\mathsf{cone}((\![\lambda x:A.b]\!)_\rho)} & \text{if } (\![\lambda x : A.b]\!)_\rho \in \mathsf{Proto} \\
\langle\!\langle M \rangle\!\rangle_{\rho\zeta} &= \mathsf{SN(M)} & \text{if } M \in \mathsf{Data} \\
\langle\!\langle M \rangle\!\rangle_{\rho\zeta} &= \{\emptyset\} & \text{otherwise}
\end{array}$$

where for every $M \in$ Pseudo,

$$\mathcal{E}_{\rho\zeta}(M) = \{(N,Q) \in \mathsf{Pseudo} \times \mathbf{A} \mid\ \vdash N : (\![M]\!)_\rho,\ N \in \langle\!\langle M \rangle\!\rangle_{\rho\zeta},\ Q \in a(N)\}$$

[6] This is not only true for dependent types but also for higher-order polymorphism as it occurs in $\lambda\omega$.

The following lemma is easily established by induction on the structure of M.

Lemma 26 *Let* $M, N \in$ Pseudo. *Let* (ρ, ζ) *and* (ρ', ζ') *be two valuations.*

- *If* $\rho x = \rho' x$ *and* $\zeta x = \zeta' x$ *for every* $x \in \mathsf{FV}(M)$, *then* $\langle\!\langle M \rangle\!\rangle_{\rho\zeta} = \langle\!\langle M \rangle\!\rangle_{\rho'\zeta'}$.
- $\langle\!\langle M[N/x] \rangle\!\rangle_{\rho\zeta} = \langle\!\langle M \rangle\!\rangle_{\rho(x:=(\![N]\!)_\rho),\zeta(x:=\langle\!\langle N \rangle\!\rangle_{\rho\zeta})}$

As a consequence of Lemma 26 and of the subject reduction property, we conclude that $\langle\!\langle . \rangle\!\rangle_{\rho\zeta}$ is invariant under reduction on legal terms.

Corollary 27 *For every valuation* (ρ, ζ) *and terms* M, N *such that* $M \rightarrow_{mix} N$ *and* $(\![M]\!)_\rho, (\![N]\!)_\rho \in$ Proto, *we have* $\langle\!\langle M \rangle\!\rangle_{\rho\zeta} = \langle\!\langle N \rangle\!\rangle_{\rho\zeta}$.

In order to prove the main theorem, we must establish that the model behaves as expected. It requires a standard soundness argument. In the sequel, we call a context a finite list of variables $\Delta = y_1, \ldots, y_n$ such that for $i = 1, \ldots, n$, $y_i \notin \mathsf{FV}(\epsilon y_j)$ $(\forall j \leq i)$. One can check that for every term M, $\mathsf{FV}(M)$ can be ordered into a context.

Definition 28 *Let* Δ *be a context. A valuation* (ρ, ζ) satisfies Δ *(notation* $(\rho, \zeta) \models \Delta$*) if for every* $x \in \Delta$,

(i) $\vdash \rho x : (\![\epsilon x]\!)_\rho$,
(ii) $\rho x \in \langle\!\langle \epsilon x \rangle\!\rangle_{\rho\zeta}$,
(iii) $\langle\!\langle x \rangle\!\rangle_{\rho\zeta} \in a((\![x]\!)_\rho)$.

We say that $\models M : A$ *if for every valuation* (ρ, ζ) *satisfying* $\mathsf{FV}(M) \cup \mathsf{FV}(A)$,

(i) $\vdash (\![M]\!)_\rho : (\![A]\!)_\rho$,
(ii) $(\![M]\!)_\rho \in \langle\!\langle A \rangle\!\rangle_{\rho\zeta}$,
(iii) $\langle\!\langle M \rangle\!\rangle_{\rho\zeta} \in a((\![M]\!)_\rho)$,

Fact 29 *Let* (ρ, ζ) *be a valuation satisfying* Δ. *Let* $x \notin \Delta$ *and* $x \notin \mathsf{FV}(\epsilon y)$ *for all* $y \in \Delta$. *Then for every* $C \in a(x)$, $\rho(x := x), \zeta(x := C)$ *satisfies* $\Delta \cup \{x\}$.

As $a(x) \neq \emptyset$, valuations can always be extended to a larger context while preserving satisfaction. We can now prove the main technical result of this paper.

Proposition 30 (Soundness) $\vdash M : A \quad \Rightarrow \quad \models M : A$.

Proof: by induction on the length of derivations.

- *Axiom:* if $\vdash s_1 : s_2$ is an axiom, then it is easy to show $\models s_1 : s_2$.
- *Start:* assume $\vdash x : A$ is deduced from $\vdash A : s$ by a start rule. Then $\epsilon x = A$. Assume (ρ, ζ) satisifies $\mathsf{FV}(A) \cup \{x\}$. By definition of satisfaction, $\vdash \rho x : (\![A]\!)_\rho$, $\rho x \in \langle\!\langle A \rangle\!\rangle_{\rho\zeta}$ and $\langle\!\langle x \rangle\!\rangle_{\rho\zeta} \in a(\rho x)$, so we are done.

- *Function symbol:* assume $\vdash f t_1 \cdots t_n : \tau$ is deduced by a function rule from $\vdash t_i : \sigma_i$ for $i = 1, \ldots, n$ where f is a function symbol of datum $(\sigma_1, \ldots, \sigma_n) \to \tau$. Assume $(\rho, \zeta) \models \mathsf{FV}(f t_1 \cdots t_n)$.
  $\vdash (\!| f t_1 \cdots t_n |\!)_\rho : \tau$ follows immediately from the induction hypothesis.
  Next one has to prove that $(\!| f t_1 \cdots t_n |\!)_\rho \in \langle\!\langle \tau \rangle\!\rangle_{\rho\zeta}$. This is an immediate consequence of Lemma 22.
  Finally, we need to prove $\langle\!\langle f t_1 \cdots t_n \rangle\!\rangle_{\rho\zeta} \in a((\!| f t_1 \cdots t_n |\!)_\rho)$. This is easy because $(\!| f t_1 \cdots t_n |\!)_\rho \notin \mathsf{Proto}$.
- *Product:* assume $\vdash \Pi x : A.B : s_3$ is deduced by a formation rule from $\vdash A : s_1$ and $\vdash B : s_2$. Let (ρ, ζ) be a valuation such that$(\rho, \zeta) \models \mathsf{FV}(\Pi x : A.B)$.
  We prove $\vdash (\!| \Pi x : A.B |\!)_\rho : s_3$. By induction hypothesis, $\vdash (\!| A |\!)_\rho : s_1$. By fact 29,
  $$\rho(x := x), \zeta(x := C) \models \mathsf{FV}(\Pi x : A.B) \cup \{x\}$$
  for every $C \in a(x)$. Hence $\vdash (\!| B |\!)_{\rho,(x:=x)} : s_2$ by induction hypothesis. By the product rule, $\vdash \Pi x : (\!| A |\!)_\rho . (\!| B |\!)_{\rho,(x:=x)} : s_3$. As $\Pi x : (\!| A |\!)_\rho . (\!| B |\!)_{\rho,(x:=x)} = (\!| \Pi x : A.B |\!)_\rho$, we conclude (i) holds.
  Next we show $(\!| \Pi x : A.B |\!)_\rho \in \langle\!\langle s_3 \rangle\!\rangle_{\rho\zeta}$. By definition of $\langle\!\langle . \rangle\!\rangle_{\rho\zeta}$, it is equivalent to show that $(\!| \Pi x : A.B |\!)_\rho$ is strongly normalising (we already know that (i) holds). By induction hypothesis, $(\!| A |\!)_\rho \in \langle\!\langle s_1 \rangle\!\rangle_{\rho\zeta} \subseteq \mathsf{SN}$ and $(\!| B |\!)_{\rho'} \in \langle\!\langle s_2 \rangle\!\rangle_{\rho'\zeta'} \subseteq \mathsf{SN}$ for every valuation (ρ', ζ') satisfying $\mathsf{FV}(B)$. Let $C \in a(x)$. Then $\rho(x := x), \zeta(x := C) \models \mathsf{FV}(\Pi x : A.B) \cup \{x\}$. Hence $(\!| B |\!)_{\rho(x:=x)} \in \mathsf{SN}$ and $(\!| \Pi x : A.B |\!)_\rho \in \mathsf{SN}$.
  Finally, we show $\langle\!\langle \Pi x : A.B \rangle\!\rangle_{\rho\zeta} \in a((\!| \Pi x : A.B |\!)_\rho)$. By (i), we know that $(\!| \Pi x : A.B |\!)_\rho \in \mathsf{Type}$, so we have to prove that $\langle\!\langle \Pi x : A.B \rangle\!\rangle_{\rho\zeta}$ is a $(\!| \Pi x : A.B |\!)_\rho$-saturated set. As $(\!| A |\!)_\rho$ is a type, it follows by induction hypothesis that $\langle\!\langle A \rangle\!\rangle_{\rho\zeta}$ is a $(\!| A |\!)_\rho$-saturated set. Besides, $(\!| B |\!)_{\rho(x:=x)}$ is a type and by the substitution lemma, $(\!| B |\!)_{\rho(x:=N)}$ is a type whenever $\vdash N : \epsilon x$. Hence $\langle\!\langle B \rangle\!\rangle_{\rho(x:=N),\zeta(x:=Q)}$ is a $(\!| B |\!)_{\rho(x:=N)}$-saturated set whenever $\rho(x := N), \zeta(x := Q) \models \mathsf{FV}(B)$ (equivalently for every $(N, Q) \in \mathcal{E}_{\rho\zeta}(A)$). We conclude $\langle\!\langle \Pi x : A.B \rangle\!\rangle_{\rho\zeta}$ is a $(\!| \Pi x : A.B |\!)_\rho$-saturated set.
- *Application:* assume $\vdash M\,N : B[N/x]$ is deduced from $\vdash M : \Pi x : A.B$ and $\vdash N : A$ by an application rule. Let (ρ, ζ) be a valuation satisfying $\mathsf{FV}(M) \cup \mathsf{FV}(B[N/x])$.
  First, we show that $\vdash (\!| MN |\!)_\rho : (\!| B[N/x] |\!)_\rho$. Consider the valuation (ρ', ζ') defined by
  $$\rho' y = \begin{cases} \rho y \text{ if } y \in \mathsf{FV}(M) \cup \mathsf{FV}(B[N/x]) \\ y \;\; \text{otherwise} \end{cases}$$
  and
  $$\zeta' y = \begin{cases} \zeta y \text{ if } y \in \mathsf{FV}(M) \cup \mathsf{FV}(B[N/x]) \\ C_y \text{ otherwise} \end{cases}$$
  where C_y is an arbitrary element of $a(y)$. Then
  $$(\rho', \zeta') \models \mathsf{FV}(MN) \cup \mathsf{FV}(\Pi x : A.B)$$
  By induction hypothesis, we have

- $\vdash (\![M]\!)_{\rho'} : (\![\Pi x : A.B]\!)_{\rho'}$;
- $\vdash (\![N]\!)_{\rho'} : (\![A]\!)_{\rho'}$.

Hence $\vdash (\![MN]\!)_{\rho'} : (\![B]\!)_{\rho',(x:=x)}[(\![N]\!)_{\rho'}/x]$. In other words, $\vdash (\![MN]\!)_{\rho'} : (\![B[N/x]]\!)_{\rho'}$. As ρ and ρ' coincide on $\mathsf{FV}(M) \cup \mathsf{FV}(B[N/x])$, we conclude that (i) holds.

Next, we show that $(\![MN]\!)_\rho \in \langle\!\langle B[N/x]\rangle\!\rangle_{\rho\zeta}$. Note that it is equivalent to show $(\![MN]\!)_{\rho'} \in \langle\!\langle B[N/x]\rangle\!\rangle_{\rho'\zeta'}$ where (ρ', ζ') is defined as above. By induction hypothesis, we know that $\vdash (\![N]\!)_{\rho'} : (\![A]\!)_{\rho'}$, $(\![N]\!)_{\rho'} \in \langle\!\langle A\rangle\!\rangle_{\rho'\zeta'}$ and $\langle\!\langle N\rangle\!\rangle_{\rho'\zeta'} \in a((\![N]\!)_{\rho'})$. Hence, $((\![N]\!)_{\rho'}, \langle\!\langle N\rangle\!\rangle_{\rho'\zeta'}) \in \mathcal{E}_{\rho'\zeta'}(A)$. By induction hypothesis, $(\![M]\!)_{\rho'} \in \langle\!\langle \Pi x : A.B\rangle\!\rangle_{\rho'\zeta'}$. Hence

$$(\![MN]\!)_{\rho'} \in \langle\!\langle B\rangle\!\rangle_{\rho'(x:=(\![N]\!)_{\rho'}),\zeta'(x:=\langle\!\langle N\rangle\!\rangle_{\rho'\zeta'})}$$

By Lemma 26, $\langle\!\langle B[N/x]\rangle\!\rangle_{\rho'\zeta'} = \langle\!\langle B\rangle\!\rangle_{\rho'(x:=(\![N]\!)_{\rho'}),\zeta'(x:=\langle\!\langle N\rangle\!\rangle_{\rho'\zeta'})}$. So we are done.

Finally, we prove that $\langle\!\langle MN\rangle\!\rangle_{\rho\zeta} \in a((\![MN]\!)_\rho)$. There are two cases two distinguish. If $(\![MN]\!)_\rho \notin$ Proto, then $a((\![MN]\!)_\rho) = \{\{\emptyset\}\}$ and $\langle\!\langle MN\rangle\!\rangle_{\rho\zeta} = \{\emptyset\}$, so we are done. Otherwise, $(\![M]\!)_\rho \in$ Proto. By induction hypothesis, $\langle\!\langle M\rangle\!\rangle_{\rho\zeta} \in a((\![M]\!)_\rho)$ and $\langle\!\langle N\rangle\!\rangle_{\rho\zeta} \in a((\![N]\!)_\rho)$. Hence $(\langle\!\langle M\rangle\!\rangle_{\rho\zeta})_{(\![N]\!)_\rho}\langle\!\langle N\rangle\!\rangle_{\rho\zeta} \in a((\![MN]\!)_\rho)$.

- *abstraction:* assume $\vdash \lambda x : A.t : \Pi x : A.B$ is deduced by an abstraction rule from $\vdash t : B$ and $\vdash \Pi x : A.B : s$. Let (ρ, ζ) be a valuation satisfying $\mathsf{FV}(\lambda x : A.t) \cup \mathsf{FV}(\Pi x : A.B)$. We prove $\vdash (\![\lambda x : A.t]\!)_\rho : (\![\Pi x : A.B]\!)_\rho$.
  By induction hypothesis, $\vdash (\![\Pi x : A.B]\!)_\rho : s$. By Fact 29 we find that $\rho(x := x), \zeta(x := C) \models \mathsf{FV}(t)$ for every $C \in a(x)$. Hence $\vdash (\![t]\!)_{\rho(x:=x)} : (\![A]\!)_{\rho(x:=x)}$. As x is not free in A, we have $(\![A]\!)_{\rho(x:=x)} = (\![A]\!)_\rho$. We can apply the abstraction rule to conclude.
  Next we prove that $(\![\lambda x : A.t]\!)_\rho \in \langle\!\langle \Pi x : A.B\rangle\!\rangle_{\rho\zeta}$. This amounts to showing that for every $(N, Q) \in \mathcal{E}_{\rho\zeta}(A)$, we have

$$(\![\lambda x : A.t]\!)_\rho\ N \in \langle\!\langle B\rangle\!\rangle_{\rho(x:=N),\zeta(x:=Q)}$$

  By definition of saturated sets, this follows from

$$(\![t]\!)_{\rho(x:=N)} \in \langle\!\langle B\rangle\!\rangle_{\rho(x:=N),\zeta(x:=Q)}$$

  which is a direct consequence of the induction hypothesis.
  Finally we prove $\langle\!\langle \lambda x : A.t\rangle\!\rangle_{\rho\zeta} \in a((\![\lambda x : A.t]\!)_\rho)$. There are two cases to distinguish. If $(\![\lambda x : A.t]\!)_\rho \notin$ Proto, this is an easy consequence of the definitions. Otherwise, we have to prove that for every $B \in \mathsf{cone}((\![\lambda x : A.t]\!)_\rho)$ and $c \in a(B)$, $\langle\!\langle t\rangle\!\rangle_{\rho(x:=B),\zeta(x:=c)} \in a((\![\lambda x : A.t]\!)_\rho B)$. By the generation lemma, it follows that $\vdash B : (\![A]\!)_\rho$, hence $(\rho(x := B), \zeta(x := c))$ satisfies $\mathsf{FV}(t)$. The result is a consequence of the induction hypothesis.
- *expansion/reduction:* assume $\vdash M : B$ is deduced from $\vdash M : A$ and $\vdash B : s$ using the expansion/reduction rule. Let (ρ, ζ) be a valuation satisfying $\mathsf{FV}(M) \cup \mathsf{FV}(B)$. As before, we can extend the valuation into a new valuation (ρ', ζ') such that (ρ', ζ') satisfies $\mathsf{FV}(M) \cup \mathsf{FV}(B) \cup \mathsf{FV}(A)$ and coincides with (ρ, ζ) on $\mathsf{FV}(M) \cup \mathsf{FV}(B)$.

To prove $\vdash (\!|M|\!)_{\rho'} : (\!|B|\!)_{\rho'}$, note that $(\!|A|\!)_{\rho'} \to (\!|B|\!)_{\rho'}$ or $(\!|B|\!)_{\rho'} \to (\!|B|\!)_{\rho'}$.
Besides, it follows from the induction hypothesis that:
- $\vdash (\!|M|\!)_{\rho'} : (\!|A|\!)_{\rho'}$;
- $\vdash (\!|B|\!)_{\rho'} : s$.

We conclude by the conversion rule.
To prove $(\!|M|\!)_{\rho} \in \langle\!\langle B\rangle\!\rangle_{\rho\zeta}$; we just apply Corollary 27.
Finally, $\langle\!\langle M\rangle\!\rangle_{\rho\zeta} \in a((\!|M|\!)_{\rho})$ is immediate from the induction hypothesis.□

Corollary 31 $\vdash M : A \quad \Rightarrow \quad M \in \mathsf{SN}$.

Proof: for every derivation $\vdash M : A$, consider the valuation (ρ, ζ) such that $\rho(x) = x$ for every $x \in V$ and $\zeta(x) = \mathsf{max}(x)$ where max is defined on pseudo-terms by $\prec$-induction:

- if $M \in \mathsf{Type}$, $\mathsf{max}(M) = \mathsf{SN}(M)$;
- if $M \in \mathsf{Proto}$, $\mathsf{max}(M) = (\lambda x : a(B).\mathsf{max}(M\ B))_{B \in \mathsf{cone}(M)}$;
- otherwise, $\mathsf{max}(M) = \{\emptyset\}$.

Then $(\rho, \zeta) \models \mathsf{FV}(M) \cup \mathsf{FV}(A)$. It follows from Proposition 30 that $M \in \langle\!\langle A\rangle\!\rangle_{\rho,\zeta}$. As $\langle\!\langle A\rangle\!\rangle_{\rho,\zeta} \subseteq \mathsf{SN}$, we conclude.

5 Applications of the main theorem

5.1 Strong normalisation results

As stated in Corollary 11, Theorem 10 has several important consequences.

For $\mathcal{R}$-terminating extensions of the λ-cube, we know from [3] that subject reduction holds; so we are left to prove that the systems are stratified. To do so, notice that, if M is a prototype, then $M : A$ with A a kind and kinds are of the form[7]:

- $*$,
- $\Pi x : A.B$ where A and B are kinds,
- $\Pi x : A.B$ where B is a kind and A is a type.

One can define a measure ν on kinds as follows:

- $\nu(*) = 1$,
- $\nu(\Pi x : A.B) = \nu(A) + \nu(B) + 1$ if A and B are kinds,
- $\nu(\Pi x : A.B) = \nu(B) + 1$ if B is a kind and A is a type.

Note that the measure is preserved by conversion. By uniqueness of types, this yields a measure μ on prototypes: define $\mu(M) = n$ if $\vdash M : A$ and $\nu(A) = n$ for some A. Extending μ to all pseudo-terms by letting $\mu(P) = 0$ if $P \notin \mathsf{Proto}$, we obtain the following result. For every P, Q,

$$P \prec Q \Rightarrow \mu(P) < \mu(Q)$$

[7] Below we are implicitely assuming that algebraic data live in $*$ as in [3]; it is easy to adapt the proof to the other case.

Hence the systems of the algebraic λ-cube are stratified. A similar technique applies to algebraic higher-order logic.

For $\mathcal{R}$-canonical and $\mathcal{R}$-left-linear extensions of the calculus of constructions with universes, the proof is more involved and requires a quasi-normalisation argument, as developed in [20]. The quasi-normalisation theorem shows that every type has a weak head normal form. This enables us to give a measure on types. As before, we can invoke uniqueness of types to turn this measure into a measure μ for prototypes with the property that $P \prec Q \Rightarrow \mu(P) < \mu(Q)$ for all pseudo-terms P, Q. Note that in this case it is crucial to know subject reduction and confluence of reduction on normal terms before the strong normalisation proof so we must restrict ourselves to confluent and left-linear rewriting systems. For such systems, the combined reduction is confluent on the set of pseudo-terms of the algebraic type system (this follows from [22]).

5.2 Confluence results

As noticed in [10], the combined reduction relation $\rightarrow_{mix}$ of an algebraic type system is in general not confluent on the set of pseudo-terms. However, it is straightforward to check that $\rightarrow_{mix}$ is locally confluent on pseudo-terms. Using Newman's Lemma, one can lift Theorem 10 to $\mathcal{R}$-canonical algebraic type systems.

Proposition 32 *Every $\mathcal{R}$-canonical algebraic type system with the subject reduction property is strongly normalising and confluent w.r.t. $\rightarrow_{mix}$.*

The results of Corollary 11 can all be lifted to $\mathcal{R}$-canonical algebraic type systems.

6 Conclusion

We have introduced in the unified framework of algebraic type systems a large class of algebraic-functional languages which includes all the systems considered in the literature so far. In this general framework, we have been able to address modularity questions. We have given a general criterion for algebraic type systems to be strongly normalising and shown that all the usual algebraic type systems satisfy this criterion. One nice aspect of the proof is that it gives a uniform treatment of all the usual algebraic type systems and emphasizes the fact that proving strong normalisation for algebraic type systems is not essentially more difficult than proving strong normalisation for pure type systems. It would be interesting to extend the present work to more powerful type systems: possible extensions to be considered are first-order inductive types (i.e. inductive types generated by first-order signatures, see for example [23]) or congruence types (an extension of algebraic type systems in which data come equipped with an elimination principle, see [8]). However, we feel more enclined to focus on two important problems which have remained unsolved so far:

- subject reduction: it is an open problem whether algebraic type systems have subject reduction. This is a serious gap in the theory of algebraic type systems. Even for systems with subject reduction, such as the algebraic Calculus of Constructions, the situation is unsatisfactory because the proof of subject reduction is long and intricate. One possible approach to solve the problem would be to consider a labelled syntax for algebraic type systems in which all the usual properties of functional pure type systems (especially subject reduction, unicity of types and classification) hold and use these properties to prove strong normalisation of the labelled syntax (for stratified systems). Then, assuming the labelled syntax to be strongly normalising, one would transfer these results to the traditional syntax by proving the equivalence between the labelled and traditional syntaxes. This approach, introduced by T. Altenkirch to prove strong normalisation for the Calculus of Constructions with $\beta\eta$-reduction ([1]), is currently investigated by P-A. Mellies and the first author.
- modular proofs: our approach to prove strong normalisation is uniform in the sense that algebraic type systems are treated simultaneously with pure type systems. Yet in practice, one would like to know that an algebraic type system is strongly normalising if its underlying pure type system is. Note that such a result would require a purely syntactic proof as no assumption is made on the algebraic type system. See [7] for some preliminary work in this direction.

Another interesting direction for future research is to study the strength of the criterion for pure(and algebraic) type systems. Although every pure type system of interest is stratified, one can easily find pure type systems which are strongly normalising without being stratified. The easiest example is probably obtained by adding to the polymorphic λ-calculus a new sort Δ and an axiom $\Delta : *$. It would be instructive to compare our criterion with other strong normalisation criteria for pure type systems. It is easy to prove that any pure type system which can be embedded in the calculus of constructions with universes is stratified. The converse is not true: consider the pure type system with set of sorts $\mathbf{N}$ and with axioms $i + 1 : i$ (and no rules). This is a stratified pure type system, yet it cannot be embedded in the calculus of constructions with universes. However, we might hope that every stratified pure type system with finitely many sorts can be embedded in the calculus of constructions with universes.

Acknowledgements

Special thanks to the anonymous referees for suggesting significant improvements to the paper. This work was partially supported by the Esprit BRA project "TYPES" (Types for Proofs and Programs).

References

1. T. Altenkirch. *Constructions, inductive types and strong normalisation.* PhD thesis, Laboratory for the Foundations of Computer Science, University of Edinburgh, 1994.
2. F. Barbanera and M. Fernandez. Combining first and higher order rewrite systems with type assignment systems. In M.Bezem and Groote [9], pages 60–74.
3. F. Barbanera, M. Fernandez, and H. Geuvers. Modularity of strong normalisation and confluence in the algebraic λ-cube. In *Proceedings of LICS'94*, pages 406–415. IEEE Press, 1994.
4. H.P. Barendregt. Lambda calculi with types. In S. Abramsky, D. M. Gabbay, and T.S.E. Maibaum, editors, *Handbook of Logic in Computer Science*, volume 2, pages 117–309. Oxford Science Publications, 1992.
5. G. Barthe. Combining dependent type theories with equational term-rewriting. Manuscript, 1995.
6. G. Barthe. η-reduction and algebraic rewriting in λ-calculus. Manuscript, 1995.
7. G. Barthe. Towards modular proofs of termination for algebraic type systems. Manuscript, submitted for publication, 1995.
8. G. Barthe and H. Geuvers. Congruence types. Presented at CSL'95. Submitted for publication in the proceedings, 1995.
9. M. Bezem and J-F. Groote, editors. *Proceedings of TLCA'93*, volume 664 of *Lecture Notes in Computer Science.* Springer-Verlag, 1993.
10. V. Breazu-Tannen. Combining algebra and higher-order types. In *Proceedings of LICS'88*, pages 82–90. IEEE Press, 1988.
11. V. Breazu-Tannen and J. Gallier. Polymorphic rewriting conserves algebraic strong normalisation. *Theoretical Computer Science*, 83:3–28, 1990.
12. M. Fernandez. *Modèles de calcul multiparadigmes fondés sur la réécriture.* PhD thesis, Université Paris-Sud Orsay, 1993.
13. M. Fernandez and J-P.Jouannaud. Modularity of termination of term-rewriting systems revisited. In *Recent Trends in Data Type Specification*, volume 906 of *Lecture Notes in Computer Science*, pages 255–272. Springer-Verlag, 1994.
14. J. Gallier. On Girard's "candidats de réducibilité". In P. Odifreddi, editor, *Logic and Computer Science*, pages 123–203. Academic Press, 1990.
15. H. Geuvers and M.J. Nederhof. A modular proof of Strong Normalization for the Calculus of Constructions, *Journal of Functional Programming* 1, 2 (1991), 155–189.
16. H. Geuvers. *Logics and type systems.* PhD thesis, University of Nijmegen, 1993.
17. H. Geuvers. A short and flexible proof of strong normalisation for the calculus of constructions. In P. Dybjer, B. Nordström, and J. Smith, editors, *Proceedings of TYPES'94*, volume 996 of *Lecture Notes in Computer Science*, pages 14–38. Springer-Verlag, 1995.
18. J-Y. Girard. *Interprétation fonctionelle et élimination des coupures dans l'arithmétique d'ordre supérieur.* PhD thesis, Université Paris 7, 1972.
19. J-P. Jouannaud and M. Okada. Executable higher-order algebraic specification languages. In *Proceedings of LICS'91*, pages 350–361. IEEE Press, 1991.
20. Z. Luo. *Computation and Reasoning: A Type Theory for Computer Science.* Number 11 in International Series of Monographs on Computer Science. Oxford University Press, 1994.
21. A. Middeldorp. *Modular properties of term-rewriting systems.* PhD thesis, Department of Computer Science, Vrije Universiteit, Amsterdam, 1990.

22. F. Müller. Confluence of the lambda calculus with left-linear algebraic rewriting. *Information Processing Letters*, 41:293–299, 1992.
23. C. Paulin-Mohring. Inductive definitions in the system Coq. Rules and properties. In Bezem and Groote [9], pages 328–345.
24. W. Tait. A realisability interpretation of the theory of species. In R. Parikh, editor, *Logic Colloquium 73*, volume 453 of *Lectures Notes in Mathematics*, pages 240–251, 1975.
25. J. Terlouw. Strong normalisation in type systems: a model-theoretical approach. In *Dirk van Dalen Festschrift*, pages 161–190. University of Utrecht, 1993. To appear in Annals of Pure and Applied Logic.
26. Y. Toyama. On the Church-Rosser property for the direct sum of term rewriting systems. *Journal of the ACM*, 34(1):128–143, 1987.

Collapsing Partial Combinatory Algebras

Inge Bethke[1,2] and Jan Willem Klop[1,3]

[1] CWI, P.O. Box 94079, 1090 GB Amsterdam, The Netherlands
[2] University of Utrecht, Department of Philosophy, P.O. Box 80126, 3508 TC Utrecht, The Netherlands
[3] Vrije Universiteit, Department of Mathematics and Computer Science, de Boelelaan 1081a, 1081 HV Amsterdam, The Netherlands

Abstract. Partial combinatory algebras occur regularly in the literature as a framework for an abstract formulation of computation theory or recursion theory. In this paper we develop some general theory concerning homomorphic images (or collapses) of pca's, obtained by identification of elements in a pca. We establish several facts concerning final collapses (maximal identification of elements). 'En passant' we find another example of a pca that cannot be extended to a total one.

1 Introduction

A *partial combinatory algebra (pca)* is a structure $\mathfrak{A} = \langle A, s, k, \cdot \rangle$ where A is a set, $\cdot$ is a partial binary operation (application) on A, and k, s are two elements of A such that

1. $\forall a, a' \in A \ \ (k \cdot a) \cdot a' = a$,
2. $\forall a, a' \in A \ \ (s \cdot a) \cdot a' \downarrow$,
3. $\forall a, a', a'' \in A \ \ ((s \cdot a) \cdot a') \cdot a'' = \begin{cases} (a \cdot a'') \cdot (a' \cdot a'') & \text{if } (a \cdot a'') \cdot (a' \cdot a'') \downarrow, \\ \text{undefined} & \text{otherwise,} \end{cases}$
4. $k \neq s$.

Here $M \downarrow$ means the expression M is defined, and $M = N$ means both expressions are defined and equal. Another useful notation is to write $M \uparrow$ if M is undefined. It is common to omit $\cdot$ and associate unparenthesized expressions to the left. In working with expressions that may or may not be defined, it is useful to write $M \simeq N$ to mean that if either M or N is defined, then both are defined and equal. These notational conventions allow us to replace clause 3 by

$$\forall a, a', a'' \in A \ \ saa'a'' \simeq aa''(a'a'') \ .$$

Total pca's (*ca*'s), where application is a total operation on the carrier set, are extensively studied in the context of models of λ-calculus and Combinatory Logic (CL) (cf. e.g. [Bar84], [HS86]); nontotal pca's (*nca*'s), where application is not defined everywhere, are a little less well-off in this respect. They figure in the semantics of programming languages (see the forthcoming book by Mitchell [Mit9?]) as well as in the formalization of constructive mathematics (see [Bee85], [TvD88]). In fact, they are the models of a 'minimal axiomatic basis for theories of operators', as stated in [TvD88]. An early approach to treat abstract computation theory was given by the notion of Wagner [Wag69] and Strong [Str68], URS (Uniform Reflexive Structure). More recently, the notion of Effective Applicative Structure, EAS, has been introduced by Asperti and Ciabattoni [AC95]); they show that this notion is in fact equivalent to PCA.

Let us briefly indicate why a study of pca's falls in the scope of higher-order algebra, logic and term rewriting - the subject of the present conference. The connection with term rewriting, via CL and λ-calculus, is obvious since pca's admit abstraction $[x]M$; in fact they were 'designed' just for that purpose. The connection with higher-order algebra is less clear, also due to the fact that there is no sharp definition of this notion. Meinke [Mei95] bases his survey of higher-order algebra on type theories. Indeed, it is shown that the finite type hierarchy HEO can be built over an arbitrary pca (Bethke [Bet91]); also Mitchell [Mit9?] generalizes the construction of HRO to arbitrary pca's. Furthermore, pca's play a role in the construction of per models for realizability. See also Streicher [Str91].

While nca's thus have enjoyed quite some attention as a tool in abstract computation theory, amazingly little is known about their structural properties. Thus, it was even an open question in [Swan79] whether an nca can always be extended to a ca. A negative answer is given in [Klo82] and [Bet87]. Dually to extending pca's, one may ask what behaviour pca's exhibit under homomorphic images. To be more precise, given a pca $\mathfrak{A} = \langle A, s, k, \cdot \rangle$ and some elements a, a' of A, one may ask whether there exists a homomorphic image $\phi(\mathfrak{A})$ such that $\phi(a) = \phi(a')$.

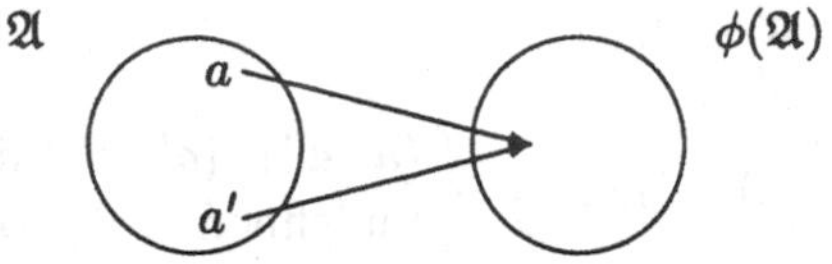

We shall call such a homomorphic image a *collapse*. There exist several investigations into collapses of ca's (cf. e.g. [Jac75], [JZ85], [BI93]). Here the leading question is whether, given λ-terms M and N, the equation $M = N$ can be added consistently to the λ-calculus. Considerations of collapses of nca's seem to be rare. In fact, we do not know of any. In the present note, we address this last theme.

Instead of considering collapses, one can also study certain well-behaved congruence relations. As it turns out, there exists a natural 1-1 correspondence between these relations and collapses: every such congruence induces a collapse and vice versa. We establish this fact in Sect. 2.

We use the correspondence between well-behaved congruence relations and collapses in Sect. 3 to show that there is at least one major difference between nca's and ca's with respect to their class of collapses: nca's always have a final collapse $\phi_{\text{fin}}(\mathfrak{A})$ which combines all possible identifications.

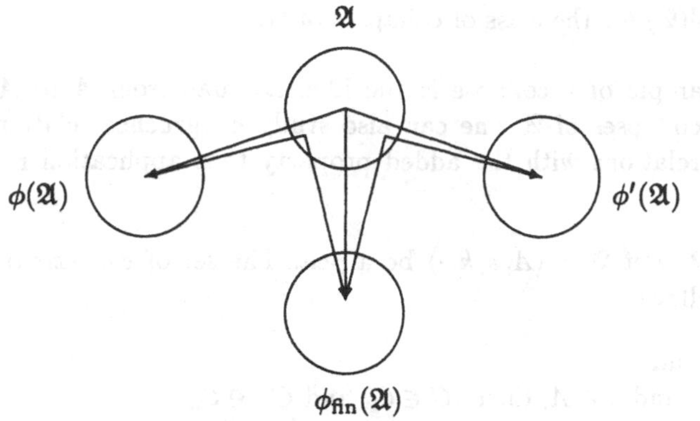

For ca's, such a final collapse does not need to exist. We provide a counterexample.

Not every pca allows for additional identifications. In Sect. 4, we give two examples of these ***irreducible*** pca's: the well-known total graph models as well as the nontotal pca of natural numbers with partial recursive function application cannot be collapsed any further.

In Sect. 5, we concentrate on extensional collapses, i.e. collapses that identify elements displaying identical applicative behaviour. We provide a condition on nca's that guarantees the existence of extensional collapses. In fact, if an nca meets this condition, then its final collapse is extensional. As an application which may be of independent interest, we show that the paradigmatic nca of strongly normalizing CL-terms has an extensional final collapse.

2 Collapses of PCA's

A homomorphism is a structure-preserving map from one algebra to another. For partial algebras, there is one basic notion for a homomorphism which simultaneously generalizes the notions of homomorphisms between total algebras and relational structures respectively. However, since its defining property is relatively weak, we select a proper subclass of homomorphisms throughout this paper. An extensive survey of the model theory of partial algebras can be found in [Bur82].

Definition 1. Let $\mathfrak{A} = \langle A, s, k, \cdot \rangle$ and $\mathfrak{B} = \langle B, s', k', \cdot' \rangle$ be pca's.

1. A *closed homomorphism* of $\mathfrak{A} = \langle A, s, k, \cdot \rangle$ into $\mathfrak{B} = \langle B, s', k', \cdot' \rangle$ is a mapping $\phi : A \to B$ such that
   (a) $\phi(s) = s'$, $\phi(k) = k'$, and
   (b) $\phi(a \cdot a') \simeq \phi(a) \cdot' \phi(a')$ for all $a, a' \in A$.
   If ϕ is surjective, then ϕ is a *closed epimorphism*, and if ϕ is bijective, then ϕ is an *isomorphism*.
2. ϕ is a *collapse* of $\mathfrak{A}$ if ϕ is a closed epimorphism of $\mathfrak{A}$ onto some pca $\mathfrak{B}$.

We write Col($\mathfrak{A}$) for the class of collapses of $\mathfrak{A}$.

A trivial example of a collapse is the identity map from A to A. Instead of considering collapses of $\mathfrak{A}$, one can also study congruence relations on A, i.e. equivalence relations with the added property that application relates related elements.

Definition 2. Let $\mathfrak{A} = \langle A, s, k, \cdot \rangle$ be a pca. The set of *contexts* over $\mathfrak{A}$, $C_{\mathfrak{A}}$, is defined as follows.

1. $\Box \in C_{\mathfrak{A}}$, and
2. if $C \in C_{\mathfrak{A}}$ and $a \in A$, then $aC \in C_{\mathfrak{A}}$ and $Ca \in C_{\mathfrak{A}}$.

If C is a context, then $C[a]$ denotes the expression obtained from C by replacing $\Box$ by a.

Definition 3. Let $\mathfrak{A} = \langle A, s, k, \cdot \rangle$ be a pca and E be an equivalence relation on A.

1. E is called *proper* if $\langle s, k \rangle \notin E$.
2. E is said to be a *congruence* if for all $\langle a, a' \rangle \in E$ and $C \in C_{\mathfrak{A}}$, if either $C[a]$ or $C[a']$ is defined, then both are defined and $\langle C[a], C[a'] \rangle \in E$.

We write Con($\mathfrak{A}$) for the set of proper congruence relations on A.

A trivial example of a proper congruence on A is the diagonal $\{\langle a, a \rangle \mid a \in A\}$. However, there may be more complex ones. In particular, every collapse corresponds to a congruence relation, namely the one that relates identified elements.

Definition 4. Let $\mathfrak{A} = \langle A, s, k, \cdot \rangle$ be a pca and $\phi \in \mathrm{Col}(\mathfrak{A})$. Put

$$E_\phi = \{\langle a, a' \rangle \in A \times A \mid \phi(a) = \phi(a')\} .$$

Proposition 5. *Let $\mathfrak{A} = \langle A, s, k, \cdot \rangle$ be a pca and $\phi \in \mathrm{Col}(\mathfrak{A})$. Then $E_\phi \in \mathrm{Con}(\mathfrak{A})$.*

Proof. E_ϕ is clearly an equivalence relation on A and is proper, since $\phi(s) = s' \neq k' = \phi(k)$. To prove that E_ϕ is a congruence, let $\langle a, a' \rangle \in E_\phi$ and $C \in C_{\mathfrak{A}}$. Then

$$\phi(C[a]) \simeq C'[\phi(a)] \simeq C'[\phi(a')] \simeq \phi(C[a'])$$

for some context C'. Hence $C[a]$ is defined if and only if $C[a']$ is defined, and if they are both defined, then $\langle C[a], C[a'] \rangle \in E_\phi$. $\Box$

Given any congruence relation E on A, we may construct a pca $\mathfrak{A}/E$ of $\mathfrak{A}$ called the *quotient of* $\mathfrak{A}$ *modulo* E. The intuitive idea behind $\mathfrak{A}/E$ is that we identify related elements of A.

Definition 6. Let $\mathfrak{A} = \langle A, s, k, \cdot \rangle$ be a pca and $E \in \mathrm{Con}(\mathfrak{A})$. We form the quotient

$$\mathfrak{A}/E = \langle A/E, [s]_E, [k]_E, \cdot_E \rangle$$

by taking the collection $A/E = \{[a]_E \mid a \in A\}$ of equivalence classes $[a]_E = \{a' \mid \langle a, a' \rangle \in E\}$ equipped with the application operation

$$[a]_E \cdot_E [a']_E = \begin{cases} [aa']_E & \text{if } aa' \downarrow \\ \text{undefined otherwise} & . \end{cases}$$

Proposition 7. *Let* $\mathfrak{A} = \langle A, s, k, \cdot \rangle$ *be a pca. For* $E \in \mathrm{Con}(\mathfrak{A})$, $\lambda a \in A.[a]_E \in \mathrm{Col}(\mathfrak{A})$.

Proof. We first show that $\cdot_E$ is well-defined. To this end, let $[a]_E = [a']_E$ and $[b]_E = [b']_E$. Then $\langle a, a' \rangle, \langle b, b' \rangle \in E$. Let $C \equiv \Box b$, $C' \equiv a'\Box$. As E is a congruence, it follows that $C[a] \downarrow$ iff $C[a'] \downarrow$, and $C'[b] \downarrow$ iff $C'[b'] \downarrow$. Thus

$$ab \downarrow \leftrightarrow a'b \downarrow \leftrightarrow a'b' \downarrow \ .$$

Hence $[a]_E[b]_E \downarrow$ iff $[a']_E[b']_E \downarrow$. Now assume $[a]_E[b]_E \downarrow$. Then $\langle ab, a'b \rangle, \langle a'b, a'b' \rangle \in E$. So $\langle ab, a'b' \rangle \in E$, i.e. $[ab]_E = [a'b']_E$. Thus $[a]_E[b]_E = [a']_E[b']_E$.
$\mathfrak{A}/E$ meets the first three conditions on pca's, since $\mathfrak{A}$ is a pca; it meets the last condition, since E is proper. Hence $\mathfrak{A}/E$ is a pca. Clearly, $\lambda a \in A.[a]_E$ is a closed epimorphism of $\mathfrak{A}$ onto $\mathfrak{A}/E$. □

If we, as is standard, identify isomorphic pca's, we can in fact pass in this way from collapses to proper congruence relations and back, and end up were we have started. This is a special case of the well-known First Homomorphism Theorem of universal algebra (see e.g. [Grä79]). Thus, given collapses ϕ, ϕ' of $\mathfrak{A}$, let us write $\phi \cong \phi'$ if the homomorphic images of $\mathfrak{A}$ under ϕ and ϕ' are isomorphic.

Theorem 8. *Let* $\mathfrak{A} = \langle A, s, k, \cdot \rangle$ *be a pca. Then*

1. $\lambda a \in A.[a]_{E_\phi} \cong \phi$ *for all* $\phi \in \mathrm{Col}(\mathfrak{A})$, *and*
2. $E_{\lambda a \in A.[a]_E} = E$ *for all* $E \in \mathrm{Con}(\mathfrak{A})$.

Proof. To prove 1., define the surjection $\psi : A/E_\phi \to \phi(A)$ by $\psi([a]_{E_\phi}) = \phi(a)$. As

$$[a]_{E_\phi} = [a']_{E_\phi} \leftrightarrow \langle a, a' \rangle \in E_\phi \leftrightarrow \phi(a) = \phi(a'),$$

it follows that ψ is well-defined and bijective, and since ϕ is a closed homomorphism, ψ is a closed homomorphism too. So ψ is an isomorphism. For 2., note that

$$E_{\lambda a \in A.[a]_E} = \{\langle a, a' \rangle \in A \times A \mid [a]_E = [a']_E\} = E \ .$$

□

3 Final Collapses of PCA's

A pca $\mathfrak{A}$ has always an *initial* collapse, i.e. a collapse ϕ such that for any collapse ϕ' there is a unique homomorphism ψ with $\psi \circ \phi = \phi'$. This initial collapse is just the identity on A that does not identify any elements. Nca's, however, also have a *final* collapse, i.e. a collapse ϕ such that for any collapse ϕ' there is a unique homomorphism ψ with $\psi \circ \phi' = \phi$. Such a final collapse then identifies all elements that can be identified. The crucial observation is the following.

Proposition 9. *Let $\mathfrak{A} = \langle A, s, k, \cdot \rangle$ be an nca and E be an equivalence relation on A. Then E is proper provided E is a congruence relation.*

Proof. Assume E is a congruence and suppose that $\langle s, k \rangle \in E$. Pick $a, a' \in A$ such that $aa' \uparrow$ and let $C \equiv \Box kaa'$. Then $skaa' \downarrow$ iff $kkaa' \downarrow$. As $kkaa' = ka'$, it follows that $skaa' = ka'(aa')$. Hence $aa' \downarrow$. Contradiction. □

In dealing with nca's, we can therefore forget about properness and concentrate on congruence only. As it turns out, the union of all congruences is again a congruence.

Definition 10. Let $\mathfrak{A} = \langle A, s, k, \cdot \rangle$ be an nca. Put

$$E_{\mathrm{fin}} = \{\langle a, a' \rangle \in A \times A \mid \forall C \in \mathcal{C}_{\mathfrak{A}}\ C[a] \downarrow \text{ if and only if } C[a'] \downarrow\} .$$

Lemma 11. *Let $\mathfrak{A} = \langle A, s, k, \cdot \rangle$ be an nca. Then*

1. $E_{\mathrm{fin}} \in \mathrm{Con}(\mathfrak{A})$,
2. $E_{\mathrm{fin}} = \bigcup \mathrm{Con}(\mathfrak{A})$.

Proof. 1. E_{fin} is clearly a congruence relation. Hence $E_{\mathrm{fin}} \in \mathrm{Con}(\mathfrak{A})$ by Proposition 9.
2. From 1. it follows that $E_{\mathrm{fin}} \subseteq \bigcup \mathrm{Con}(\mathfrak{A})$. For the other inclusion, let $\langle a, a' \rangle \in \bigcup \mathrm{Con}(\mathfrak{A})$. Then $\langle a, a' \rangle \in E$ for some $E \in \mathrm{Con}(\mathfrak{A})$. Thus, since E is a congruence, $\langle a, a' \rangle \in E_{\mathrm{fin}}$. □

Theorem 12. *Every nca $\mathfrak{A} = \langle A, s, k, \cdot \rangle$ has a final collapse.*

Proof. We shall prove that $\lambda a \in A.[a]_{E_{\mathrm{fin}}}$ is final. To this end, let ϕ be any collapse of $\mathfrak{A}$ onto some pca $\mathfrak{B} = \langle B, s', k', \cdot' \rangle$ and put $\psi(b) = [a]_{E_{\mathrm{fin}}}$ where $\phi(a) = b$. Observe that ψ is well-defined. For, if $\phi(a) = b = \phi(a')$, then $\langle a, a' \rangle \in E_\phi \subseteq E_{\mathrm{fin}}$ and hence $[a]_{E_{\mathrm{fin}}} = [a']_{E_{\mathrm{fin}}}$. Clearly ψ is a homomorphism. And as $\psi(\phi(a)) = [a]_{E_{\mathrm{fin}}}$ for all $a \in A$, it follows that $\psi \circ \phi = \lambda a \in A.[a]_{E_{\mathrm{fin}}}$. Now let ψ' be such that $\psi' \circ \phi = \lambda a \in A.[a]_{E_{\mathrm{fin}}}$. Then $\psi(\phi(a)) = \psi'(\phi(a))$ for all $a \in A$. Hence $\psi(b) = \psi'(b)$ for all $b \in B$. So $\psi = \psi'$. □

For ca's, such a final collapse does not need to exist. To see this, we recall a well-known result from [Jac75]. *Extensional combinatory logic*, ECL, is an equational theory consisting of expressions of the form $M = N$ where M and N

are terms constructed as usual from variables, the two constants S and K, and a binary application operator $\cdot$ which we do not write. The axioms and rules of inference of ECL are those of equational logic together with the axioms

$$Kxy = x \quad Sxyz = xz(yz)$$

and the rule

$$\frac{Mx = Nx}{M = N}$$

where the variable x occurs in neither M nor N. Closed terms modulo provable equality form a ca in the following way: We let

$$\mathfrak{A}_{\mathrm{ECL}} = \langle T^0/\mathrm{ECL}, [S]_{\mathrm{ECL}}, [K]_{\mathrm{ECL}}, \cdot \rangle$$

where T^0 is the set of closed terms (i.e. the set of terms without any variable),

$$T^0/\mathrm{ECL} = \{[M]_{\mathrm{ECL}} \mid M \in T^0\},$$

$$[M]_{\mathrm{ECL}} = \{N \in T^0 \mid \mathrm{ECL} \vdash M = N\}$$

and

$$[M]_{\mathrm{ECL}} \cdot [N]_{\mathrm{ECL}} = [MN]_{\mathrm{ECL}} \; .$$

In [Jac75], Jacopini - using a slightly different terminology - proved that $[\Omega]_{\mathrm{ECL}}$, where

$$\Omega \equiv S(SKK)(SKK)(S(SKK)(SKK)),$$

can be identified with any other element in this ca. This means in particular that $\mathfrak{A}_{\mathrm{ECL}}$ has collapses ϕ and ϕ' such that $\phi([\Omega]_{\mathrm{ECL}}) = \phi([S]_{\mathrm{ECL}})$ and $\phi'([\Omega]_{\mathrm{ECL}}) = \phi'([K]_{\mathrm{ECL}})$. It follows that $\mathfrak{A}_{\mathrm{ECL}}$ lacks a final collapse. For, suppose $\mathfrak{A}_{\mathrm{ECL}}$ has a final collapse onto some pca $\mathfrak{B} = \langle B, s', k', \cdot' \rangle$. Then there are homomorphisms ψ and ψ' such that $\psi \circ \phi = \psi' \circ \phi'$. So

$$s' = \psi(\phi([S]_{\mathrm{ECL}})) = \psi(\phi([\Omega]_{\mathrm{ECL}})) = \Omega' = \psi'(\phi'([\Omega]_{\mathrm{ECL}})) = \psi'(\phi'([K]_{\mathrm{ECL}})) = k'$$

where $\Omega' \equiv s'(s'k'k')(s'k'k')(s'(s'k'k')(s'k'k'))$. This constitutes a contradiction with the fact that the homomorphic image of $\mathfrak{A}_{\mathrm{ECL}}$ under the final collapse meets the last condition on pca's.

Theorem 13. *Not every ca has a final collapse.*

4 Irreducible PCA's

Not every pca allows for further identifications. For example, the codomain of every final collapse has reached its maximal degree of identifications. We shall call such a pca, where the only collapse is the trivial initial one, *irreducible*.

Definition 14. Let $\mathfrak{A} = \langle A, s, k, \cdot \rangle$ be a pca. $\mathfrak{A}$ is *irreducible* if $E_\phi \subseteq \{\langle a, a \rangle \mid a \in A\}$ for every collapse ϕ of $\mathfrak{A}$.

There are prominent pca's which share this property. We give two examples.

Example 1. The first example uses only elementary properties of sets, and is directly taken from Engeler [Eng81]. It is in fact a notational variant of one of several ca's first described in Plotkin [Plo72] which in turn are nearly the same as the better known $P\omega$ construction of Scott [Sco76].

Let A be any nonempty set, and let B be the least set containing A and all ordered pairs consisting of a finite subset $\beta \subseteq B$ and an element $b \in B$. Assume that elements of A are distinguishable from ordered pairs. Let D_A be the power set of B, and define the total application operation on D_A by

$$xy = \{b \in B \mid (\beta, b) \in x \text{ for some } \beta \subseteq y\} \ .$$

Choose

$$s = \{(\alpha, (\beta, (\gamma, b))) \mid b \in \alpha\gamma(\beta\gamma)\},$$

and

$$k = \{(\alpha, (\beta, b)) \mid b \in \alpha\} \ .$$

Then $\mathfrak{D} = \langle D_A, s, k, \cdot \rangle$ is a ca.

To prove that $\mathfrak{D}$ is irreducible, let E_ϕ be any collapse of $\mathfrak{D}$ and let $\langle x, y \rangle \in E_\phi$. Assume $x \neq y$. Say, $b \in x$ and $b \notin y$ for some $b \in B$. Define

$$z = \{(\{b\}, b') \mid b' \in k\} \ .$$

Then $z \in D_A$. Now let C be the context $z\Box$. Since E_ϕ is a congruence, $\langle zx, zy \rangle \in E_\phi$. Observe that $zx = k$ and $zy = \emptyset$. Hence $\langle k, \emptyset \rangle \in E_\phi$, and therefore $\langle kss, \emptyset ss \rangle \in E_\phi$. That is, also $\langle s, \emptyset \rangle \in E_\phi$. It follows that $\langle s, k \rangle \in E_\phi$. Thus E_ϕ is improper. This is a contradiction. So $x = y$; whence $\langle x, y \rangle \in \{\langle x, x \rangle \mid x \in D_A\}$.

Remark. The argument given above extends to the family of $P\omega$-models which consists of coded versions of $\mathfrak{D}$. At first sight, this may seem to contradict the remarkable result of Baeten and Boerboom in their 1979 paper *Ω can be anything it shouldn't be* (cf. [BB79]). The authors, however, do not consider collapses. Rather they show that, given an arbitrary closed λ-term M, there exists a member of the $P\omega$-family which identifies M and Ω.

Example 2. As second example we consider the nca of natural numbers with partial recursive function application. More specifically, we define a nontotal application operation on the natural numbers $\mathbb{N}$ by

$$nm = \{n\}(m)$$

where $\{n\}$ is the partial recursive function with Gödel number n. We may let k be any Gödel number of the recursive function which, given some argument x, returns a Gödel number of the constant function returning x. The natural number s is slightly more complicated: we let s be a Gödel number of the recursive function

$$f(x) = n_x$$

where n_x is a Gödel number of the recursive function

$$g(y) = m_{x,y}$$

with $m_{x,y}$ a Gödel number of the partial recursive function

$$h(z) \simeq \{\{x\}(z)\}(\{y\}(z)) \ .$$

The existence of this function is easiest to explain using Turing machines, or some other model of computation. Then $\mathfrak{N} = \langle \mathbb{N}, s, k, \cdot \rangle$ is an nca.

To prove that $\mathfrak{N}$ is irreducible, let E_ϕ be any collapse of $\mathfrak{N}$ and let $\langle x, y \rangle \in E_\phi$. Assume $x \neq y$. It is now not hard to imagine a partial recursive function g with Gödel number z, say, such that $g(x)\uparrow$ and $g(y)\downarrow$. Then $C \equiv z\Box$ is a context with $C[x]\uparrow$ and $C[y]\downarrow$. Thus E_ϕ is not a congruence. This is a contradiction. So $x = y$; whence $\langle x, y \rangle \in \{\langle x, x \rangle \mid x \in \mathbb{N}\}$.

5 Extensional Collapses of PCA's

In this final section, we shall consider collapses that identify elements which display identical applicative behaviour.

Definition 15. Let $\mathfrak{A} = \langle A, s, k, \cdot \rangle$ be a pca.

1. $\mathfrak{A}$ is *extensional* if for all $a, a' \in A$,

$$(\forall a'' \in A \; aa'' \simeq a'a'') \rightarrow a = a' \ .$$

2. $\mathfrak{A}$ has an *extensional collapse* if $\mathfrak{A}$ has a collapse onto some extensional pca.

Proposition 16. *Let $\mathfrak{A} = \langle A, s, k, \cdot \rangle$ be a pca and let ϕ be an extensional collapse of $\mathfrak{A}$. Then*

$$\{\langle a, a' \rangle \in A \times A \mid \forall a'' \in A \; aa'' \simeq a'a''\} \subseteq E_\phi \ .$$

Proof. Suppose ϕ is a collapse onto the extensional pca $\mathfrak{B} = \langle B, s', k', \cdot' \rangle$. Let $a, a' \in A$ be such that $aa'' \simeq a'a''$ for all $a'' \in A$ and let $b \in B$. Say, $b = \phi(a'')$. Then

$$\phi(a)b \simeq \phi(a)\phi(a'') \simeq \phi(aa'') \simeq \phi(a'a'') \simeq \phi(a')\phi(a'') \simeq \phi(a')b \ .$$

Hence $\phi(a) = \phi(a')$, since $\mathfrak{B}$ is extensional. Therefore $\langle a, a' \rangle \in E_\phi$. □

Not every pca has an extensional collapse. Observe, for example, that the two pca's considered in Example 1 and 2 are not extensional. As they are both irreducible, it follows that they do not have an extensional collapse.

Theorem 17. *Not every pca has an extensional collapse.*

For nca's, there exists a simple condition such that the final collapse is extensional.

Theorem 18. *Let $\mathfrak{A} = \langle A, s, k, \cdot \rangle$ be an nca. Its final collapse is extensional if and only if*

$$(\dagger) \quad \forall a, a' \in A\ (\forall C \in \mathcal{C}_{\mathfrak{A}}\, \forall a'' \in A\, (C[aa''] \downarrow \leftrightarrow C[a'a''] \downarrow) \rightarrow \ \langle a, a' \rangle \in E_{\text{fin}}) \ .$$

Proof. Suppose (†) holds. To prove that $\mathfrak{A}/E_{\text{fin}}$ is extensional, let $[a]_{E_{\text{fin}}}, [a']_{E_{\text{fin}}} \in A/E_{\text{fin}}$ be such that

$$[a]_{E_{\text{fin}}}[a'']_{E_{\text{fin}}} \simeq [a']_{E_{\text{fin}}}[a'']_{E_{\text{fin}}}$$

for every $[a'']_{E_{\text{fin}}} \in A/E_{\text{fin}}$. Now let $a'' \in A$, C be any context and assume one of $C[aa'']$ and $C[a'a'']$ is defined, say $C[aa''] \downarrow$. Then $aa'' \downarrow$ and hence $[aa'']_{E_{\text{fin}}} = [a'a'']_{E_{\text{fin}}}$. So $\langle aa'', a'a'' \rangle \in E_{\text{fin}}$ and therefore $C[a'a''] \downarrow$. Thus $\langle a, a' \rangle \in E_{\text{fin}}$ by (†), i.e.

$$[a]_{E_{\text{fin}}} = [a']_{E_{\text{fin}}} \ .$$

For the other direction, assume $\mathfrak{A}/E_{\text{fin}}$ is extensional and let $a, a' \in A$ be such that $C[aa''] \downarrow$ if and only if $C[a'a''] \downarrow$ for all contexts C and all $a'' \in A$. Then, in particular,

$$[a]_{E_{\text{fin}}}[a'']_{E_{\text{fin}}} \simeq [a']_{E_{\text{fin}}}[a'']_{E_{\text{fin}}}$$

for every $[a'']_{E_{\text{fin}}} \in A/E_{\text{fin}}$. Hence $[a]_{E_{\text{fin}}} = [a']_{E_{\text{fin}}}$, since $\mathfrak{A}/E_{\text{fin}}$ is extensional. So $\langle a, a' \rangle \in E_{\text{fin}}$. □

We shall apply this result in the next and final example of this paper where we prove that the final collapse of the nca of closed, strongly normalizing CL-terms is extensional. In the example, we employ fundamental definitions and notions of term rewrite systems. Extensive surveys of term rewriting can be found in [Klo92] and [DJ90].

Example 3. Reduction in CL is generated by the rules

1. $SLMN \rightarrow LN(MN)$

2. $KLM \to L$

for all CL-terms L, M, N. Here 'generated' means:

3. if $L \to M$ then $C[L] \to C[M]$

for every context C. Contexts are defined as in Definition 2 with element a changed into CL-term L.

We write $L \equiv M$ if L and M are identical terms. The transitive-reflexive closure of the rewrite relation $\to$ is denoted by $\twoheadrightarrow$. If $L \twoheadrightarrow M$, we say that L *reduces* to M. The equivalence relation generated by $\to$ is called *convertibility* and written as $=$.

A term of the form $SLMN$ or KLM is a *redex*; its *contractum* is $LN(MN)$ or L, respectively. A term not containing such redexes is a *normal form (nf)* and has a nf if it reduces to one. A reduction of L is a sequence of terms $L \equiv L_1 \to L_2 \to L_3 \to \cdots$. Reductions may be infinite. If every reduction of L terminates eventually (in a normal form), then L is said to be *strongly normalizing*. We let SN be the set of all strongly normalizing CL-terms, and SN^0 be the set of all closed, strongly normalizing CL-terms. Observe that $\omega \equiv S(SKK)(SKK) \in \mathrm{SN}^0$; however, $\Omega \equiv \omega\omega \notin \mathrm{SN}$.

The rewrite system CL is orthogonal and has therefore nice properties such as confluence. Another pleasantness is:

(∗) Let $L \notin \mathrm{SN}$ and $L \to M$ be such that $M \in \mathrm{SN}$. Then the redex contracted in the reduction step must contain a proper subterm N with $N \notin \mathrm{SN}$ that is erased in the step $L \to M$

(cf. Exercise 3.1.13 of [Klo92]). From this we obtain the following proposition.

Proposition 19. *Let C be a context and $L, M \in \mathrm{SN}$.*

1. *If $L \to M$, then $C[L] \in \mathrm{SN}$ if and only if $C[M] \in \mathrm{SN}$.*
2. *If $L \twoheadrightarrow M$, then $C[L] \in \mathrm{SN}$ if and only if $C[M] \in \mathrm{SN}$.*
3. *If $L = M$, then $C[L] \in \mathrm{SN}$ if and only if $C[M] \in \mathrm{SN}$.*

Proof. 1. If $L \to M$, then $C[L] \to C[M]$. Hence $C[M] \in \mathrm{SN}$ if $C[L] \in \mathrm{SN}$. For the other direction, assume $C[M] \in \mathrm{SN}$ and suppose $C[L] \notin \mathrm{SN}$. By (∗) there must be a subterm N of L with $N \notin \mathrm{SN}$. This is of course impossible, since $L \in \mathrm{SN}$.
2. Follows from 1.
3. If $L = M$, then by confluence, $L \twoheadrightarrow N \twoheadleftarrow M$ for some term N. Moreover, $N \in \mathrm{SN}$, since $L, M \in \mathrm{SN}$. Therefore $C[L] \in \mathrm{SN}$ iff $C[N] \in \mathrm{SN}$ iff $C[M] \in \mathrm{SN}$ by 2. □

Closed, strongly normalizing terms modulo convertibility form an nca in the following way: We let

$$\mathfrak{A}_{\mathrm{SN}} = \langle \{[M]_{\mathrm{SN}} \mid M \in \mathrm{SN}^0\}, [S]_{\mathrm{SN}}, [K]_{\mathrm{SN}}, \cdot \rangle$$

where

$$[M]_{\mathrm{SN}} = \{N \in \mathrm{SN}^0 \mid M = N\}$$

and

$$[M]_{\mathrm{SN}} \cdot [N]_{\mathrm{SN}} = \begin{cases} [MN]_{\mathrm{SN}} & \text{if } MN \in \mathrm{SN}, \\ \text{undefined otherwise} & . \end{cases}$$

Observe that application is well-defined. For, if $M = M'$ and $N = N'$, then $MN \in \mathrm{SN}$ iff $M'N \in \mathrm{SN}$ iff $M'N' \in \mathrm{SN}$ by Proposition 19.3. By a similar argument, $\mathfrak{A}_{\mathrm{SN}}$ satisfies conditions 1. and 3. on pca's. Moreover, $SLM \in \mathrm{SN}$ if $L, M \in \mathrm{SN}$. Hence also condition 2. is met. Finally, $[S]_{\mathrm{SN}} \neq [K]_{\mathrm{SN}}$, since $S \not\equiv K$. So $\mathfrak{A}_{\mathrm{SN}}$ is an nca.

To prove that the final collapse of $\mathfrak{A}_{\mathrm{SN}}$ is extensional, we invoke Theorem 18. That is, we shall prove that for all $L, M \in \mathrm{SN}^0$, if

$$\forall C \in \mathcal{C}_{\mathfrak{A}_{\mathrm{SN}}} \forall N \in \mathrm{SN}^0 \, (C[[L]_{\mathrm{SN}}[N]_{\mathrm{SN}}] \downarrow \leftrightarrow C[[M]_{\mathrm{SN}}[N]_{\mathrm{SN}}] \downarrow),$$

then $\langle [L]_{\mathrm{SN}}, [M]_{\mathrm{SN}} \rangle \in E_{\mathrm{fin}}$. If we denote the set of contexts built from the hole $\Box$ and closed, strongly normalizing terms by $\mathcal{C}_{\mathrm{SN}}$, the requirement for Theorem 18 boils down to the following: for all $L, M \in \mathrm{SN}^0$, if

$$(\ddagger) \quad \forall C \in \mathcal{C}_{\mathrm{SN}} \forall N \in \mathrm{SN}^0 \, (C[LN] \in \mathrm{SN} \leftrightarrow C[MN] \in \mathrm{SN}),$$

then $C[L] \in \mathrm{SN}$ iff $C[M] \in \mathrm{SN}$ for all $C \in \mathcal{C}_{\mathrm{SN}}$. We start with an intuitive description of the proof.

We first recall the notion of *descendants* of a specific occurrence of a subterm L of M under a reduction $M \twoheadrightarrow N$: we underline the given occurrence of L in M (and nothing else) and perform the reduction $M \twoheadrightarrow N$. Then we look for the set of all underlined subterms of N. These subterms (actually subterm occurrences) are the descendants of L. We moreover say that L is *activated* in this reduction if $N \equiv C[L^*P]$ for some context C and some term P where L^* is a descendant of L.

Now suppose $(\ddagger)$ holds and $C[L] \notin \mathrm{SN}$, i.e. $C[L]$ has an infinite reduction. Observe that by Proposition 19.2 we may assume that L is a normal form. This means that the infinite reduction is sustained by just one source: the 'material' present in the context C. In the course of the infinite reduction, L will be multiplied in several descendants and the only contribution of L to sustaining the infinite reduction is that a descendant of L, L^*, is activated such that L^*P eventually will develop into a redex and will be contracted.

Indeed, if no descendant of L ever would be activated, all activity would be due to the context. In that case we also have an infinite reduction after replacing L by M.

Given the fact that $C[L]$ has an infinite reduction, we want to construct an infinite reduction of $C[M]$. This is done by gradually replacing all descendants of L by M, in the following manner: as soon as a descendant of L is activated, we replace it by M. Because of $(\ddagger)$, this replacement does not loose the possibility of an infinite reduction. Performing this infinite reduction in the so obtained new context, we again wait until the first of the remaining descendants of L is activated and replace it again by M. This procedure is repeated ad infinitum. In each step of the procedure, we gain some finite piece of the reduction of $C[M]$; if the procedure stops because no more descendants of L exist, or will be activated, then we gain an infinite reduction of $C[M]$.

In the following, we make this intuitive description more precise. We deviate from the practice up to now and allow for contexts with several holes. If C is a context with n holes, we write $C[L_1, \ldots, L_n]$ for the term obtained from C by replacing the holes by $L_1, \ldots, L_n$ in that order. Moreover, we write $C[L, \ldots, L] \twoheadrightarrow C'[L, \ldots, L]$ if the occurrences of L displayed in $C'[L, \ldots, L]$ are precisely the descendants of the occurrences of L displayed in $C[L, \ldots, L]$.

Proposition 20. *Let L be a normal form.*

1. *Let*
$$C[L, \ldots, L] \to \cdots \to C'[L, \ldots, L^* P[L, \ldots, L], \ldots, L]$$
be a reduction until for the first time a descendant (displayed as L^*) of one of the L's shown in $C[L, \ldots, L]$ is activated. Then for every M,
$$C[M, \ldots, M] \to \cdots \to C'[M, \ldots, MP[M, \ldots, M], \ldots, M]$$
is a reduction obtained by replacing every descendant of the L's by M.
2. *Let $C[L, \ldots, L] \twoheadrightarrow \cdots$ be an infinite reduction in which no descendant of the displayed L's ever is activated. Then for every M, $C[M, \ldots, M] \twoheadrightarrow \cdots$ is an infinite reduction obtained by replacing every descendant of the L's by M.*

Proof. Routine. □

Theorem 21. *The final collapse of $\mathfrak{A}_{\mathrm{SN}}$ is extensional.*

Proof. Let L, M be normal forms such that

$$(\ddagger) \quad \forall C \in \mathcal{C}_{\mathrm{SN}}\, \forall N \in \mathrm{SN}^0\, (C[LN] \in \mathrm{SN} \leftrightarrow C[MN] \in \mathrm{SN})\ .$$

We shall prove that $C[L] \in \mathrm{SN}$ iff $C[M] \in \mathrm{SN}$ for all $C \in \mathcal{C}_{\mathrm{SN}}$. Suppose this is not the case, say $C[L] \notin \mathrm{SN}$ and $C[M] \in \mathrm{SN}$ for some $C \in \mathcal{C}_{\mathrm{SN}}$. We shall derive a contradiction by constructing an infinite reduction of $C[M]$ as follows: Let $\mathcal{R} : C[L] \twoheadrightarrow \cdots$ be an infinite reduction. If no descendant of L ever is activated, then $\mathcal{R}' : C[M] \twoheadrightarrow \cdots$ obtained by replacing every descendant of L by M is an infinite reduction by Proposition 20.2. Otherwise we consider the initial part of $\mathcal{R}$ up to the first moment in which some descendant of L is activated:

$$C[L] \to \cdots \to C^*[L, \ldots, L^* P[L, \ldots, L], \ldots, L]\ .$$

This is the A_0B_1-edge in the diagram below. Now replace the activated descendant of L by M. Observe that this term stays infinite (i.e. is not strongly normalizing). For, either

(i) $C^*[L, \ldots, \Box, \ldots, L] \in \mathcal{C}_{\mathrm{SN}}$ and $P[L, \ldots, L] \in \mathrm{SN}$: then we can apply $(\ddagger)$, or

(ii) $C^*[L, \ldots, \Box, \ldots, L] \notin \mathcal{C}_{SN}$: then $C^*[L, \ldots, \Box, \ldots, L]$ contains a subterm that is not strongly normalizing and hence $C^*[L, \ldots, MP[L, \ldots, L], \ldots, L] \notin$ SN, or

(iii) $P[L, \ldots, L] \notin$ SN: then also $C^*[L, \ldots, MP[L, \ldots, L], \ldots, L] \notin$ SN.

In case of a final S-redex contraction, there may be another activated descendant of L. That is,

$$C^*[L, \ldots, MP[L, \ldots, L], \ldots, L] \equiv C^{\maltese}[L, \ldots, L^{\maltese}Q[L, \ldots, L], \ldots, L] .$$

In this case we replace also $L^{\maltese}$ by M. Applying (i)-(iii) a second time, we find that this new term stays also infinite. By Proposition 20.1, we have a reduction

$$C[M] \to \cdots \to C^*[M, \ldots, MP[M, \ldots, M], \ldots, M]$$

which we depict by the D_0D_1-edge in the diagram. Observe that

$$C^*[M, \ldots, MP[M, \ldots, M], \ldots, M] \equiv C^{\maltese}[M, \ldots, MQ[M, \ldots, M], \ldots, M] .$$

We now reiterate this procedure, using instead of $\mathcal{R}$ an infinite reduction

$$\mathcal{R}^* : C^*[L, \ldots, MP[L, \ldots, L], \ldots, L] \twoheadrightarrow \cdots$$

or

$$\mathcal{R}^{\maltese} : C^{\maltese}[L, \ldots, MQ[L, \ldots, L], \ldots, L] \twoheadrightarrow \cdots$$

which corresponds to the horizontal edge starting in point A_1. Note that L^* and $L^{\maltese}$ changed into M is now part of the context. If there are no descendants of L left, then

$$C^*[L, \ldots, MP[L, \ldots, L], \ldots, L] \equiv C^*[MP] \equiv C^*[M, \ldots, MP[M, \ldots, M], \ldots, M]$$

and we are done:

$$C[M] \to \cdots \to C^*[MP] \twoheadrightarrow \cdots$$

is the wanted infinite reduction. Likewise for $C^{\maltese}$. We are also done, if no descendant of L ever is activated in $\mathcal{R}^*$ or $\mathcal{R}^{\maltese}$. For, in that case we obtain an infinite reduction

$$C[M] \to \cdots \to C^*[M, \ldots, MP[M, \ldots, M], \ldots, M] \twoheadrightarrow \cdots$$

by Proposition 19.2; likewise for $C^{\maltese}$. In the remaining case, we consider the initial part of $\mathcal{R}^*$ ($\mathcal{R}^{\maltese}$) up to the first moment in which a descendant of the remaining descendants of L is activated. This is the A_1B_2-edge in the diagram. Employing Proposition 19.1, we gain the edge D_1D_2. In this way, we proceed ad infinitum.

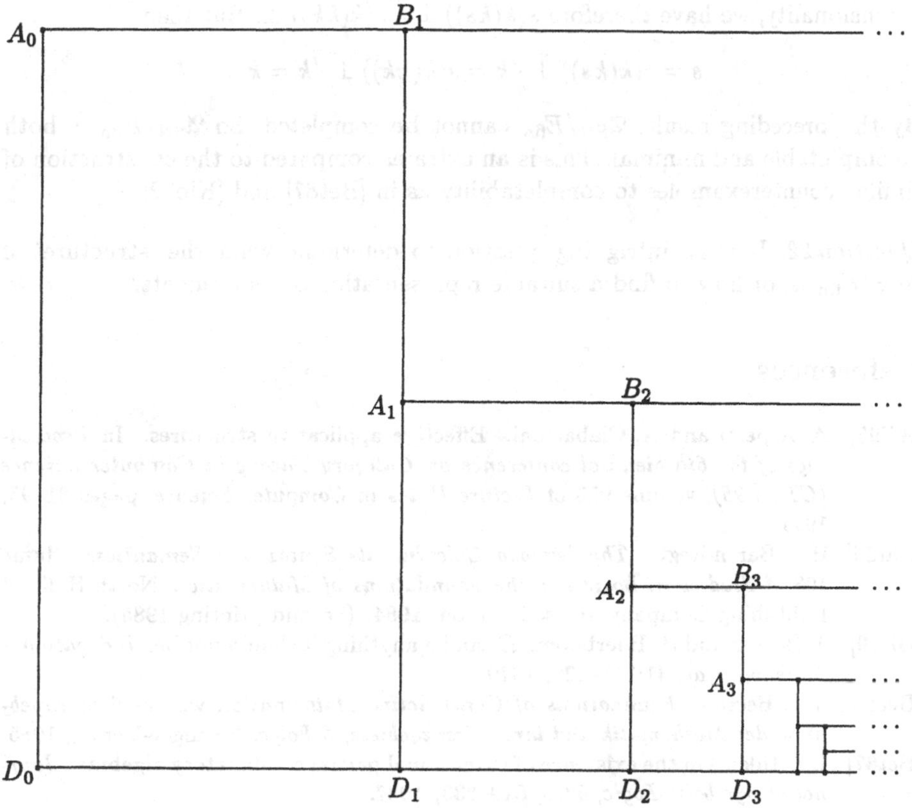

□

Remark. Let $\mathfrak{A} = \langle A, s, k, \cdot\rangle$ be a pca. We call $\ker(\mathfrak{A})$, the *kernel* of $\mathfrak{A}$, the subset of A containing all elements generated by k and s. So $\ker(\mathfrak{A})$ is defined by:

1. $k, s \in \ker(\mathfrak{A})$, and
2. if $a, a' \in \ker(\mathfrak{A})$ and $aa' \downarrow$, then $aa' \in \ker(\mathfrak{A})$.

In case $\ker(\mathfrak{A}) = A$, we call $\mathfrak{A}$ a *minimal* pca. Note that the nca $\mathfrak{A}_{\mathrm{SN}}/E_{\mathrm{fin}}$ is in fact minimal.

As observed in [Bet87], extensional nca's cannot be *completed* to a ca by adding some elements and completing the application operation. For, suppose $\mathfrak{A} = \langle A, s, k, \cdot\rangle$ is an extensional nca and $\mathfrak{A}'$ is some completion of $\mathfrak{A}$. Choose $a, a' \in A$ such that $aa' \uparrow$ and put $\bot \equiv s(ka)(ka')$. Observe that $\bot\, a'' \uparrow$ for every $a'' \in A$, and hence $s(k(kk))\,\bot\, a'' \uparrow$ and $s(k(ks))\,\bot\, a'' \uparrow$ for every $a'' \in A$. By extensionality, we have therefore $s(k(ks)) \perp = s(k(kk)) \perp$. But then

$$s = s(k(ks)) \perp \cdot' k = s(k(kk)) \perp \cdot' k = k \ .$$

By the preceding result, $\mathfrak{A}_{\mathrm{SN}}/E_{\mathrm{fin}}$ cannot be completed. So $\mathfrak{A}_{\mathrm{SN}}/E_{\mathrm{fin}}$ is both incompletable and minimal. This is an extra as compared to the construction of similar counterexamples to completability as in [Bet87] and [Klo82].

Question 22. It is an intriguing question to determine what the 'structure' of $\mathfrak{A}_{\mathrm{SN}}/E_{\mathrm{fin}}$ is, or how to find a suitable representation of its elements.

References

[AC95] A. Asperti and A. Ciabattoni. Effective applicative structures. In *Proceedings of the 6th biennial conference on Category Theory in Computer Science (CTCS'95)*, volume 953 of *Lecture Notes in Computer Science*, pages 81–95, 1995.

[Bar84] H.P. Barendregt. *The Lambda Calculus, its Syntax and Semantics*, volume 103 of *Studies in Logic and the Foundations of Mathematics*. North-Holland Publishing Company, revised edition, 1984. (Second printing 1985).

[BB79] J. Baeten and B. Boerboom. *Ω* can be anything it should not be. *Indagationes Mathematicae*, 41:111–120, 1979.

[Bee85] M.J. Beeson. *Foundations of Constructive Mathematics*, volume 6 of *Ergebnisse der Mathematik und ihrer Grenzgebiete, 3.Folge*. Springer-Verlag, 1985.

[Bet87] I. Bethke. On the existence of extensional partial combinatory algebras. *Journal of Symbolic Logic*, 52(3):819–833, 1987.

[Bet91] I. Bethke. Finite type structures within combinatory algebras. *Annals of Pure and Applied Logic*, 55:101–123, 1991.

[BI93] A. Berarducci and B. Intrigila. Some new results on easy lambda-terms. *Theoretical Computer Science*, 121:71–88, 1993.

[Bur82] P. Burmeister. Partial algebras - survey of a unifying approach towards a two-valued model theory for partial algebras. *Algebra Universalis*, 15:306–358, 1982.

[DJ90] N. Dershowitz and J.-P. Jouannaud. Rewrite systems. In J. van Leeuwen, editor, *Formal Methods and Semantics, Handbook of Theoretical Computer Science, Volume B*, chapter 6, pages 243–320. MIT Press, 1990.

[Eng81] E. Engeler. Algebras and combinators. *Algebra Universalis*, 13:389–392, 1981.

[Grä79] G. Grätzer. *Universal algebra. Second edition*. Springer-Verlag, 1979.

[HS86] J.R. Hindley and J.P. Seldin. *Introduction to Combinators and λ-calculus*, volume 1 of *London Mathematical Society Student Texts*. Cambridge University Press, 1986.

[Jac75] G. Jacopini. A condition for identifying two elements of whatever model of combinatory logic. In C. Böhm, editor, *λ-Calculus and Computer Science Theory*, volume 37 of *Lecture Notes in Computer Science*, pages 213–219. Springer-Verlag, 1975.

[JZ85] G. Jacopini and M. Venturini Zilli. Easy terms in the lambda-calculus. *Fundamenta Informaticae*, VIII(2):225–233, 1985.

[Klo82] J.W. Klop. Extending partial combinatory algebras. *Bulletin of the European Association for Theoretical Computer Science*, 16:472–482, 1982.

[Klo92] J.W. Klop. Term rewriting systems. In D. Gabbay S. Abramsky and T. Maibaum, editors, *Handbook of Logic in Computer Science, Volume II.* Oxford University Press, 1992.

[Mei95] K. Meinke. A survey of higher-order algebra. Technical Report U.U.D.M. 1995:39, ISSN 1101-3591, University of Uppsala, 1995.

[Mit9?] J.C. Mitchell. *Introduction to Programming Language Theory.* Forthcoming.

[Plo72] G.D. Plotkin. A set-theoretical definition of application. Technical Report Memorandum MIP-R-95, School of Artificial Intelligence, University of Edinburgh, 1972.

[Sco76] D.S. Scott. Data types as lattices. *SIAM J. Comput.*, 5:522–587, 1976.

[Str91] T. Streicher. *Semantics of type theory : correctness, completeness and independence results.* Birkhäuser Verlag, 1991.

[Str68] H.R. Strong. Algebraically generalized recursive function theory. *IBM J. Research and Development*, 12:465–475, 1968.

[Swan79] Open problems; Swansea λ-calculus meeting; 21 September 1979. *Bulletin of the European Association for Theoretical Computer Science*, 10:136–140, 1980.

[TvD88] A.S. Troelstra and D. van Dalen. *Constructivism in Mathematics. An Introduction. Volume II*, volume 123 of *Studies in Logic and the Foundations of Mathematics.* North-Holland Publishing Company, 1988.

[Wag69] E. Wagner. Uniformly reflexive structures. *Trans. American Math. Society*, 144:1–41, 1969.

A Complete Proof System for Nested Term Graphs

Stefan Blom

Department of Computer Science, Vrije Universiteit Amsterdam,
de Boelelaan 1081, 1018 HV Amsterdam, The Netherlands

Abstract. Nested Term Graphs are syntactic representations of cyclic term graphs. Via a simple translation they contain μ-terms as a subset. There exists a characterization of the μ-terms that unwind to the same tree, presented as a complete proof system. This paper gives a similar characterization for Nested Term Graphs. The semantics of tree unwinding is presented via bisimulations.

1 Introduction

Sharing and cycles are essential features in a program development system [10]. In fact, they occur at the source level, after parsing, in the intermediate program representation and during program execution. However, traditional computational models such as term rewriting systems [8] and λ-calculus [3] do not allow to reason about them directly. A natural way to capture sharing and cycles is to rewrite graphs instead of terms. Since the graphs considered here have locally the same structure as terms, we will call them term graphs as is nowadays common practice [11]. A term graph can be represented as sets of nodes and edges [4, 12, 5], as terms with loop-back operators [7, 13, 6] or as sets of recursive equations [1]. In this paper we will follow the latter approach.

Actually we will consider not only 'flat' systems of recursion equations, but also allow nesting of the recursion construct in the same way as it was done in [2] for the λ-calculus. This is inspired by the simple and elegant μ-calculus in which many recursive behaviors can be expressed. Clearly the μ-calculus can also be viewed as a simple form of a calculus for recursion expressions where nesting is present. However the μ-calculus is not as expressive as one would wish, since it does not allow 'subterm sharing' (In the terminology of [1], the μ-calculus admits 'vertical sharing', but not 'horizontal sharing'.) On the other hand there is a simple complete proof system for the μ-calculus [1], where complete refers to the semantics of μ-terms obtained by infinite tree unwinding, or what is equivalent, to bisimilarity. This proof system is very reminiscent of the complete proof system in [9] for recursive expressions in the area of communicating processes. In the present paper we generalize μ-terms to our nested term graphs (NTG's) and also construct a complete proof system for these NTG's. Here the semantics is again given by bisimilarity.

Historically, recursive or iterative expressions have been studied widely and deeply [7, 13, 6]. There are two major points of difference between this 'classical' work and the present paper. First, the μ-calculus and the present generalized system of NTG's employ the feature of bound variables as in λ-calculus. This has both advantages and disadvantages; a disadvantage is the need for a careful treatment of renaming just as in λ-calculus, an advantage is the greater ease of expressing what one wants. The same situation arises in λ-calculus versus the variable-free system of combinatory logic, and we will refrain from further discussing this issue here. An advantage of the approach in [7, 13, 6] is that the treatment can be more algebraic since there are only first order operators like Kleene star or, dagger. By contrast, our treatment can be called 'higher-order', where this phrase is meant to indicate the presence of bound variables, thus going beyond the first-order framework. Another reason for nesting is, that nesting allows application of term rewriting on a same syntactic level, just as the μ-calculus does. After a minor syntactic change the framework of [7, 13, 6] allows for this too, but the way pointers are represented causes a lot of very small very simple steps when one has to unwind the graph to expose a redex. While the notion of term graph rewriting in [4, 12, 5] has more expressive power in general than term rewriting, term rewriting is easier to reason about and the term graph rewriting systems of main interest to this line of research can be expressed as simple extensions of term rewriting systems. The second major difference between our work and that in [7, 13, 6] is the following. As semantics for term graphs and μ-terms infinite tree unwinding is used. In [1] it was shown that bisimulation gives the same equivalence as infinite tree unwinding. The work on terms with loop-back operators uses bisimulation as semantics when soundness is considered, but in many cases proves completeness only with respect to graph isomorphism. Completeness with respect to bisimulation is proven for restricted versions of the theory. These restricted versions are applicable to term graphs, but only after encoding of the problem.

There are many NTG's representing the same graph and many graphs with the same semantics. The natural question to ask is: which NTG's have the same semantics? This paper answers the question in the form of a complete proof system. We will use bisimulation instead of unwinding as the basic notion of equality on NTG's. Finally, we want to remark that our work can be seen as a first step in developing the proof theory of the fundamental programming constructs let and letrec.

The rest of the paper is organized as follows: In section 2 we introduce the notion of nested term graph and related notions such as substitution on NTG's. In section 3 we introduce the notion of equality on NTG's: bisimulation. Actually bisimulation is defined on term graphs so we must translate NTG's to term graphs. The translation is called the underlying graph. After this we are ready to specify the proof system in section 4 and prove its soundness (section 5) and completeness (section 6).

2 Syntax

We will adopt the convention that $\mathcal{V}$ is a set of variables with typical members α, $\beta, \gamma, \delta, \alpha_1, \alpha_2, \ldots$ and that $\mathcal{F}$ is a set of function symbols $f, g, h, F, G, H, \ldots$ each with a given fixed arity. By arity(f) we mean the arity of the function symbol f, constants are function symbols with arity 0. There is a special constant $\bullet$, the ***black hole***.

Definition 1. A ***nested term graph*** can be:

- α, for each $\alpha \in \mathcal{V}$,
- $f(t_1, \ldots, t_n)$ for each n-ary $f \in \mathcal{F}$ and NTG's $t_1, \ldots, t_n$ or
- $\langle s | \alpha_1 = t_1, \ldots, \alpha_n = t_n \rangle$, for NTG's $s, t_1, \ldots, t_n$ and variables $\alpha_1, \ldots, \alpha_n$ which are pairwise different.

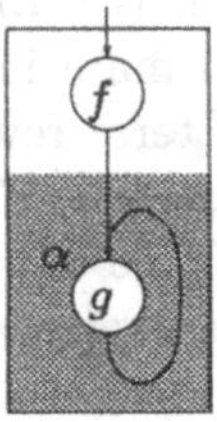

Fig. 1. The picture of $\langle f(\alpha) | \alpha = g(\alpha) \rangle$.

We will now introduce some terminology and notation. Some of the terminology has its roots in the pictures we can draw of NTG's. An example of such a picture is given in Fig. 1.

- Instead of $\langle s | E \rangle$ we will often write s^E.
- Given a nested term graph of the form $\langle s | \alpha_1 = t_1, \ldots, \alpha_n = t_n \rangle$, we refer to the entire NTG as a ***box***, to the s as the ***external part***, and to the set of equations $\alpha_1 = t_1, \ldots, \alpha_n = t_n$ as the ***internal part*** or the ***environment***. The symbols $E, F, \ldots$ are normally used to match environments.
- A NTG without box constructs is a ***term***.
- A nested term graph is ***guarded*** if it is of the form $f(t_1, \ldots, t_n)$.
- A nested term graph of the form α^E where all right-hand sides of the equations in E are of the form $f(\alpha_1, \ldots, \alpha_n)$ is called ***flat***.
- We say that an occurrence of a variable is a ***free occurrence***, if the occurrence is not inside a box, where the variable occurs as the left-hand side of an equation in the environment. If a variable occurs as the left-hand side of an equation in the internal part, then that occurrence is a ***binding occurrence*** and all free occurrences of the variable in the external part and in the right-hand sides of the environment are ***bound*** by this left-hand side occurrence.

– For any nested term graph x or environment x we define $\text{var}(x)$ to be the set of variables occurring in x and $\text{free}(x)$ to be the set of variables having at least one free occurrence in x. We define $\text{def}(\{\alpha_1 = t_1, \ldots, \alpha_n = t_n\})$ to be the set of variables the environment defines: $\{\alpha_1, \ldots, \alpha_n\}$.

Example 1. (free and bound variables)

$$\langle f(\alpha_1, \langle f(\alpha_2, \beta_1) | \alpha_3 = f(\alpha_4, \gamma), \beta_2 = \bullet \rangle) | \alpha_5 = f(\alpha_6, \beta_3) \rangle$$

Here the subscripts denote the number of the occurrence. The only occurrence of γ is a free occurrence, the 1st occurrence of α is bound to the 5^{th}occurrence, the $2^{\text{nd}}$and 4^{th}occurrences of α are bound to the 3^{rd}occurrence, the 1^{st}occurrence of β is bound to the $2^{\text{nd}}$and the 3^{rd}occurrence of β is free.

Definition 2. When t is obtained from s by replacing in s a binding occurrence of a variable and all occurrences bound to that occurrence by a variable that does not occur is s, we say that $s =_\alpha t$. We denote by the same $=_\alpha$ the transitive reflexive closure of the relation and we name it *α-conversion*. A nested term graph is maximally α-converted if no variable occurs both free and as a binder and no variable occurs twice as a binder. Unless stated otherwise we will not distinguish between two α-convertible nested term graphs.

Definition 3. A context $C[.]$ is a NTG with one occurrence of a constant replaced by $\Box$. By $C[s]$ we mean $C[.]$ with the box replaced by s.

Example 2. Following the definition these are contexts:

$$\langle \Box | \alpha = f(\alpha) \rangle \quad \langle \alpha | \alpha = f(\Box) \rangle \quad f(\Box)$$

and these are not:

$$\langle \Box | \alpha = f(\Box) \rangle \quad \langle \alpha | \Box = f(\alpha) \rangle$$

Definition 4. A *substitution* is a partial function σ from $\mathcal{V}$ to the set of NTG's. If the NTG the substitution is applied to satisfies the condition that the set of variables occurring in the left-hand sides in the NTG is disjoint from the set of variables the substitution is defined for and from the set of variables occurring free in the images of variables under the substitution then the substitution extends to NTG's by the following rules:

$$\alpha\sigma = \begin{cases} \sigma(\alpha) \text{ , if } \alpha \in D_\sigma \\ \alpha \quad \text{ , otherwise} \end{cases}$$

$$f(t_1, \ldots, t_n)\sigma = f(t_1\sigma, \ldots, t_n\sigma)$$

$$\langle s | \alpha_1 = t_1, \ldots, \alpha_n = t_n \rangle\sigma = \langle s\sigma | \alpha_1 = t_1\sigma, \ldots, \alpha_n = t_n\sigma \rangle \ .$$

Because NTG's are taken modulo α-conversion the condition can always be satisfied.

3 Bisimulation

We want to consider NTG's equivalent if they unwind to the same rational tree. On term graphs there is the equivalent notion of bisimilarity ([1]). This notion is useful because it is decidable and it will give us the strategy for the completeness proof later on. To define bisimilarity on NTG's we first associate a term graph with every NTG. This term graph, called the underlying graph, forgets about boxes. The underlying graph is defined using an i-successor relation on occurrences of function and variable symbols.

Definition 5. We define i-successor on occurrences of function and variable symbols as follows:

- $x \to_0 y$ if x is the left-hand side of an equation and y is the head symbol of the same equation.
- $x \to_0 y$ if x is a bound variable and y is the left-hand side of the equation x is bound to.
- $x \to_0 x$ if x is an occurrence of the black hole $\bullet$.

We put $x \sim_0 y$ if x and y are free occurrences of the same variable.

The relation $\sim$ on occurrences of both variables and function symbols is the equivalence relation generated by the union of $\to_0$ and $\sim_0$.

We say $x \to_i y$ $(i > 0)$ if x is the occurrence of a function symbol and y is the head symbol of the ith argument of that symbol. We denote the equivalence class that a symbol x belongs to by $\overline{x}$.

The head symbol is the same as the left most symbol of any NTG and is denoted by $\mathrm{lm}(t)$ for any NTG t. The head symbol of a sub NTG t of $C[t]$ is denoted by $\mathrm{lm}(t \subset C[t])$.

Definition 6. The *underlying graph* of a nested term graph t , notation $\mathrm{graph}(t)$, is a term graph.

The nodes are the equivalence classes of the equivalence relation $\sim$. The class a symbol x belongs to is denoted by $\overline{x}$. If a class contains an occurrence of a function symbol that function symbol is the label. If a class contains an occurrence of a free variable that variable is the label. Otherwise $\bullet$ is the label.

If $x \to_i y$ $(i > 0)$ then $\overline{y}$ is the ith argument of $\overline{x}$. We will denote this by $\overline{x} \to_i \overline{y}$.

The relations $\to_i$ $(i > 0)$ are now defined on both occurrences of symbols in a NTG and on nodes of the underlying graphs. This overloading is harmless. Note that $\to_0$ is only defined on occurrences. The underlying graph is actually a term graph.

Example 3. Figure 2 illustrates the $\to_i$ relations and the equivalence classes in the construction of the underlying graph of $\langle f(\alpha) | \alpha = g(\langle \alpha | \alpha = f(\beta) \rangle), \beta = f(\alpha) \rangle$. The resulting underlying graph is drawn in Fig. 3.

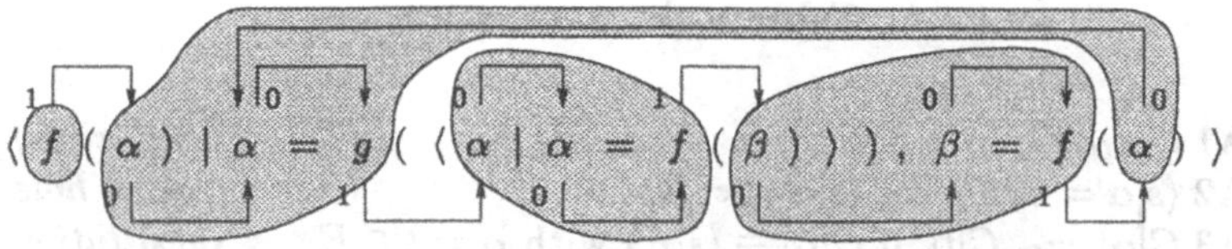

Fig. 2. The construction of an underlying graph

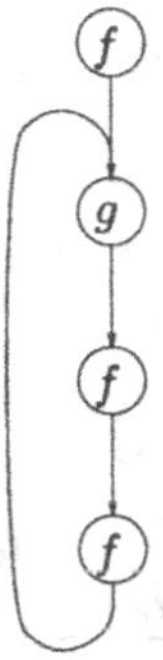

Fig. 3. An example of an underlying graph

Definition 7. For two graphs g and h a relation $R \subset \text{nodes}(g) \times \text{nodes}(h)$ is a *bisimulation* if

$$(\text{root}(g), \text{root}(h)) \in R$$

and

$$\forall (a, b) \in R : \text{label}(a) = \text{label}(b) \wedge \forall a', b', i : (a \rightarrow_i a' \wedge b \rightarrow_i b') \Rightarrow ((a', b') \in R) \ .$$

If such a relation exists we say g and h are *bisimilar* ($g \leftrightarrow h$). We say that two NTG's s and t are *bisimilar*, denoted $s \leftrightarrow t$ if $\text{graph}(s) \leftrightarrow \text{graph}(t)$.

4 The Proof System

We assume that the expressions occurring in the proof system are maximally α-converted, i.e. all variables occurring as left-hand sides are pairwise different, also if they occur in different environments, and there is no overlap between free and bound variables. The equations we will reason about in the proof system will be written $s =_{\mathsf{P}} t$ to distinguish the equal sign used from the equal sign in the syntax of NTG's. If there is no confusion we will sometimes write $s = t$.

Some remarks now on how to read the system. Reflexivity, symmetry and transitivity are obvious. The black hole rule is there to express that defining $\alpha = \alpha$ is the same as explicitly saying it is undefined. Since the undefinedness

Table 1. the proof system

A1	$t =_{\mathsf{p}} t$	*reflexivity*
A2	$\langle s \mid \alpha = \alpha, E\rangle =_{\mathsf{p}} \langle s \mid \alpha = \bullet, E\rangle$	*black hole*
A3	$C[\alpha] =_{\mathsf{p}} C[t]$, if $C[\alpha] \equiv \langle s \mid E\rangle$ with $\alpha = t \in E$	*substitution*
A4	$\langle s \mid\rangle =_{\mathsf{p}} s$	*box removal*
A5	$f(s_1,\ldots,s_n)^E =_{\mathsf{p}} f(s_1^E,\ldots,s_n^E)$	*unwinding*
A6	$s^{E,F} =_{\mathsf{p}} s^E$, if $\mathrm{def}(F) \cap \mathrm{free}(s^E) = \emptyset$	*garbage collection*
R1	$\dfrac{s =_{\mathsf{p}} t}{t =_{\mathsf{p}} s}$	*symmetry*
R2	$\dfrac{s =_{\mathsf{p}} t \quad t =_{\mathsf{p}} u}{s =_{\mathsf{p}} u}$	*transitivity*
R3	$\dfrac{s =_{\mathsf{p}} t}{C[s] =_{\mathsf{p}} C[t]}$	*context*
R4	$\dfrac{\begin{array}{c} s_1 =_{\mathsf{p}} t_1\{s_1/\alpha_1,\ldots,s_n/\alpha_n\} \\ \vdots \\ s_n =_{\mathsf{p}} t_n\{s_1/\alpha_1,\ldots,s_n/\alpha_n\}\end{array}}{s_1 =_{\mathsf{p}} \langle \alpha_1 \mid \alpha_1 = t_1,\ldots,\alpha_n = t_n\rangle}$, for guarded $t_1,\ldots,t_n$	*folding*

is of a special kind we use a special symbol for it $\bullet$. Substitution requires some care. We don't want e.g.

$$\langle\langle \alpha \mid \beta = A\rangle \mid \alpha = F(\beta)\rangle =_{\mathsf{p}} \langle\langle F(\beta) \mid \beta = A\rangle \mid \alpha = F(\beta)\rangle$$

to be derivable as it is clearly wrong. The cause of the trouble is that there are both free and bound occurrences of β. The maximal α-conversion convention saves us here. Box removal and garbage collection are used to get rid of old unused definitions. Unwinding or distribution expresses the equivalence of shared definitions and identical local definitions. The context rule needs to be read carefully. Maximal α-conversion doesn't mean that variables that occur free in s or t in the premise are still free in the conclusion. I.e. deriving

$$\langle \alpha \mid \alpha = A\rangle =_{\mathsf{p}} \langle\langle \alpha \mid\rangle \mid \alpha = A\rangle$$

from

$$\alpha =_{\mathsf{p}} \langle \alpha \mid\rangle$$

is legal. The context rule is used to replace any subterm by an equal one. The folding rule states that a system of guarded recursive equations has a solution of a special form.

Definition 8. We say that s and t are semantically equal iff s and t are bisimilar:

$$\models s =_{\mathsf{p}} t \iff s \leftrightarrows t\ .$$

Table 2. The formal proof from Ex. 4

1	$\langle f(\alpha)|\alpha = \langle \beta|\beta = f(\alpha)\rangle\rangle =_p f(\langle \alpha|\alpha = \langle \beta|\beta = f(\alpha)\rangle\rangle)$	unwinding
2	$\langle \alpha|\alpha = \langle \beta|\beta = f(\alpha)\rangle\rangle =_p \langle\langle \beta|\beta = f(\alpha)\rangle|\alpha = \langle \beta|\beta = f(\alpha)\rangle\rangle$	substitution
3	$\langle \beta|\beta = f(\alpha)\rangle =_p \langle f(\alpha)|\beta = f(\alpha)\rangle$	substitution
4	$\langle f(\alpha)|\beta = f(\alpha)\rangle =_p \langle f(\alpha)|\rangle$	garbage collection
5	$\langle f(\alpha)|\rangle =_p f(\alpha)$	box removal
6	$\langle \beta|\beta = f(\alpha)\rangle =_p f(\alpha)$	transitivity, 3, 4, 5
7	$\langle\langle \beta|\beta = f(\alpha)\rangle|\alpha = \langle \beta|\beta = f(\alpha)\rangle\rangle =_p \langle f(\alpha)|\alpha = \langle \beta|\beta = f(\alpha)\rangle\rangle$	context, 6
8	$\langle f(\alpha)|\alpha = \langle \beta|\beta = f(\alpha)\rangle\rangle =_p f(\langle \alpha|\alpha = \langle \beta|\beta = f(\alpha)\rangle\rangle)$	unwinding
9	$\langle \alpha|\alpha = \langle \beta|\beta = f(\alpha)\rangle\rangle =_p f(\langle \alpha|\alpha = \langle \beta|\beta = f(\alpha)\rangle\rangle)$	transitivity, 2, 7, 8
10	$\langle \alpha|\alpha = \langle \beta|\beta = f(\alpha)\rangle\rangle =_p \langle \delta|\delta = f(\delta)\rangle$	folding, 9
11	$f(\langle \alpha|\alpha = \langle \beta|\beta = f(\alpha)\rangle\rangle) =_p f(\langle \delta|\delta = f(\delta)\rangle)$	context, 10
12	$\langle f(\alpha)|\alpha = \langle \beta|\beta = f(\alpha)\rangle\rangle =_p f(\langle \delta|\delta = f(\delta)\rangle)$	transitivity, 1, 11
13	$\langle \alpha|\alpha = f(\alpha)\rangle =_p \langle f(\alpha)|\alpha = f(\alpha)\rangle$	substitution
14	$\langle f(\alpha)|\alpha = f(\alpha)\rangle =_p f(\langle \alpha|\alpha = f(\alpha)\rangle)$	unwinding
15	$\langle \alpha|\alpha = f(\alpha)\rangle =_p f(\langle \alpha|\alpha = f(\alpha)\rangle)$	transitivity, 13, 14
16	$f(\langle \alpha|\alpha = f(\alpha)\rangle) =_p \langle \alpha|\alpha = f(\alpha)\rangle$	symmetry, 15
17	$\langle f(\alpha)|\alpha = \langle \beta|\beta = f(\alpha)\rangle\rangle =_p \langle \alpha|\alpha = f(\alpha)\rangle$	transitivity, 12, 16

We say that an equation $s = t$ is provable or derivable:

$$\vdash s =_p t$$

if there is a list of equations $E_1, E_2, \ldots E_n$, where $E_n \equiv s = t$, such that each equation E_i is either an instantiation of an axiom in the proof system or the result of an instantiation of a rule where the instantiations of the premises of the rule are present in the list prior to E_i itself. Such a list will be called a derivation from now on.

The system is not minimal. As an example the reflexivity axiom is derived from the other axioms and rules:

$$1 \ \langle t|\rangle = t \ \text{A4}$$
$$2 \ t = \langle t|\rangle \ \text{R1}, 1$$
$$3 \ t = t \quad \text{R2}, 2, 1$$

Example 4. In table 2 the formal proof of

$$\vdash \langle f(\alpha)|\alpha = \langle \beta|\beta = f(\alpha)\rangle\rangle =_p \langle \alpha|\alpha = f(\alpha)\rangle$$

is given. The proof doesn't use any particular strategy.

5 Soundness

This section is rather technical and uses some notions from rewriting theory: If $\rightarrow$ is a relation $\twoheadrightarrow$ is the transitive reflexive closure. A normal form is an element a such that there is no b with $a \rightarrow b$. An element x is terminating in a relation $\rightarrow$ (notation $SN^{\rightarrow}(x)$) if there is no sequence $x \equiv x_0 \rightarrow x_1 \rightarrow x_2 \rightarrow \ldots$.

Theorem 9 (soundness). *If $\vdash s =_p t$ then $\models s =_p t$.*

For the proof we will make use of a characterization of bisimulation on the syntactic level. We will give a sufficient condition for a relation on the occurrences of two NTG's. Such a relation will be called an extensible relation or a syntax level bisimulation.

Definition 10. If R is a relation on occurrences of symbols in two NTG's s and t we define $x\ R^{\to\circ}\ y$ iff $\exists x', y' : x \twoheadrightarrow_0 x' \wedge y \twoheadrightarrow_0 y' \wedge x'\ R\ y'$

A relation R on occurrences of symbols in two NTG's s and t is extensible if it satisfies the following conditions:

- $\operatorname{lm}(s)\ R^{\to\circ}\ \operatorname{lm}(t)$
- If $x\ R\ y$, $x \twoheadrightarrow_0 x'$ and $y \twoheadrightarrow_0 y'$ then $x'\ R^{\to\circ}\ y'$
- If $x\ R\ y$ then $SN^{\to\circ}(x)$ iff $SN^{\to\circ}(y)$
- If $x\ R\ y$, x and $y \twoheadrightarrow_0$ normal forms then x and y are occurrences of the same symbol.
- If $x\ R\ y$, $x \to_i x'$ and $y \to_i y'$ $(i > 0)$ then $x'\ R^{\to\circ}\ y'$.

Lemma 11. *If R is an extensible relation between s and t then $s \leftrightarrow t$.*

Proof. For any relation R we define $C_1\ R^*\ C_2$ if $\exists x, y : x\ R\ y$, $\overline{x} = C_1$ and $\overline{y} = C_2$. If R is extensible we have that R^* is a bisimulation. Note that $R^* = (R^{\to\circ})^*$. This means that $root(s)\ R\ root(t)$, because $root(s) = \overline{\operatorname{lm}(s)}$, $root(t) = \overline{\operatorname{lm}(t)}$ and $\operatorname{lm}(s)\ R^{\to\circ}\ \operatorname{lm}(t)$. If $C_1\ R^*\ C_2$ then $C_1 = \overline{x}$ and $C_2 = \overline{y}$ for x and y with $x\ R\ y$. If x and y have no $\twoheadrightarrow_0$ normal forms then the label of C_1 is $\bullet$ just as the label of C_2. If x and y have normal forms x' and y' respectively then $x'\ R^{\to\circ}\ y'$ which means that $x'\ R\ y'$. This last statement means that x' and y' must be occurrences of the same symbol. This means that they are either free occurrences of a variable or occurrence of the same function symbol. If $C_1 \to_i D_1$ and $C_2 \to_i D_2$ $(i > 0)$ then there are $x'' \in D_1$ and $y'' \in D_2$ such that $x' \to_i x''$ and $y' \to_i y''$. This means that $x''\ R^{\to\circ}\ y''$ so $\overline{x''}\ R^*\ \overline{y''}$ or $D_1\ R^*\ D_2$ as required.

With the syntactic characterization in hand we will now prove soundness.

Proof of Thm. 9. We will do this by induction on the length of the derivation. Suppose the last step was:

A1 The identity relation is a bisimulation.
A2 The underlying graphs of left- and right-hand sides are exactly the same.
A3 Substitution has three cases. The first one is

$$\langle C[\alpha] | \alpha = s, E \rangle =_p \langle C[s] | \alpha = s, E \rangle \ .$$

We define R on occurrences as follows: $x\ R\ y$ if x and y are matching occurrences outside of the hole or matching occurrences of the s sub terms. This R satisfies the conditions of the lemma. See Ex. 5 for an instance of this construction. The second case is

$$\langle t | \gamma = C[\alpha], \alpha = s, E \rangle =_p \langle t | \gamma = C[s], \alpha = s, E \rangle \ .$$

This case is almost identical to the previous one. The third case is

$$\langle t|\alpha = C[\alpha]E\rangle =_{\mathrm{p}} \langle t|\alpha = C[C[\alpha]], E\rangle \ .$$

For this last case we have to modify R slightly: We still have $x\ R\ y$ for matching occurrences in t and E, but we add matching occurrences in the $C[.]$ part and the right hand side α occurrences.

A4 It is trivial that $\langle s|\rangle \leftrightarrow s$.

A5 The natural syntactic relation on occurrences works.

A6 The symbol occurrences of s^E form a natural subset of those of $s^{E,F}$ this gives a natural 'identity' relation R on the occurrences of s^E and $s^{E,F}$. When lifted this relation is a bisimulation.

R1 The definition of bisimilarity is symmetric.

R2 Given bisimulations R_1 for $s \leftrightarrow t$ and R_2 for $t \leftrightarrow u$ the composition $R = R_1 \circ R_2$ is a bisimulation:
- We have $\overline{\mathrm{lm}(s)}\ R_1\ \overline{\mathrm{lm}(t)}$ and $\overline{\mathrm{lm}(t)}\ R_2\ \overline{\mathrm{lm}(u)}$ so $\overline{\mathrm{lm}(s)}\ R\ \overline{\mathrm{lm}(u)}$.
- If $C_1\ R\ C_3$ then there is a C_2 such that $C_1\ R_1\ C_2$ and $C_2\ R_2\ C_3$. Because R_1 and R_2 are bisimulations we know that label(C_1) = label(C_2) and label(C_2) = label(C_3) and so label(C_1) = label(C_3).
- If $\overline{x}\ R\ \overline{y}$, $x \rightarrow_i x'$ and $y \rightarrow_i y'$ then there is a C such that $\overline{x}\ R_1\ C$ and $C\ R_2\ \overline{y}$. because the labels of $\overline{x}$ and C are the same C can be written as $C = \overline{z}$ with $z \rightarrow_i z'$. We then know $\overline{x'}\ R_1\ \overline{z'}$ and $\overline{z'}\ R_2\ \overline{y'}$ so $\overline{x'}\ R\ \overline{y'}$

R3 Given a bisimulation R between s and t and a context $C[.]$ we can define S on occurrences as $x\ S\ x$ if x is an occurrence in $C[.]$ and xSy if x is an occurrence in s and y one is t such that $\overline{x}\ R\ \overline{y}$. The relation S is extensible so $C[s] \leftrightarrow C[t]$.

R4 We are given that

$$\models s_1 =_{\mathrm{p}} t_1\{s_1/\alpha_1, \ldots, s_n/\alpha_n\}, \ldots, s_n =_{\mathrm{p}} t_n\{s_1/\alpha_1, \ldots, s_n/\alpha_n\} \ .$$

Let $R_1, \ldots, R_n$ be bisimulations that witness these equalities. We must now construct a bisimulation between s_1 and $\langle \alpha_1|\alpha_1 = t_1, \ldots, \alpha_n = t_n\rangle$.
On the syntactic level we can do this by defining a relation R on the symbol occurrences of s_1, ..., s_n, $t_1\{s_1/\alpha_1, \ldots, s_n/\alpha_n\}$, ..., $t_n\{s_1/\alpha_1, \ldots, s_n/\alpha_n\}$ and $\langle \alpha_1|\alpha_1 = t_1, \ldots, \alpha_n = t_n\rangle$: for any two symbols occurrences x in s_i and y in $t_i\{s_1/\alpha_1, \ldots, s_n/\alpha_n\}$ such that $\overline{x}\ R_i\ \overline{y}$ we have $x\ R\ y$. Because of maximal α-conversion the NTG $t_i\{s_1/\alpha_1, \ldots, s_n/\alpha_n\}$ is formed by replacing the free occurrences of α_j in t_i by s_j. For any occurrence x in $t_i\{s_1/\alpha_1, \ldots, s_n/\alpha_n\}$ that is not a replacement we have $x\ R\ y$, where y is the same occurrence in the t_i sub-NTG in $\langle \alpha_1|\alpha_1 = t_1, \ldots, \alpha_n = t_n\rangle$. For any replacement occurrence x we have $x\ R\ y$, where y is the matching occurrence in the s_j (assuming that x is in a replacement of α_j. See Ex. 5 for an example of this construction.
If we take the transitive closure of this R then restrict it to occurrences in s_1 and $\langle \alpha_1|\alpha_1 = t_1, \ldots, \alpha_n = t_n\rangle$ and then lift it to the level of nodes, the relation is a bisimulation. To prove this we will use the lemma again. By the fact that R_1 is a bisimulation:

$$\mathrm{lm}(s_1)\ R\ \mathrm{lm}(t_1\{s_1/\alpha_1, \ldots, s_n/\alpha_n\}) \ .$$

Because t_1 is guarded:

$$\mathrm{lm}(t_1\{s_1/\alpha_1,\ldots,s_n/\alpha_n) \; R \; \mathrm{lm}(t_1 \subset \langle\alpha_1|\alpha_1 = t_1,\ldots,\alpha_n = t_n\rangle) \ .$$

By definition of $\rightarrow_0$

$$\mathrm{lm}(\langle\alpha_1|\alpha_1 = t_1,\ldots,\alpha_n = t_n\rangle) \rightarrow_0^2 \mathrm{lm}(t_1 \subset \langle\alpha_1|\alpha_1 = t_1,\ldots,\alpha_n = t_n\rangle) \ .$$

□

Example 5. In Fig. 4 the auxiliary relation described in the proof of soundness is shown for $\langle\alpha|\alpha = f(\beta), \beta = f(\alpha)\rangle$ and $\langle f(\beta)|\alpha = f(\beta), \beta = f(\alpha)\rangle$.

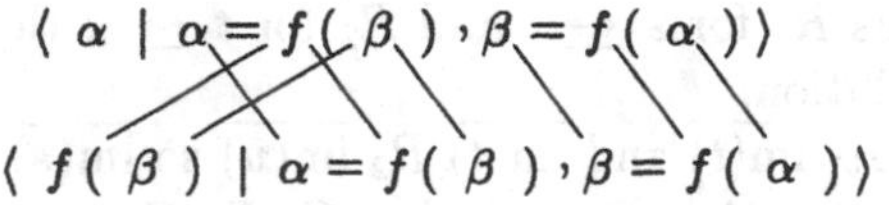

Fig. 4. soundness of substitution

In Fig. 5 the same is done for the constructed relation for the case of soundness. This picture looks rather complicated because of the many lines between

$$\langle\alpha|\alpha = f(f(\alpha))\rangle \text{ and } f(\langle f(\alpha)|\alpha = f(f(\alpha))\rangle)$$

and between

$$\langle f(\alpha)|\alpha = f(f(\alpha))\rangle \text{ and } f(\langle\alpha|\alpha = f(f(\alpha))\rangle) \ .$$

This jungle of lines comes from the fact that between these pairs we use a syntax level bisimulation generated by a graph level bisimulation. The three line types (solid, dotted and dashed) are used to distinguish between lines starting in the top NTG from symbols belonging to different nodes in the underlying graph.

6 Completeness

To prove completeness we first show that any NTG may be proven equal to a NTG of a special form: a flat NTG. We then prove completeness for term graphs only and conclude that because of soundness we also have proven completeness for arbitrary NTG's. The proof uses a derived axiom:

$$\langle s|\alpha = \beta, E\rangle =_{\mathbf{p}} \langle s\{\beta/\alpha\}|E\{\beta/\alpha\}\rangle \text{ if } \alpha \not\equiv \beta \ .$$

The idea is that we perform a simultaneous substitution of all occurrences of α by β and then apply garbage collection. This is the same as a sequence of substitution steps followed by garbage collection combined by using transitivity. The formal proof is an induction on the number of occurrences of α and is left to the reader.

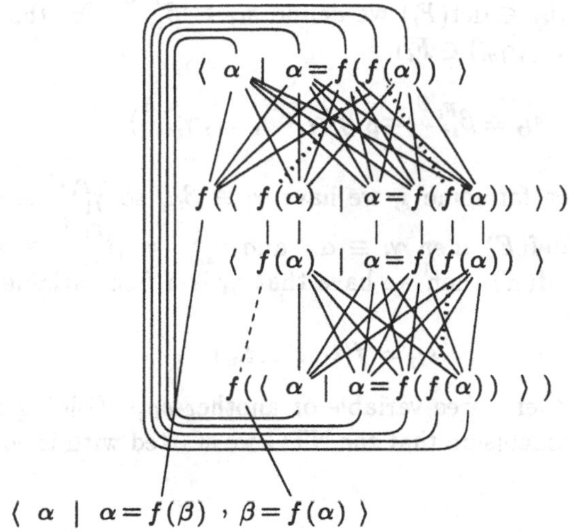

Fig. 5. soundness of folding

Theorem 12 (completeness). *If* $\models s =_{\mathsf{P}} t$ *then* $\vdash s =_{\mathsf{P}} t$.

Lemma 13. *Every NTG can be proven equal to a flat NTG.*

Proof. We will prove this statement by structural induction.

- If the given NTG is a variable α then $\alpha =_{\mathsf{P}} \langle\alpha|\rangle$.
- If the given NTG is of the form $f(t_1,\ldots,t_n)$ we know that

$$f(t_1,\ldots,t_n) =_{\mathsf{P}} f(\alpha_1^{E_1},\ldots,\alpha_n^{E_n}) \ .$$

Define $E = \cup_{i=1}^{n} E_i$ and we have that

$$\langle\alpha|\alpha = f(\alpha_1,\ldots,\alpha_n), E\rangle =_{\mathsf{P}}$$

$$=_{\mathsf{P}} f(\langle\alpha_1|\alpha = f(\alpha_1,\ldots,\alpha_n), E\rangle,\ldots,\langle\alpha_n|\alpha = f(\alpha_1,\ldots,\alpha_n), E\rangle) =_{\mathsf{P}}$$

$$=_{\mathsf{P}} f(\alpha_1^{E_1},\ldots,\alpha_n^{E_n}) \ .$$

- Given a box construct $\langle s|\alpha_1 = t_1,\ldots,\alpha_n = t_n\rangle$ we know by induction that this NTG is provably equal to one of the form $\langle\beta_0^{F_0}|\alpha_1 = \beta_1^{F_1},\ldots,\alpha_n = \beta_n^{F_n}\rangle$. This is provably equal to

$$\langle\alpha_0|\alpha_0 = \beta_0^{F_0},\alpha_1 = \beta_1^{F_1},\ldots,\alpha_n = \beta_n^{F_n}\rangle \equiv \alpha_0^{E} .$$

In this NTG we could have for some β_i that $\beta_i \notin \mathrm{def}(F_i)$. If this is the case we transform $\alpha_i = \beta_i^{F_i}$ into $\alpha_i = \beta_i$ and use either the $\alpha = \beta$ rule or the $\alpha = \alpha$ rule followed by $\bullet = \langle\beta_i|\beta_i = \bullet\rangle$. This process is terminating so we may assume that for each β_i, $\beta_i \in \mathrm{def}(F_i)$.

For each variable $\beta_{ij} \in \operatorname{def}(F_i)$ we define $s_{ij} \equiv \beta_{ij}^{F_i E}$. For this s_{ij} we can prove $(\beta_{ij} = f(\gamma_1, \ldots, \gamma_m) \in F_i)$:

$$s_{ij} \equiv \beta_{ij}^{F_i E} =_{\mathrm{P}} f(\gamma_1^{F_i E}, \ldots, \gamma_m^{F_i E}) \ .$$

If $\gamma_i \in \operatorname{def}(F_i)$ then for some j_i we have $\gamma_i \equiv \beta_{ij_i}$ so $\gamma_1^{F_i E} \equiv s_{ij_i}$. If $\gamma_i \notin \operatorname{def}(F_i)$ and $\gamma_i \in \operatorname{def}(E)$ then $\gamma_i \equiv \alpha_{i_i}$ and $\gamma_1^{F_i E} = \beta_{i_i 0}^{F_{i_i} E} \equiv s_{i_i 0}$. If $\gamma_i \notin \operatorname{def}(F_i)$ and $\gamma_i \notin \operatorname{def}(E)$ then we have that γ_i is a free variable. This gives rise to an equation

$$s_{ij} = f(t_1, \ldots, t_m)$$

where each t_i is either a free variable or another $s_{i'j'}$. Folding over the set of s_{ij}'s yields the conclusion that the NTG we started with is equal to a flat NTG.

□

Example 6. We will give a detailed derivation showing that

$$\langle f(\langle \alpha | \alpha = f(\alpha, \beta), \beta = \gamma \rangle, \delta) | \delta = \delta \rangle$$

is equal to some flat NTG. In that derivation the strategy of the proof is used. The level of indentation indicates the depth of recursion and the NTG between brackets on each level of indentation is the NTG we are proving equal to a term graph at that point.

$[\langle f(\langle \alpha | \alpha = f(\alpha, \beta), \beta = \gamma \rangle, \delta) | \delta = \delta \rangle]$
  $[f(\langle \alpha | \alpha = f(\alpha, \beta), \beta = \gamma \rangle, \delta)]$
    $[\langle \alpha | \alpha = f(\alpha, \beta), \beta = \gamma \rangle]$
      $[\alpha]\ \alpha = \langle \alpha | \rangle$
      $[f(\alpha, \beta)]$
        $[\alpha]\ \alpha = \langle \alpha | \rangle$
        $[\beta]\ \beta = \langle \beta | \rangle$
        $f(\alpha, \beta) = f(\langle \alpha | \rangle, \langle \beta | \rangle) = \langle \gamma | \gamma = f(\alpha, \beta) \rangle$
      $[\gamma]\ \gamma = \langle \gamma | \rangle$
      $\langle \alpha | \alpha = f(\alpha, \beta), \beta = \gamma \rangle =$
        $= \langle \delta | \delta = \langle \alpha | \rangle, \alpha = \langle \gamma' | \gamma' = f(\alpha, \beta) \rangle, \beta = \langle \gamma | \rangle \rangle =$
        $= \langle \delta | \delta = \alpha, \alpha = \langle \gamma' | \gamma' = f(\alpha, \beta) \rangle \beta = \langle \gamma | \rangle \rangle =$
        $= \langle \alpha | \alpha = \langle \gamma' | \gamma' = f(\alpha, \beta) \rangle \beta = \langle \gamma | \rangle \rangle =$
        $= \langle \alpha | \alpha = \langle \gamma' | \gamma' = f(\alpha, \gamma) \rangle \rangle$
      Giving the equation:
      $\langle \langle \gamma' | \gamma' = f(\alpha, \gamma) \rangle | \alpha = \langle \gamma' | \gamma' = f(\alpha, \gamma) \rangle \rangle =$
        $= f(\langle \langle \gamma' | \gamma' = f(\alpha, \gamma) \rangle | \alpha = \langle \gamma' | \gamma' = f(\alpha, \gamma) \rangle \rangle, \gamma)$
      So after folding
      $\langle \alpha | \alpha = f(\alpha, \beta), \beta = \gamma \rangle = \langle \alpha | \alpha = f(\alpha, \gamma) \rangle$
    $[\delta]\ \delta = \langle \delta | \rangle$

$f(\langle\alpha|\alpha = f(\alpha,\beta), \beta = \gamma\rangle, \delta) = f(\langle\alpha|\alpha = f(\alpha,\gamma)\rangle, \langle\delta|\rangle) =$
$= \langle\beta|\beta = f(\alpha,\delta), \alpha = f(\alpha,\gamma)\rangle$
$[\delta]\ \delta = \langle\delta|\rangle$
$\langle f(\langle\alpha|\alpha = f(\alpha,\beta), \beta = \gamma\rangle, \delta)|\delta = \delta\rangle =$
$= \langle\epsilon|\epsilon = \langle\beta|\beta = f(\alpha,\delta), \alpha = f(\alpha,\gamma)\rangle, \delta = \langle\delta|\rangle\rangle =$

$$= \langle\epsilon|\,\overbrace{\epsilon = \langle\beta|\,\underbrace{\beta = f(\alpha,\delta), \alpha = f(\alpha,\gamma)}_{E_1}\rangle, \delta = \langle\delta'|\underbrace{\delta' = \bullet}_{E_2}\rangle}^{E}\rangle$$

Leading to the recursion equations
$\langle\beta^{E_1}|E\rangle = f(\langle\alpha^{E_1}|E\rangle, \langle\delta'^{E_2}|E\rangle)$
$\langle\alpha^{E_1}|E\rangle = f(\langle\alpha^{E_1}|E\rangle, \gamma)$
$\langle\delta'^{E_2}|E\rangle = \bullet$
Folding yields:
$\langle f(\langle\alpha|\alpha = f(\alpha,\beta), \beta = \gamma\rangle, \delta)|\delta = \delta\rangle = \langle\alpha|\alpha = f(\beta,\delta), \beta = f(\beta,\gamma), \delta = \bullet\rangle$

Proof of Thm. 12. Given $\models s = t$ we know that there are flat NTG's s' and t' such that $\vdash s = s'$ and $\vdash t = t'$, so $\models s' = t'$. If we can show $\vdash s' = t'$, it follows that $\vdash s = t$.

A flat NTG has the special property that there is a one to one correspondence between the nodes of the graph and the variables occurring. The given bisimulation R between s' and t' can therefore be seen as a relation on the variables occurring. Using that observation for every pair $(\alpha,\beta) \in R$ we define a variable $\gamma_{(\alpha,\beta)}$. If α is a free variable then $\gamma_{(\alpha,\beta)} \equiv \alpha$. If α is bound and $\alpha = f(\alpha_1,\ldots,\alpha_n)$ and $\beta = f(\beta_1,\ldots,\beta_n)$ are the definitions of α and β then we also define a term

$$t_{(\alpha,\beta)} \equiv f(\gamma_{(\alpha_1,\beta_1)},\ldots,\gamma_{(\alpha_n,\beta_n)}) \ .$$

Due to the fact that R is a bisimulation, the variables in $t_{(\alpha,\beta)}$ are all defined. Assume $s' \equiv \alpha_0^E$ and $t' \equiv \beta_0^F$. We define

$$r = \langle\gamma_{(\alpha_0,\beta_0)}|\gamma_{(\alpha,\beta)} = t_{(\alpha,\beta)} \text{ for } (\alpha,\beta) \in R \text{ with } t_{(\alpha,\beta)} \text{ defined}\rangle$$

For all α defined in s', we can derive:

$$\alpha^E = f(\alpha_1^E,\ldots,\alpha_n^E)$$

which, defining $u_{(\alpha,\beta)} \equiv \alpha^E$, can be read as:

$$u_{(\alpha,\beta)} = f(u_{(\alpha_1,\beta_1)},\ldots,u_{(\alpha_n,\beta_n)}) = t_{(\alpha,\beta)}\{u_{(\alpha_i,\beta_i)}/\gamma_{(\alpha_i,\beta_i)}\}_{i=1}^n$$

for each β such that $\alpha R \beta$. This β is defined as $\beta = f(\beta_1,\ldots,\beta_n)$. So by folding:

$$\vdash s' = r \ .$$

In the same way one can derive $r = t'$ and so

$$\vdash s' = t' \ .$$

□

Example 7. We will now prove the equality of two flat NTG's using the strategy of the previous proof. We have that

$$\models s \equiv \alpha_0^{\{\alpha_0=f(\alpha_0,\beta,\alpha_1),\alpha_1=f(\alpha_1,\beta,\alpha_0)\}} = \alpha_0^{\{\alpha_0=f(\alpha_1,\beta,\alpha_0),\alpha_1=f(\alpha_0,\beta,\alpha_1)\}} \equiv t$$

via the bisimulation $R = \{(\alpha_0, \alpha_0), (\alpha_0, \alpha_1), (\alpha_1, \alpha_0), (\alpha_1, \alpha_1), (\beta, \beta)\}$.

We define

$$r \equiv \alpha_{00}^{\{\alpha_{00}=f(\alpha_{01},\beta,\alpha_{10}),\alpha_{01}=f(\alpha_{00},\beta,\alpha_{11}),\alpha_{10}=f(\alpha_{11},\beta,\alpha_{00}),\alpha_{11}=f(\alpha_{10},\beta,\alpha_{01})\}} .$$

Putting $E = \{\alpha_0 = f(\alpha_0, \beta, \alpha_1), \alpha_1 = f(\alpha_1, \beta, \alpha_0)\}$ we can derive

$$\alpha_0^E = f(\alpha_0^E, \beta, \alpha_1^E)$$

and

$$\alpha_1^E = f(\alpha_1^E, \beta, \alpha_0^E) .$$

If we define

$$s_{ij} = \alpha_i^E \text{ for } i, j = 0, 1$$

we can read the derived equations as

$$\begin{aligned} s_{00} &= f(s_{01}, \beta, s_{10}) \\ s_{01} &= f(s_{00}, \beta, s_{11}) \\ s_{10} &= f(s_{11}, \beta, s_{00}) \\ s_{11} &= f(s_{10}, \beta, s_{01}) \end{aligned}$$

and fold then into

$$s_{00} = \alpha_{00}^{\{\alpha_{00}=f(\alpha_{01},\beta,\alpha_{10}),\alpha_{01}=f(\alpha_{00},\beta,\alpha_{11}),\alpha_{10}=f(\alpha_{11},\beta,\alpha_{00}),\alpha_{11}=f(\alpha_{10},\beta,\alpha_{01})\}} .$$

Thus we have derived $s = r$. In a similar way $t = r$ can be derived. Hence $s = t$.

7 Acknowledgments

The author thanks Zena Ariola for making possible his stay at the University of Oregon, where part of the work on the final revision of this paper was done. The author thanks Zena Ariola and Jan Willem Klop for commenting on drafts of this paper.

References

1. Z. M. Ariola and J. W. Klop. Equational Term Graph Rewriting. *To appear in Fundamenta Informaticae. Extended version: CWI Report CS-R9552*, 1995.
2. Z.M. Ariola and J.W. Klop. Cyclic Lambda Graph Rewriting. In *Proceedings, Ninth Annual IEEE Symposium on Logic in Computer Science*, pages 416–425, Paris, France, July 1994. IEEE Computer Society Press.
3. H. P. Barendregt. *The Lambda Calculus: Its Syntax and Semantics.* North-Holland, Amsterdam, 1984.
4. H.P. Barendregt, M.J.C.D. van Eekelen, J.R.W. Glauert, J.R. Kennaway, M.J. Plasmeijer, and M.R. Sleep. Term Graph Reduction. In *Proc. Conference on Parallel ARchitecture and Languages Europe (PARLE)*, pages 141–158, Eindhoven, The Netherlands, 1987. Springer-Verlag LNCS 259.
5. E. Barendsen. *Types and Computations in Lambda Calculi and Graph Rewrite Systems.* PhD thesis, University of Nijmegen, 1995.
6. S.L. Bloom and Z. Esik. *Iteration Theories.* EATCS Monograph. Springer, 1993.
7. C.C. Elgot. Monadic Computation and Iterative Algebraic Theories. In H. E. Rose and J. C. Shepherdson, editors, *Logic Colloquium '73 (Bristol, England)*, number 80 in Studies in Logic and the Foundations of Mathematics, pages 175–230. North Holland, 1975.
8. J. W. Klop. Term Rewriting Systems. In S. Abramsky, D. Gabbay, and T. Maibaum, editors, *Handbook of Logic in Computer Science*, volume II, pages 1–116. Oxford University Press, 1992.
9. R. Milner. A complete inference system for a class of regular behaviours. *JCSS*, 28:227–247, 1984.
10. S. L. Peyton Jones. *The implementation of Functional Programming Languages.* Prentice-Hall International, Englewood Cliffs, N.J., 1987.
11. M.R. Sleep, M.J. Plasmeijer, and M.C.D.J. van Eekelen, editors. *Term Graph Rewriting: Theory and Practice.* John Wiley & Sons, 1993.
12. S. Smetsers. *Graph Rewriting and Functional Languages.* PhD thesis, University of Nijmegen, 1993.
13. G. Ştefănescu. The Algebra of Flownomials Part 1: Binary Flownomials; Basic Theory. Technical report, Technical University Munich, November 1994.

R^n- and G^n-Logics

Claus Hintermeier, Hélène Kirchner[1] and Peter D. Mosses[2]

[1] CRIN-CNRS & INRIA-Lorraine,
BP 239,
F-54506 Vandœuvre-lès-Nancy Cedex, France
email: hkirchne@loria.fr
[2] BRICS, University of Aarhus,
Ny Munkegade, bldg. 540
DK-8000 Aarhus C, Danmark
email: pdmosses@brics.dk

Abstract. This paper proposes a simple, set-theoretic framework providing expressive typing, higher-order functions and initial models at the same time. Building upon Russell's ramified theory of types, we develop the theory of R^n-logics, which are axiomatisable by an order-sorted equational Horn logic with a membership predicate, and of G^n-logics, that provide in addition partial functions. The latter are therefore more adapted to the use in the program specification domain, while sharing interesting properties, like existence of an initial model, with R^n-logics. Operational semantics of R^n-/G^n-logics presentations is obtained through order-sorted conditional rewriting.

1 Motivations

The general goal of this work is to give a simple, set-theoretic framework providing expressive typing, higher-order functions and initial models at the same time. The decision to use set-theoretic interpretations is taken mainly because of their simplicity, and their intuitive appeal for formal software specification. Higher-order functions and highly expressive types including polymorphism, dependent and higher-order types are frequently used concepts that we want to handle in a uniform way. We also want to provide a concise and sufficiently simple deduction system, easily implemented by rewriting.

Algebraic specification techniques model types as sets and subtypes as subsets, called sorts and subsorts, respectively. However, in conventional algebraic frameworks, sort expressions are generally restricted to constants, functions are first-order and sort assertions are static and unconditional. From the algebraic approach, we want to keep the initial semantics which provides a unique model up to isomorphism for classes of models, and rewrite techniques for operational semantics.

Defining function graphs as sets of argument/value pairs for each function is a classical set-theoretic technique to give semantics to functions. This is the case in Russell's ramified theory of types from which we started our work. However, self-applicable functions, as in the untyped λ-calculus, are not possible in that theory; our approach will include references to sets (i.e. names) in order to cope with this problem.

2 Introduction to R^n-Logics and G^n-logics

An R^n-logic is an equational Horn logic with membership predicate $\in$. The parameter n, which is a natural number ≥ 0, gives a bound on the nesting depth of the sets used in interpretations. Analogous to Whitehead and Russell [37], we assign orders $i \in [0..n]$ to variables and terms, so that a term of order 0 is interpreted as an individual value and a term of order 1 or greater as a set. Moreover formulas are restricted to stratified ones, i.e. $t \in t'$ is a valid formula only if t is of one order lower than t' and $t = t'$ is an admissible equation only if all its instances are order-preserving, i.e. left and right hand side are of the same order. This prevents Russell's paradox. Furthermore, the syntax of terms does not include the empty set as a predefined constant, in order to avoid negation in the considered Horn clause fragment. Instead we have the restriction that all sets represented by terms are non-empty.

The difference from Russell's ramified theory of types [37] is the consideration of non-term-generated models. Our choice of models avoids Gödel's second theorem which proves the incompleteness of deduction systems like the one in Principia Mathematica [37]. We get a complete deduction system by using non-standard axioms of choice and extensionality. This goes along with an extension of the signature by choice functions, which are deterministic in our framework. The essential use of choices here is the possibility to express that there may be other objects than those represented by terms in a particular model. Hence, given a set of individuals, we define sets together with choices as follows: a choice of order 0 is a term representing a (possibly non-standard) individual. A set of order 1 is a set of choices of order 0 and individuals. A choice of order $k \in [1..n-1]$ is a term representing a set of order k. A set of order $k \in [2..n]$ is a set of choices and sets of order $k-1$.

Therefore, the underlying idea for the sort structure is to start with Russell's ramified theory of types up to order n, which is basically many-sorted. Assume $\{s_0, \ldots, s_n\}$ is the set of sorts. Then s_0 is the sort of individuals represented by terms and for $i \in [1..n]$, s_i is the sort of terms representing sets of sets of ... (i times) of individuals. Therefore, sets in s_i are called sets of *order* i. Now, we add supersorts $s'_0, \ldots, s'_n$ for $s_0 \ldots, s_n$, respectively, such that s'_0 is the sort of all individuals not necessarily represented by a term, s'_i, $i \in [1..n]$, is the sort of all sets of sets of ... (i times) of individuals, also not necessarily represented by a term. This allows us for example to reason about real numbers as individuals although it is impossible to represent all of them as terms. Choice functions are thus defined from s_i to s'_{i-1}, $i \in [1..n]$. Let us call this intermediate theory *simple R^n-logic*. The interpretation of the sort structure for R^n-logic is illustrated in Figure 1.

Simple G^n-logics are defined analogously, except that functions are not necessarily total. When f is declared as a function e.g. from individuals to individuals

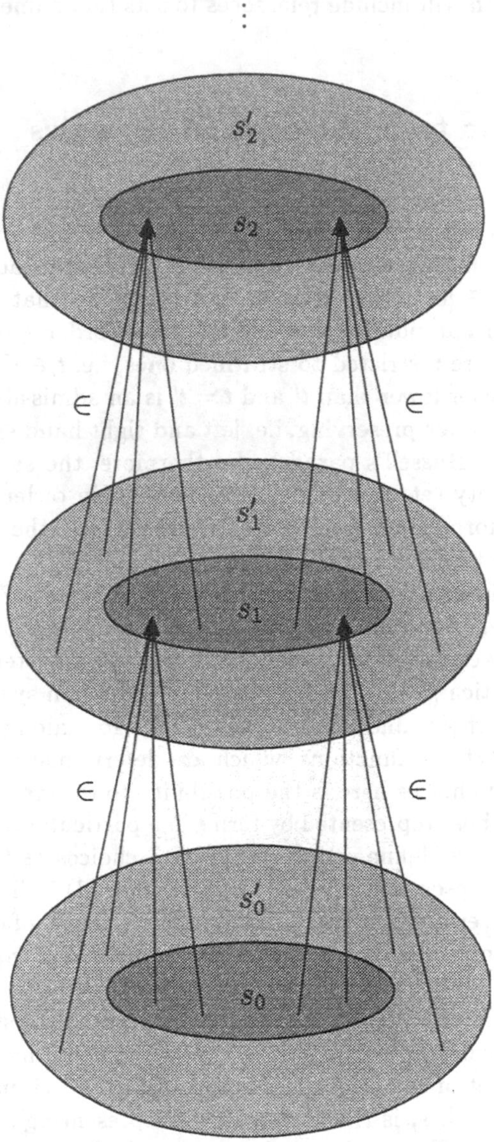

Fig. 1. Interpretation of sorts in R^n-logic

in G^n-logics, this does not imply that f is completely defined over all individuals; however, if $f(t)$ is defined for some individual term t, then $f(t)$ has to be an individual. G^n-logics seem to be more natural for program specifications than R^n-logics. However, they are also a bit more complex since we have to handle definedness of a term t using an additional predicate $Ex\ t$.

The difference between simple and full R^n-/G^n-logics is that the latter include *references to sets*, which are individuals. As the last step of the construction, we add sorts for references to objects (i.e. to individuals, sets of individuals, etc.) represented by terms in one of the sorts $s_0, \ldots, s_n$. The intuition for references is that they are names for the objects they refer to. References in our framework are mainly useful for the construction of function graphs: it is possible to define a graph of a higher-order function as a set of individual pairs of references to arguments and results.

The basic result in this paper is the existence of a sound and complete deduction system for full R^n-/G^n-logics. This is not in contradiction to the general incompleteness of higher-order predicate logics, since we use a particular non-standard model notion, similar to Henkin models [9]. Furthermore, this category of models together with standard order-sorted homomorphisms contains an initial object for each presentation, since we use a Horn clause fragment and all operations are deterministic.

3 Illustration of R^n-and G^n-logics

Let us give some examples of specifications in R^n- and G^n-logics, before proceeding to the formal definitions. To start with, the example of polymorphic ordered lists illustrates the expressiveness of R^n- and G^n-logics, as it involves types depending on functions.

Recall that s_0, s_1 and s_2 are the sorts of individuals, sets of individuals, and sets of such sets, respectively. Let the signature contain the operators *lists* (ordered lists), *nil*, *cons* (ordered list constructors), *insert* (element insertion), *orders* (orders over elements) and *pair* with the following declarations of domains and co-domains:

$$\begin{array}{llll} lists & : s_1, s_1 \to s_1 & nil & : s_1 \to s_0 \\ cons & : s_0, s_0 \to s_0 & orders & : s_1 \to s_2 \\ insert & : s_0, s_0 \to s_0 & pair & : s_0, s_0 \to s_0. \end{array}$$

Let additionally x, y, l be variables of sort s_0, and i, o be variables of sort s_1. Let ϕ be the conjunction $o \in orders(i) \wedge x \in i \wedge y \in i \wedge l \in lists(i, o)$ and ϕ' be $\phi \wedge cons(y, l) \in lists(i, o)$. The axioms are:

$$\begin{array}{r} o \in orders(i) \Rightarrow nil(o) \in lists(i,o) \\ o \in orders(i) \wedge x \in i \Rightarrow cons(x, nil(o)) \in lists(i,o) \\ pair(x,y) \in o \wedge \phi' \Rightarrow cons(x, cons(y,l)) \in lists(i,o) \\ \phi \Rightarrow insert(x,l) \in lists(i,o) \end{array}$$

$$pair(x,y) \in o \wedge \phi' \Rightarrow insert(x, cons(y,l)) = cons(x, cons(y,l))$$
$$pair(y,x) \in o \wedge \phi \Rightarrow insert(x, cons(y,l)) = cons(y, insert(x,l)).$$

The reader should not be alarmed by the amount of detail in the above axioms: here, for simplicity, we are using the bare R^n-logic specifications, without introducing any of the syntactic sugar that would be needed for large-scale use in practical applications.

Some simple consequences of the above specification in R^n-logic, taking the set of natural numbers N for i and assuming $leq = \{pair(m,n) \mid m \leq n\} \in orders(N)$, include:

$$cons(1, cons(3, nil(leq))) \in lists(N, leq)$$
$$insert(2, cons(1, cons(3, nil(leq)))) = cons(1, cons(2, cons(3, nil(leq))))$$

whereas $cons(3, cons(1, nil(leq))) \in lists(N, leq)$ is *not* a consequence.

In R^n-logics functions are total, so all well-sorted terms are required to have values. One may however restrict ones attention to those terms whose values belong to some particular sets. For example, above we may be interested only in those terms whose values belong to $lists(N, leq)$; we may regard $cons(3, cons(1, nil(leq)))$ as an error term.

In G^n-logics, however, functions may be partial, and the values of error terms may be left undefined. This provides a canonical way of distinguishing errors: we do not have to identify some particular sets of interest. Taking the specification above in G^n-logics, we get that $cons(3, cons(1, nil(leq)))$ is undefined (in the standard model of the specification, at least).

R^n-/G^n-logics offer the possibility to express and manipulate functions via their graphs. Suppose $f : s_i \rightarrow s_j$. Then we may specify the graph $g :\rightarrow s_1$ of f by:

$$pair(ref(x), ref(y)) \in g \iff f(x) = y$$

where $ref : s_k \rightarrow s'_0$ is the function that maps each value in s_k to the corresponding reference, and $pair : s'_0, s'_0 \rightarrow s_0$.

Our logics also allow us to specify self-applicable functions with set theoretic semantics. Assume we want to define domain restrictions $restrict(f, s)$ for functions, often written $f|_s$. Now, our type system does not prevent us writing $restrict(restrict, s)$, where s is a set of function graphs defined in the same logic, since the graph of restrict may be defined as a simple set of individuals as shown above.

Let us conclude these illustrations with another familiar example: *maplist*, an operator that takes an operator and a list as arguments, and applies the operator to each item in the list, making a list of the results. Here, having already specified ordered lists, we consider only the case where the applied operator belongs to *monotones*, the set of monotone increasing functions on the ordered items, so that the resulting list is also ordered. The signature for ordered lists is extended as follows:

$$maplist : s_1, s_0 \rightarrow s_0 \qquad monotones : s_1, s_1 \rightarrow s_2.$$

We use the same variables as in the specification of ordered lists above, together with m of sort s_1. The axioms are:

$$\begin{array}{l} o \in orders(i) \wedge m \in monotones(i,o) \Rightarrow maplist(m, nil(o)) = nil(o) \\ o \in orders(i) \wedge m \in monotones(i,o) \wedge x \in i \wedge l \in lists(i,o) \wedge \\ cons(x,l) \in lists(i,o) \wedge pair(x,y) \in m \Rightarrow \\ \qquad maplist(m, cons(x,l)) = cons(y, maplist(m,l)) \end{array}$$

Notice that the domain of m and the set of list items i have to be the same. When i is properly included in the domain of m, we may either increase i to match (thereby increasing the set of lists) or make use of $maplist(restrict(m,i),l)$.

4 R^n-Logics

Presentations in simple and full R^n-logics have a signature and Horn clause axioms, including conditional membership formulas and conditional equalities, which allow to express, among other things, polymorphic and dependent types. Let us give their formal definition, using a fragment of conventional first-order order-sorted logic [30]:

Definition 1. An *R^n-signature* Σ is an order-sorted signature $(\mathbf{S}, \leq_{\mathbf{S}}, \mathbf{F}, \mathbf{R})$, such that:

- $\mathbf{S}$ is a non-empty set of sorts, $\mathbf{S} = \{s_0, \ldots, s_n\} \cup \{s_{0r}, \ldots, s_{nr}\} \cup \{s'_0, \ldots, s'_n\}$,
- $\leq_{\mathbf{S}}$ is an ordering relation on $\mathbf{S}$ defined by: for all $i \in [0..n], s_i \leq_{\mathbf{S}} s'_i$ and $s_{ir} \leq_{\mathbf{S}} s'_0$,
- $\mathbf{F}$ is a set of function symbols. Any f in $\mathbf{F}$ with an arity k has a set of ranks $f : s_1, \ldots, s_k \rightarrow s$ with $s_1, \ldots, s_k, s$ in $\mathbf{S}$. If $n > 0$, $\mathbf{F}$ contains the following functions:
  - *choose* with ranks $\{(choose : s_i \rightarrow s'_{i-1}) \mid i \in [1..n]\}$,
  - *ref* with ranks $\{(ref : s_i \rightarrow s_{ir}) \mid i \in [0..n]\}$,
  - *deref* with ranks $\{(deref : s_{ir} \rightarrow s_i) \mid i \in [0..n]\}$,

  All other functions f have ranks of the form $(f : s''_1, \ldots, s''_q \rightarrow s'')$, where the sorts s'', s''_i belong to $\{s_0, \ldots, s_n, s_{0r}, \ldots, s_{nr}\}$ for all $i \in [1..q]$, so that each well-sorted term has a unique least sort (Σ is called *regular* in this case).
- $\mathbf{R}$ is a set of relation symbols. Any p in $\mathbf{R}$ with an arity k has a set of ranks $p : s_1, \ldots, s_k$ with $s_1, \ldots, s_k$ in $\mathbf{S}$. If $n > 0$, $\mathbf{R}$ contains the relation $\in$ with ranks $\{(\in : s'_{i-1}\ s_i) \mid i \in [1..n]\}$.

A Σ-term (or literal) is ground if it does not contain any variable.

An *R^n-presentation* $\mathbf{P}$ is a set of Σ-Horn clauses, written $G \Rightarrow L$, where G is the premiss (or the body) and L is the conclusion (or the head). A clause $G \Rightarrow$ with an empty conclusion is called a goal clause by analogy with logic programming. All variables of sorts $s'_0, \ldots, s'_n$ in $\mathbf{P}$ occur only on the left of a membership relation '$\in$', '*choose*' only appears as top operator of a left argument of '$\in$' and all equalities occurring in clauses of $\mathbf{P}$ are sort preserving, which means that, for each instance, the least sort of the left-hand side is the same as the least sort of the right-hand side.

The sort structure of R^n-signatures is illustrated in Figure 2.

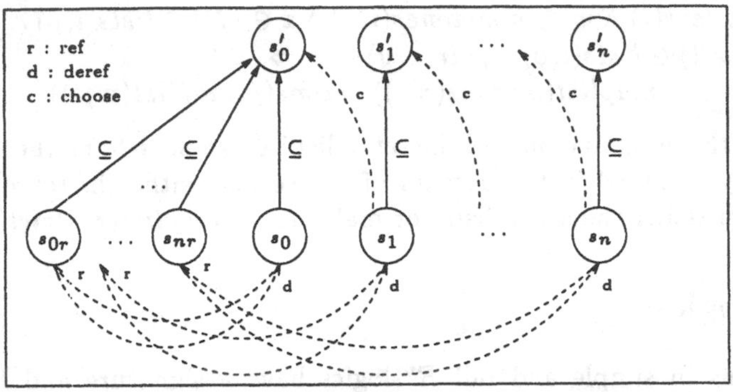

Fig. 2. The Sort Structure of R^n-signatures

First of all, remark the nature of the terms in the different sorts: the lower sorts s_m for m in $[0..n]$ should contain only terms without the symbol *choose*, the ones in s'_{m-1} those of s_m with one *choose* symbol on the top. In addition, s_{mr} contains only terms of s_m with *ref* additionally on its top, whenever $m \in [0..n]$. Regularity of the signature implies that for all $i, j \in [0..n]$, $i \neq j$, there is no ground term both of sort s_i and s_j.

The signature restrictions are relatively strong, because of the clear separation of sets of different order. In fact, the sort preservation of equalities results in static typing like in many-sorted logics. The choice functions are introduced in order to get a characteristic element for each set represented by a term. They also play a role in constructing the initial model.

Some difficulty arises with the treatment of these choice functions, which classically have a non-deterministic behaviour, but which we need to keep deterministic for technical reasons (Order-sorted equational Horn logic does not handle non-deterministic functions!). However, the definition of R^n-presentations does not allow us to define the result of a choice. Hence, the non-determinism has to move to the model level. Remark that these choice functions correspond with Hilbert's ϵ symbol [10, 17] and with the ϵ operator of HOL (cf. [8]).

References are particularly useful when we regard functions as sets, since their introduction allows for arbitrary arguments for functions in s'_0. From a set theoretic point of view, there may be objections to the use of references as elements different from sets. However, similar to individuals, we regard them as *a priori* given objects, just as terms are. The appropriate intuition is to think of references as purely syntactic, finite objects (just like terms), although the set they represent might be infinite.

R^n-models are a special case of first-order Σ-models defined for order-sorted equational Horn logic with non-overloaded semantics [34].

Definition 2. Let Σ be an R^n-signature, **X** a set of **S**-sorted variables and **P** an R^n-presentation. An *R^n-model* **A** of **P** is a Σ-model of **P**, where the carrier $C^{\mathbf{A}}$ contains a non-empty set of individuals ${s'_0}^{\mathbf{A}}$ in the sense of [37], such that:

- s_0 is interpreted as the set of individuals represented by a term,
- for $i \in [1..n]$, $s_i^{\mathbf{A}}$ contains only non-empty sets of order i represented by a term,
- for $i \in [0..n]$, $s_{ir}^{\mathbf{A}}$ is a set of symbolic, finite names contained in ${s'_0}^{\mathbf{A}}$,
- for $i \in [1..n]$, ${s'_i}^{\mathbf{A}}$ is the set of all sets of order i, plus all choices of order $i+1$ whenever $i < n$,
- $choose^{\mathbf{A}}$ gives a unique element from any set in $C^{\mathbf{A}}$ given as argument (choice function),
- *ref* is a bijective function over $s_i^{\mathbf{A}}$, $i \in [0..n]$, taking a set of order i and returning its unique name in $s_{ir}^{\mathbf{A}}$,
- *deref* is its inverse function on $s_{ir}^{\mathbf{A}}$, and
- $\in$ is interpreted as the membership relation on all sets in $C^{\mathbf{A}}$.

The notation $(\Sigma, \mathbf{P}, \mathbf{X}) \models_{R^n} \phi$ means that the formula ϕ is true in all R^n-models of **P**.

Figure 3 now shows the set of deduction rules necessary to perform deduction in R^n-logics. Here, u^m, t^m stand for terms of sorts s_m, x^k, x^m, x^{kr} for variables of sorts s_k, s_m and s_{kr}, respectively. It is not difficult to prove their soundness:

Theorem 3. *Let* **P** *be an R^n-presentation. Assume that there exists at least one ground term of sort s_m for each $m \in [0..n]$. The deduction rules of Figure 3 are sound with respect to deduction in R^n-models.*

Figure 3 actually shows deduction rule schemes, which should be understood for all $m \in [1..n]$ and $k \in [0..n]$. Remark that **Choice** is the axiom of choice and **Ext** is a non-standard version of the axiom of extensionality, which can also be given as hereditary Harrop formula, although going out of the syntax of R^n-logics:

$$\textbf{ClassicExt}\ (\forall {x'}^{m-1}, {x'}^{m-1} \in y^m \Rightarrow {x'}^{m-1} \in z^m) \wedge$$
$$(\forall {x'}^{m-1}, {x'}^{m-1} \in z^m \Rightarrow {x'}^{m-1} \in y^m) \Rightarrow (y^m = z^m)$$

The intuition behind **Ext** is the freeness of *choose*, the choice function, which we keep without equational axioms by requiring that it does not occur in the conclusion of a clause in an R^n-presentation. Remark that **Ext** cannot be formulated as a Horn clause of the following form:

$$\textbf{HornExt}\quad choose(y^m) \in z^m \wedge choose(z^m) \in y^m \Rightarrow y^m = z^m.$$

This is simply not a valid clause, i.e. satisfied by all R^n-logic models, since it does not hold for any fixed interpretation of *choose*, as the following example shows:

1. R^n-Deduction Rules from Order Sorted Equational Horn Logics:

Reflexivity $\quad \dfrac{}{x = x}$ if $x \in \mathbf{X}$

Axioms $\quad \dfrac{}{G \Rightarrow L}$ if $G \Rightarrow L \in \mathbf{P}$

Substitutivity $\quad \dfrac{G \Rightarrow L}{\sigma(G) \Rightarrow \sigma(L)}$ if $\sigma \in \text{SUBST}_\Sigma(\mathbf{X})$

Cut $\quad \dfrac{G \wedge L' \Rightarrow L \quad G' \Rightarrow L'}{G \wedge G' \Rightarrow L}$

Paramodulation $\quad \dfrac{G \Rightarrow L[s] \quad G' \Rightarrow (s = t)}{G \wedge G' \Rightarrow L[t]}$

2. Specific Deduction Rules of R^n-logics (for $k \in [0..n]$ and $m \in [1..n]$):

Ref $\quad \dfrac{}{ref(deref(x^{kr})) = x^{kr}}$

Deref $\quad \dfrac{}{deref(ref(x^k)) = x^k}$

Choice $\quad \dfrac{}{choose(x^m) \in x^m}$

Ext $\quad \dfrac{choose(u^m) \in t^m \quad choose(t^m) \in u^m}{u^m = t^m}$

Fig. 3. R^n-deduction rules

Example 1. Let $n = 1$ and $\mathbf{P} = \{a \in A,\ b \in B,\ a \in C,\ b \in C\}$. Now, let $\mathbf{A}$ be the model interpreting A as $\{a\}$, B as $\{b\}$ and C as $\{a, b\}$. $\mathbf{A}$ trivially satisfies **ClassicExt**, since the premiss never gets true. However $choose^{\mathbf{A}}(C^{\mathbf{A}})$ has to coincide with either $choose^{\mathbf{A}}(A^{\mathbf{A}})$ or $choose^{\mathbf{A}}(B^{\mathbf{A}})$. But neither $A^{\mathbf{A}} = C^{\mathbf{A}}$ nor $B^{\mathbf{A}} = C^{\mathbf{A}}$. So **HornExt** is not valid in this R^n-model of $\mathbf{P}$. But, since *choose* is free, we may quantify over all possible interpretations, and therefore **Ext** is sound for deduction.

The fact that choice functions allow the formulation of non-standard extensionality as a deduction rule over Horn clauses without inductive conditions, is the key for achieving completeness of deduction for R^n-logics using the simple first-order, order-sorted, equational Horn clause calculus given in Figure 3.

The notation $(\Sigma, \mathbf{P}, \mathbf{X}) \vdash_{\mathcal{RNL}} \phi$ means that the formula ϕ is deducible from $\mathbf{P}$ using the R^n-deduction rules.

Theorem 4. *Let* $\mathbf{P}$ *be a* R^n*-presentation. Assume that there exists at least one ground term of sort* s_m *for each* $m \in [0..n]$*. The deduction rules of Figure 3 are complete: for any* $(\Sigma, \mathbf{X})$*-atom* L*, if* $(\Sigma, \mathbf{P}, \mathbf{X}) \models_{R^n} L$*, then* $(\Sigma, \mathbf{P}, \mathbf{X}) \vdash_{\mathcal{RNL}} L$*.*

The construction of the initial model $\mathbf{I}_R$ can be done inductively: for ground terms in s_0, i.e. representing individuals, and for non-standard ground terms in s'_i, $i \in [1..n]$, it is just the usual quotient construction. For terms t in s_i, $i \in [1..n]$, it is the set of all terms $u^{\mathbf{I}_R}$ such that $(\Sigma, \mathbf{P}, \mathbf{X}) \vdash_{\mathcal{RNL}} u \in t$. Using this construction, we get:

Theorem 5. *Let* $\mathbf{P}$ *be a* R^n*-presentation. Assume that there exists at least one ground term of sort* s_m *for each* $m \in [0..n]$. *Then* $\mathbf{I}_R$ *is an initial object in the category of* R^n*-models of* $\mathbf{P}$.

The proof relies on three lemmas stating the following facts:

- Equational replacement using sort-preserving equalities preserves the sort of the terms, i.e. their order.
- Deductive term models (as described above in the construction of the initial model) interpret "=" as the equality relation and "∈" as the membership relation.
- All terms occuring in a deductive term model are well-sorted.

Full details of the proofs can be found in [11].

5 G^n-Logics

In this section, we go over from R^n-logics to G^n-logics. The main difference between the two is that functions in G^n-logics are partial over the universe. This leads to complications when we define the corresponding models, since terms, that have to be interpreted in order-sorted equational Horn clause logic, might not necessarily be interpreted in G^n-logics.

G^n-logics were inspired by G-algebras [20], but can also be seen as closely related to Scott's logic of partial equality [33, 6] and to partial models in [32]. In what follows strictness is reflected by well-definedness axioms and leads to a unique notion of equality. G^n-signatures are just extensions of R^n-signatures.

Definition 6. A G^n*-signature* is an R^n-signature $\Sigma = (\mathbf{S}, \leq_{\mathbf{S}}, \mathbf{F}, \mathbf{R})$, such that $\mathbf{R}$ contains additionally the relation Ex with ranks $\{(Ex : s'_i) \mid i \in [0..n]\}$.
A G^n*-presentation* is an R^n-presentation, such that Σ is a G^n-signature.

The difference between R^n- and G^n-models relies on partiality of functions and on the existential predicate. Let us precisely state the additional requirements for G^n-models: in a G^n*-interpretation* $\mathbf{A}$,

- $\forall f \in \mathbf{F}$, if $(f : s_1, \ldots, s_q \to s)$, then $f^{\mathbf{A}}$ is a partial function from the Cartesian product $s_1^{\mathbf{A}} \times \ldots \times s_q^{\mathbf{A}}$ into $s^{\mathbf{A}}$.
- Moreover Ex is interpreted as the membership relation in $C^{\mathbf{A}}$.

The difference between R^n- and G^n-interpretations seems small at first glance, but is more fundamental. There is no more totality obligation for function symbols. However, a function f with rank $(f : s_1, \ldots, s_q \to s)$ must have $s^{\mathbf{A}}$ as codomain whenever it is defined for an element in the Cartesian product $s_1^{\mathbf{A}} \times \ldots \times s_q^{\mathbf{A}}$. So G^n-formulas have to include declarations for the domain of functions. The satisfaction relation in all G^n-interpretations is now denoted by $\models_{G^n}$.

The less restrictive definition of term interpretations prevents us to use order-sorted equational Horn clause deduction directly. In particular, **Reflexivity** is no more sound. Instead, we have to ask for well-definedness of the interpretation of a term before deducing reflexivity. Figure 4 shows the set of deduction rules for G^n-logics incorporating well-definedness formulas and therefore suitable for partial function handling.

1. G^n-Deduction Rules from Order Sorted Equational Horn Logics with Partial Functions:

PartialReflex $\dfrac{Ex\ t}{t = t}$

Axioms $\dfrac{}{G \Rightarrow L}$ if $G \Rightarrow L \in \mathbf{P}$

SubstConform $\dfrac{G \Rightarrow L}{\sigma(G) \Rightarrow \sigma(L)}$ if $\sigma \in \mathbf{P}\text{-}\mathrm{SUBST}_{\Sigma}(\mathbf{X})$

Cut $\dfrac{G \wedge L' \Rightarrow L \quad G' \Rightarrow L'}{G \wedge G' \Rightarrow L}$

Paramodulation $\dfrac{G \Rightarrow L[s] \quad G' \Rightarrow (s = t)}{G \wedge G' \Rightarrow L[t]}$

WellDef $\dfrac{\Phi[t]}{Ex\ t}$ if $\Phi[t]$ *is a* $(\Sigma, \mathbf{X})$-atom containing t

2. Specific Deduction Rules of G^n-logics (for $k \in [0..n]$ and $m \in [1..n]$):

Ref $\dfrac{}{ref(deref(x^{kr})) = x^{kr}}$

Deref $\dfrac{}{deref(ref(x^{k})) = x^{k}}$

Choice $\dfrac{}{choose(x^{m}) \in x^{m}}$

Ext $\dfrac{choose(u^{m}) \in t^{m} \quad choose(t^{m}) \in u^{m}}{u^{m} = t^{m}}$

Fig. 4. G^n-deduction rules

As a consequence, the deduction system for G^n-logics is not a conservative extension of the R^n-logics system, since the introduction of partial functions makes it necessary to use a restricted form of reflexivity and substitution. A Σ-substitution is called **P**-conform, when $Ex\ \sigma(x)$ can be proved for all x in the domain of σ. $\mathbf{P}\text{-SUBST}_\Sigma(\mathbf{X})$ stands for the set of **P**-conform Σ-substitutions whose domain is a subset of **X**. Additionally, we need to add a well-definedness rule **WellDef**.

As in the case of R^n-logics, soundness and completeness of the calculus with respect to the corresponding class of models can be stated. The notation $(\Sigma, \mathbf{P}, \mathbf{X}) \vdash_{\mathcal{GNL}} \phi$ means that the formula ϕ is deducible from **P** using the G^n-deduction rules.

Theorem 7. (soundness, completeness of deduction, initial model)
Let **P** *be a* G^n*-logic presentation. Assume that there exists at least one ground term of sort* s_m *for each* $m \in [0..n]$.
The deduction rules of Figure 4 are sound w.r.t. deduction in G^n*-models.*
The deduction rules of Figure 4 are complete:
for any $(\Sigma, \mathbf{X})$*-atom* L*, if* $(\Sigma, \mathbf{P}, \mathbf{X}) \models_{G^n} L$*, then* $(\Sigma, \mathbf{P}, \mathbf{X}) \vdash_{\mathcal{GNL}} L$*.*
Furthermore, there exists an initial object $\mathbf{I}_G$ *in the category of* G^n*-models of* **P**.

The proofs of these results can be found in [11]. The main difficulty here comes from partiality of functions. In contrast to R^n-logics, we do not need to interpret all terms in the set of all Σ-ground terms, just those terms that have to denote an element in all models.

6 Operationalisation by Rewriting Techniques

The main application for R^n-and G^n-logics is the software specification and verification domain. The major idea is to use set theoretic semantics for both types and higher-order features. We have therefore investigated operationalisation aspects for R^n-and G^n-presentations through first-order typed conditional term rewriting systems [11]. These techniques may be used for the design of a programming language in the style of OBJ-3, but using dynamic types and sort constraints with clear set theoretic semantics.

In order to transform a R^n-or G^n-presentation into a conditional term rewriting system, we can adapt a saturation procedure on equational Horn clauses, such as the ones described for instance in [2, 27]. The three main inference rules are superposition into conclusion, superposition into premisses and equality resolution. Application of these rules requires the existence of a well-founded reduction ordering [4] on terms and literals. Superposition is performed by unifying the maximum term in an equational conclusion of a clause with a subterm in another clause, then performing a paramodulation step on the instantiated clauses. In this process, relation symbols are considered as boolean functions. Other inference rules, such as subsumption by another clause and elimination of tautologies, are also added to eliminate redundant clauses. An ordered strategy is used for reducing the search space by using only maximal terms and literals with respect to the given ordering. A saturation process is a sequence of presentations $(P_0, P_1, \ldots)$, also called a derivation, where P_i is deduced from P_{i-1} by application of one inference rule. This derivation must be fair in the intuitive sense that no clause is forgotten in the process of generating consequences. P_0 is consistent if and only if the empty clause does not belong to any P_i. Moreover if P_∞ is the set of persisting clauses in this fair derivation and does not contain the empty clause, then one can construct from P_∞ a conditional term rewriting system which is terminating and confluent in the initial model of P_0. This indeed provides a way to compute in a finite and unambiguous way the normal form of any expression in P_0. The complete description of the process and its proof can be found in [11].

In R^n-logics, the saturation procedure defined for instance in [29], can be reused after replacing the unsorted unification algorithm by an order-sorted one. This is possible thanks to the sort preservation of R^n-presentations. If saturation terminates, then the set of ground instances of rules decreasing with respect to a given ordering on terms [5], forms a terminating and confluent term rewriting system on ground terms.

Concerning G^n-logics, we have to change the inference rules for saturation a little bit due to the partiality of functions, which results in a partial form of reflexivity (cf. **PartialReflex**). We omit details here (the interested reader may refer to [11]). Let us instead illustrate the saturation technique with the following example :

Example 2. Assume we want to define stacks as set St over elements of sort E. Let the R^n-presentation **P** be defined as follows:

$$\begin{aligned}
&\epsilon \in E \\
&empty \in St \\
x \in E \wedge y \in St \Rightarrow\ &pop(push(x,y)) \in St \\
x \in E \wedge y \in St \Rightarrow\ &top(push(x,y)) \in E \\
x \in E \wedge y \in St \Rightarrow\ &push(x,y) \in St \\
x \in E \wedge y \in St \Rightarrow\ &pop(push(x,y)) = y \\
x \in E \wedge y \in St \Rightarrow\ &top(push(x,y)) = x.
\end{aligned}$$

Let $pop \succ top \succ push \succ empty \succ \epsilon \succ E \succ St$ be the precedence for a lexicographic path ordering (LPO) on terms [14]. The multiset expression giving the complexity of membership formulas with respect to this LPO is defined in the same way as for equalities in [26], i.e. by ignoring the relation symbol.

Using the inference rules from [26] with an order-sorted equational constraint solver, we can then eliminate the membership formulas for $pop(push(x,y))$ and $top(push(x,y))$ (i.e. the third and fourth clauses in **P** above) by simplification with the two last equalities. The result is a saturated presentation.

Refuting $(top(pop(push(x,y))) = top(y) \Rightarrow)$ then gives $(top(y) = top(y) \Rightarrow)$, which has the identity substitution as solution.

It should be more natural to consider **P** as a presentation in G^n-logic, since the operations *pop* and *top* are partial functions not defined on the empty stack. Then the goal $(top(pop(push(x,y))) = top(y) \Rightarrow)$ has an infinite set of solutions

$\{y \mapsto push(\epsilon, \ldots push(\epsilon, empty) \ldots)\}$, but no more the identity substitution, since *top* is undefined for *empty*.

In order to get a better efficiency for membership proofs, we extend Horn clauses furthermore by assertions, which are a kind of cache mechanism for atoms derivable from the premiss of a clause using the current presentation. This is illustrated in the following example :

Example 3. Let $\mathbf{F} = \{id, 0, Nat\}$ and $\mathbf{P}$ the following set of clauses :

$$\begin{array}{ll} & 0 \in Nat \\ x \in Nat & \Rightarrow id(x) = x \\ id(0) \in Nat \Rightarrow & \end{array}$$

We start saturation (using an LPO with precedence $id \succ 0 \succ Nat$) by deducing from the first clause an assertion for the last one. The result is :

$$id(0) \in Nat \Rightarrow \quad [\![0 \in Nat]\!]$$

Here, the part added to the clause between the brackets $[\![$ and $]\!]$ is the assertion. The remainder is a usual clause. For better readability, we omit ordering conditions and constraints in our argumentation. Now, we can superpose the second clause into the previous one to get :

$$0 \in Nat \Rightarrow \quad [\![0 \in Nat]\!]$$

Now the goal $0 \in Nat$ is satisfied since the corresponding atom is already present in the assertion. Hence, we can deduce without further superposition the empty clause, which proves the inconsistency of $\mathbf{P}$. The gain of efficiency appears in bigger examples when an assertion is used several times.

A saturation calculus with assertions in developed in [11] and gives a semantics to saturation in a fragment of G^1-logics in the style of [12, 13]. Saturation is performed for order-sorted presentations with polymorphic, dynamic types and partial functions, using so-called decorated terms, in which set terms are added locally to term nodes to store typing information [11]. The possibility to mix this style of dynamic typing with the more efficient static typing, like in [1], is to be investigated.

7 Discussion

In this section, we discuss more precisely the connections of R^n-and G^n-logics with set theory and algebraic specifications, but also mention relations with higher-order logic and functional programming.

Set Theory: R^n-logics stem from naive set theory and may also be seen as a fragment of Z [35]. Let us examine which part of set theory can be easily specified.

Horn clauses are built with two logical connectives, implication and conjunction, which correspond directly with inclusion and intersection. It is therefore not surprising that R^n-and G^n-logics allow specifying set inclusion $\subseteq$ and intersection $\cap$, assuming x^m, y^m to be variables of sort s_m, z'^{m-1} be a variable of sort s'_{m-1}, for $m \in [1..n]$:

$$\begin{array}{ll}
choose(x^m) \in y^m & \Rightarrow x^m \subseteq y^m \\
z'^{m-1} \in x^m \wedge x^m \subseteq y^m & \Rightarrow z'^{m-1} \in y^m \\
z'^{m-1} \in x^m \wedge z'^{m-1} \in y^m & \Rightarrow z'^{m-1} \in x^m \cap y^m \\
z'^{m-1} \in x^m \cap y^m & \Rightarrow z'^{m-1} \in x^m \\
z'^{m-1} \in x^m \cap y^m & \Rightarrow z'^{m-1} \in y^m .
\end{array}$$

Depending on the logic, we have different behaviours. In R^n-logics, all terms are defined and therefore all intersections of non-empty sets (represented by a ground term) denote a non-empty set, since they are represented by a ground term. In G^n-logics, terms t are non-empty if they are defined, i.e. if we can derive $Ex\ t$. Hence, intersections like $int \cap list$, where int is the set of integers and $list$ is the set of lists, can be specified as empty if we avoid to define their existence and use initial semantics.

One might try to define singletons $sgl(x^m)$ by:

$$\begin{array}{ll}
z'^m = x^m & \Rightarrow z'^m \in sgl(x^m) \\
z'^m \in sgl(x^m) & \Rightarrow z'^m = x^m .
\end{array}$$

But this last definition is not admissible due to the requirement of sort preservation for the equalities in clauses. Union would be:

$$\begin{array}{ll}
z'^k \in x^m & \Rightarrow z'^k \in x^m \cup y^m \\
z'^k \in y^m & \Rightarrow z'^k \in x^m \cup y^m \\
z'^k \in x^m \cup y^m & \Rightarrow z'^k \in x^m \vee z'^k \in y^m .
\end{array}$$

The last clause is obviously not a Horn clause. However, we may define a weak union that covers the exact one:

$$\begin{array}{c}
x^m \subseteq x^m \cup y^m \\
y^m \subseteq x^m \cup y^m \\
x^m \subseteq z^m \wedge y^m \subseteq z^m \Rightarrow x^m \cup y^m \subseteq z^m .
\end{array}$$

We may conclude from this short outline that the set theory that can be described by R^n-and G^n-logics is rather weak, due to the absence of negation. Negation must be avoided as long as we want initial models. A limited amount of negation can be used by admitting goal clauses, which are Horn clauses without conclusion. The existence of initial models is then guaranteed if the presentation is consistent. In [11], we extended the completeness results for consistency tests in form of saturation procedures to R^n-and G^n-logics, as illustrated in Section 6.

Order-sorted algebras, ETL and Unified Algebras: From an algebraic point of view, R^n-and G^n-logics compare best with many and order-sorted algebras [7], already implemented via rewriting for instance in OBJ-3 [15]. However simple G^n-logics provide arbitrary terms as sorts and thus achieve a greater expressivity. Polymorphic order-sorted algebras can be seen as a fragment of simple G^1 logic.

In order to compare R^n-logics with ETL and unified algebras, we tried to encode them in our framework. ETL [19] is in fact a fragment of R^n-logics. An ETL presentation is a triple $\langle \Omega, V, E \rangle$, such that Ω is a set of function symbols (with associated arity), V is a set of unsorted variables and E is a set of Ω-Horn clauses using only equality "=" and the typing relation ":" as binary operators. The typing relation satisfies the paramodulation axiom and therefore we may use it as a new relation symbol in R^n-logics. Remark that we cannot reuse "$\in$" for this purpose, since it has more properties than ":" in ETL. Now, it is possible to construct an R^n-presentation, such that an Ω-atom is true in R^n-logic if and only if it holds in ETL [11].

Concerning unified algebras [25], the main difference is the absence of the empty set, which should take the role of the bottom element in unified algebra. As mentioned above, we cannot allow for the empty set if we want to have initial models for all presentations. Extending our formula language by goals, it is actually possible to extend R^n-and G^n-logics by a predefined constant representing the empty set. However, *choose* has then to become partial. The problem with singletons, which are necessary for the relation ":" in unified algebras, cannot be solved as easily. A work-around is the use of quasi-singletons defined by the axioms for sgl given above, after replacing the variable z'^m of sort s'_m by one of sort s_m. Then, $sgl(x)$ is a set with exactly one standard element, namely x, but arbitrarily many non-standard elements. To cope with real singletons, we would need to extend our type theory in order to avoid that *choose* becomes deterministic, which is in conflict with the sort preservation property of R^n-and G^n-logic presentations and deduction rules. The other operators of unified algebras, lower bound $\leq$, join $|$ and meet &, can be realised by set inclusion, intersection and weak union, as defined above, so that we may come quite close to unified algebras at least. Simple R^n-logics have also strong similarities with power algebras, i.e. unified algebras with set interpretations [25].

Relation to Higher-Order Logic and Algebras: The framework of R^n- and G^n-logics provides some higher-order features, since function graphs are specified as set constants, which can be passed to other functions as higher-order arguments. This can be situated in the context of higher-order logic that provides variables for subsets, relations, functions on the universe, functions defined on functions and quantification over these. An introductory survey to the literature on higher-order logic, its relation to set theory, and reduction to first-order logic can be found in [36].

During the last years, a number of papers have dealt with the extension of first-order algebraic specifications to higher-order ones. Among them are [18, 32, 22, 24, 23, 21]. We share with these approaches the objectives of integrating higher-order function space with algebraic specifications, and keeping the existence of initial (and possibly terminal) models. But we differ from universal algebras with higher-order types as developed for instance in [21] in the fact that our models are set-theoretic rather than purely algebraic, and we so provide a uniform treatment of types and higher-order functions.

Relation to λ-calculi and higher-order rewriting: As in higher-order algebras, we do not allow λ-abstractions as terms, but rather consider that $\lambda x.t$ can always be replaced by a new constant symbol f together with the axiom $f(x) = t$. This has the advantage to avoid technical problems associated with binding mechanism and to minimize the functions in the initial model, which may be crucial for limiting the search space for automated theorem proving.

Polymorphically typed λ-calculi have shown that types and subtypes provide interesting features for functional programming and some connections can be drawn also from our work to the system $F_{\leq}$ of [3] and to dynamic types proposed in [1].

Concerning the operationalisation of deduction, we have a purely first-order mechanism and avoid higher-order rewriting or unification as developed in [28] as well as combinatory reduction systems [16]. In contrast, in the theorem proving domain, HOL [8] and ISABELLE [31] are based on type theory but provide means to reason in set theory.

Relation to object-oriented languages: Through their use of sets, both for types and for higher-order functions, R^n-and G^n-logics are close to the semantics of imperative programming languages with subtyping, inheritance and polymorphism, like C++. In the latter language, two types declared to hold exactly the same objects have to be different. Furthermore, it is possible to compare function symbols. Such a test has to fail for two functions defined in exactly the same way but with different names. In analogy, we cannot derive $A = B$ from $\{a \in A, a \in B\}$ in R^n-and G^n-logics. Although R^n-and G^n-logics do not provide any built-in mechanism for features or records, extensions in this direction could be done.

Acknowledgements: We are grateful to anonymous referees for pointing out missing references in the original version of this paper. This work was partially supported by the ESPRIT Basic Research Working Group 6112 COMPASS.

References

1. M. Abadi, L. Cardelli, B. Pierce, and G. Plotkin. Dynamic typing in a statically typed language. *ACM Transactions on Programming Languages and Systems*, 13(2):237–268, Apr. 1991.
2. L. Bachmair, H. Ganzinger, C. Lynch, and W. Snyder. Basic paramodulation and superposition. In *Proceedings 11th International Conference on Automated Deduction, Saratoga Springs (N.Y., USA)*, pages 462–476, 1992.

3. P.-L. Curien and G. Ghelli. Coherence of subsumption, minimum typing and type-checking in $F_{\leq}$. *Mathematical Structures in Computer Science*, 2(1):55–91, 1991.
4. N. Dershowitz. Orderings for term-rewriting systems. *Theoretical Computer Science*, 17:279–301, 1982.
5. N. Dershowitz and M. Okada. A rationale for conditional equational programming. *Theoretical Computer Science*, 75:111–138, 1990.
6. M. Fourman and D. Scott. Sheaves and logic. In M. Fourman and C. Mulvey, editors, *Applications of Sheaves*, volume 753 of *Lecture Notes in Mathematics*, pages 302–401. Springer-Verlag, 1979.
7. J. A. Goguen and J. Meseguer. Order-sorted algebra I: equational deduction for multiple inheritance, overloading, exceptions and partial operations. *Theoretical Computer Science*, 2(105):217–273, 1992.
8. M. J. C. Gordon and T. F. Melham. *Introduction to HOL: a theorem proving environment for higher order logic.* Cambridge University Press, 1993.
9. L. Henkin. Completeness in the theory of types. *The Journal of Symbolic Logic*, 15(2):81–91, June 1950.
10. D. Hilbert and P. Bernays. *Grundlagen der Mathematik*, volume 2. Berlin. Reprinted Ann Arbor, Mich., 1944, 1939.
11. C. Hintermeier. *Déduction avec sortes ordonnées et égalités.* Thèse de Doctorat d'Université, Université Henri Poincaré – Nancy 1, Oct. 1995.
12. C. Hintermeier, C. Kirchner, and H. Kirchner. Dynamically-typed computations for order-sorted equational presentations –extended abstract–. In S. Abiteboul and E. Shamir, editors, *Proc. 21st International Colloquium on Automata, Languages, and Programming*, volume 820 of *Lecture Notes in Computer Science*, pages 450–461. Springer-Verlag, 1994.
13. C. Hintermeier, C. Kirchner, and H. Kirchner. Sort inheritance for order-sorted equational presentations. In *Recent Trends in Data Types Specification*, volume 906 of *Lecture Notes in Computer Science*, pages 319–335. Springer-Verlag, 1995.
14. S. Kamin and J.-J. Lévy. Attempts for generalizing the recursive path ordering. Unpublished manuscript, 1980.
15. C. Kirchner, H. Kirchner, and J. Meseguer. Operational semantics of OBJ-3. In *Proceedings of 15th International Colloquium on Automata, Languages and Programming*, volume 317 of *Lecture Notes in Computer Science*, pages 287–301. Springer-Verlag, 1988.
16. J. W. Klop. *Combinatory Reduction Systems.* PhD thesis, CWI, 1980.
17. A. Leisenring. *Mathematical logic and Hilbert's ϵ-symbol.* University Mathematical Series, Bedford College, London, 1969.
18. T. Maibaum and C. Lucena. Higher order data types. *International Journal of Computer and Information Sciences*, 9:31–53, 1980.
19. V. Manca, A. Salibra, and G. Scollo. Equational type logic. *Theoretical Computer Science*, 77(1-2):131–159, 1990.
20. A. Mégrelis. *Algèbre galactique — Un procédé de calcul formel, relatif aux semi-fonctions, à l'inclusion et à l'égalité.* Thèse de Doctorat d'Université, Université Henri Poincaré – Nancy 1, 1990.
21. K. Meinke. Universal algebra in higher types. *Theoretical Computer Science*, 100:385–417, 1992.
22. B. Möller. Algebraic specifications with high-order operators. In L. Meertens, editor, *Proceedings IFIP TC2 Working Conf. on Program Specification and Transformation*, pages 367–392. IFIP, Elsevier Science Publishers B. V. (North-Holland), 1987.

23. B. Möller, A. Tarlecki, and M. Wirsing. Algebraic specification with built-in domain constructions. In M. Dauchet and M. Nivat, editors, *Proceedings of CAAP'88*, Lecture Notes in Computer Science, pages 132–148. Springer-Verlag, 1988.
24. B. Möller, A. Tarlecki, and M. Wirsing. Algebraic specifications or reachable higher-order algebras. In D. Sannella and A. Tarlecki, editors, *Recent Trends in Data Type Specification*, volume 332 of *Lecture Notes in Computer Science*, pages 154–169. Springer-Verlag, 1988.
25. P. D. Mosses. Unified algebras and institutions. In *Proceedings 4th IEEE Symposium on Logic in Computer Science, Pacific Grove*, pages 304–312, 1989.
26. R. Nieuwenhuis and A. Rubio. Basic superposition is complete. In B. Krieg-Brückner, editor, *Proceedings of ESOP'92*, volume 582 of *Lecture Notes in Computer Science*, pages 371–389. Springer-Verlag, 1992.
27. R. Nieuwenhuis and A. Rubio. Theorem proving with ordering constrained clauses. In D. Kapur, editor, *Proceedings 11th International Conference on Automated Deduction, Saratoga Springs (N.Y., USA)*, volume 607 of *Lecture Notes in Computer Science*, pages 477–491. Springer-Verlag, 1992.
28. T. Nipkow. Higher-order critical pairs. In *Proc. 6th IEEE symposium on Logic in Computer Science (LICS)*, pages 342–349. IEEE Computer Society Press, Los Alamitos, July 1991.
29. P. Nivela and R. Nieuwenhuis. Saturation of first-order (constrained) clauses with the *saturate* system. In C. Kirchner, editor, *Proceedings 5th Conference on Rewriting Techniques and Applications, Montreal (Canada)*, volume 690 of *Lecture Notes in Computer Science*, pages 436–440. Springer-Verlag, 1993.
30. A. Oberschelp. Untersuchungen zur mehrsortigen Quantorenlogik. *Math. Annalen*, 145(1):297–333, 1962.
31. L. C. Paulson. The foundation of a generic theorem prover. *Journal of Automated Reasoning*, 5:363–397, 1989.
32. A. Poigné. Partial algebras, subsorting and dependent types - prerequisites of error handling in algebraic specifications. In *Proceedings of Workshop on Abstract Data Types*, volume 332 of *Lecture Notes in Computer Science*. Springer-Verlag, 1988.
33. D. Scott. Data types as lattices. *SIAM Journal of Computing*, 5(3):522–587, 1976.
34. G. Smolka, W. Nutt, J. A. Goguen, and J. Meseguer. Order-sorted equational computation. In H. Aït-Kaci and M. Nivat, editors, *Resolution of Equations in Algebraic Structures, Volume 2: Rewriting Techniques*, pages 297–367. Academic Press inc., 1989.
35. J. M. Spivey. *Understanding Z: a specification language and its formal semantics*. Cambridge University Press, 1988.
36. J. Van Benthem and K. Doets. Higher-Order Logic. In D. Gabbay and F. Guenthner, editors, *Handbook of Philosophical Logic*, volume 1, pages 275–329. Reidel Publishing Company, 1983.
37. A. N. Whitehead and B. Russell. *Principia Mathematica*, volume 1. Cambridge University Press, Cambridge, MA, 1925.

The Variable Containment Problem

Stefan Kahrs*

University of Edinburgh
Laboratory for Foundations of Computer Science
King's Buildings,
Edinburgh EH9 3JZ
United Kingdom
email: smk@dcs.ed.ac.uk

Abstract. The *essentially* free variables of a term t in some λ-calculus, $\mathrm{FV}_\beta(t)$, form the set $\{x \mid \forall u.\, t =_\beta u \Rightarrow x \in \mathrm{FV}(u)\}$. This set is significant once we consider equivalence classes of λ-terms rather than λ-terms themselves, as for instance in higher-order rewriting.
An important problem for (generalised) higher-order rewrite systems is the *variable containment problem*: given two terms t and u, do we have for all substitutions θ and contexts $C[\,]$ that $\mathrm{FV}_\beta(C[t^\theta]) \supseteq \mathrm{FV}_\beta(C[u^\theta])$? This property is important when we want to consider $t \rightarrow u$ as a rewrite rule and keep n-step rewriting decidable. Variable containment is in general *not* implied by $\mathrm{FV}_\beta(t) \supseteq \mathrm{FV}_\beta(u)$. We give a decision procedure for the variable containment problem of the second-order fragment of $\lambda^\rightarrow$. For full $\lambda^\rightarrow$ we show the equivalence of variable containment to an open problem in the theory of PCF; this equivalence also shows that the problem is decidable in the third-order case.

1 Introduction

As soon as we make the step from terms to equivalence classes of terms as the objects of our interest, the question whether a variable occurs (free) in such an object becomes a bit delicate. Should the variable occur in all terms of the class or only in some, or can we sensibly ask this question at all?

Typically, the equivalence relation $\equiv$ in question is preserved by substitution application, i.e. $t \equiv u$ implies $t^\theta \equiv u^\theta$ for arbitary substitutions θ. In particular, if $t \equiv u$ and x is free in t but not in u then $t[y/x] \equiv u[y/x] = u$ for any variable (or term) y. This suggests the following definition:

Definition 1. Let $=_e$ be a substitutive equivalence relation. The *free variables modulo* $=_e$ of a term t, $\mathrm{FV}_e(t)$, are defined as follows:

$$\mathrm{FV}_e(t) = \bigcap_{t =_e u} \mathrm{FV}(u)$$

* The research reported here was partially supported by SERC grant GR/J07303.

In the above definition I was deliberately a bit vague about basic notions such as term, substitution, and free variable, because the concept makes sense for various (typed or untyped) λ-calculi as well as first-order terms. In the following, we shall concentrate on the equivalence $=_\beta$ and call variables which are free modulo $=_\beta$ "essentially free".

For each equivalence class $[u]_\beta$ that contains a normal form $u\downarrow$, we have $\mathrm{FV}_\beta(u) = \mathrm{FV}(u\downarrow)$. Unfortunately, the set $\mathrm{FV}_\beta(t)$ is in general (for the untyped λ-calculus) not recursive, i.e. the problem $x \in \mathrm{FV}_\beta(t)$ is undecidable: the set $M_x = \{t \mid x \in \mathrm{FV}_\beta(t)\}$ is closed under β-conversion and non-trivial which already implies that M_x is not recursive (theorem 6.6.2 (ii) in [1]); moreover, $t \in M_x \iff x \in \mathrm{FV}_\beta(t)$.

In [1] the notation $x \in_\beta M$ is used instead of $x \in \mathrm{FV}_\beta(M)$ (exercise 3.5.15, notation 4.1.4) — the concept is not really new, even though Barendregt defines it only for λ-theories rather than arbitrary (substitutive) equivalence relations. Similarly, Middeldorp [15] uses the notation $V_{\mathit{fix}}([t]_R)$ to describe the set of variables that occur in every term that is R-equivalent to t.

Most typed λ-calculi studied in the literature [3] have a strongly normalising β-reduction, which implies that $\mathrm{FV}_\beta(t)$ is recursive for each typable term t: we reduce t to its β-normal form $t\downarrow$ and find the set as $\mathrm{FV}(t\downarrow)$.

Moving from terms to equivalence classes of terms is not entirely unproblematic: the property $\mathrm{FV}_\beta(t) \subseteq \mathrm{FV}_\beta(u)$ is in a sense less informative than the ordinary $\mathrm{FV}(t) \subseteq \mathrm{FV}(u)$. From the latter we can deduce $\mathrm{FV}(t^\theta) \subseteq \mathrm{FV}(u^\theta)$ and $\mathrm{FV}(C[t]) \subseteq \mathrm{FV}(C[u])$, which means that the property is a rewrite relation. But we *cannot* deduce from the former $\mathrm{FV}_\beta(t^\theta) \subseteq \mathrm{FV}_\beta([u^\theta])$ or $\mathrm{FV}_\beta(C[t]) \subseteq \mathrm{FV}_\beta(C[u])$. Example: the terms $t \equiv (x\ y)$ and $u \equiv (\lambda z.\ z\ x\ y)$ are in normal form and have the same essentially free variables $\{x, y\}$. But when we apply the substitution $\theta = [(\lambda v.\ x')/x]$ to both terms then x' is essentially free in both but y is only essentially free in u^θ; similarly, the context $C[\_] = \_\,(\lambda v.x')$ also distinguishes these terms: x is essentially free in $C[t]$ but not in $C[u]$.

Why does this matter?

The condition $\mathrm{FV}(t) \supseteq \mathrm{FV}(u)$ is typically used as a requirement for rewrite rules $t \rightarrow u$, to make sure that rewriting never introduces free variables. This property is desirable for a number of reasons:

1. Without it, the rewrite system could not be strongly normalising, because rewriting is substitutive and extra variables on the right-hand sides could be instantiated to (terms containing instances of) left-hand sides of rules, including the left-hand side of the very rule with the extra variables.
2. Without it, confluence is unlikely: if $t \rightarrow u$ and if u contains an extra variable x then also $t \rightarrow u[y/x]$ and confluence would require that u and $u[y/x]$ have a common reduct.
3. To decide the one-step rewrite relation $t \rightarrow u$, one has to decide matching problems, i.e. matching occurrences in t to left-hand sides of rules. This remains true if we allow extra variables though we have then an additional matching problem of (a subterm of) u against the instance of the right-hand side of the applied rule. However, if we consider n-step rewriting for $n > 1$,

then we have to solve unification problems if the rules have extra variables.

To make the last point clear: we can encode any unification problem as a two-step rewriting problem of a rewrite system with extra variables as we shall see shortly. By "unification problem" we mean the following.

Definition 2. The *unification problem* $t \doteq u$ is the property $\exists\sigma.\, t^\sigma = u^\sigma$.

Again, I am deliberately vague about what the terms t and u and the substitution σ range over, and how substitution application t^σ is actually defined — terms, substitutions, and their unification problems exist in a variety of formalisms.

Theorem 3. *For any first-order unification problem $t \doteq u$ there is a finite (generalised) rewrite system R and terms C, D such that $t \doteq u \iff C \to_R ; \to_R D$.*

Proof. We choose symbols C, D, F not occurring in t and u. R has two rules: $C \to F(t,u)$ and $F(x,x) \to D$. The required intermediate term E with $C \to_R E \to_R D$ must have the properties $\exists\sigma.\, E = F(t,u)^\sigma$, because we can only apply the first rule to C, and similarly $\exists\tau. E = F(x,x)^\tau$, because we can only apply the second rule backwards to D. Both conditions together are sufficient as neither C nor D are affected by substitution application. Thus, $C \to_R E \to_R D$ is equivalent to the problem $\exists\sigma.\,\exists\tau.\; F(t,u)^\sigma = F(x,x)^\tau$ which is equivalent to $\exists\sigma.\,\exists\tau.\, \tau(x) = t^\sigma \wedge \tau(x) = u^\sigma$ which in turn is equivalent to $t \doteq u$. □

The same kind of situation appears in higher-order rewriting, where it is even more significant: matching up to fourth-order is known to be decidable [7, 4, 17], but second-order unification is already undecidable [6]. The decidability of higher-order matching is still an open problem but it is often conjectured to be decidable [4].

Remark: In view of Loader's recent result [12] that (absolute) λ-definability for (arbitrary) finite models of $\lambda^{\to}$ is undecidable this conjecture is rather doubtful. Looking at the details of Loader's proof we can observe that it shows that *relative* λ-definability is already undecidable for third-order types which (see Loader's proof of Lemma 1 in [12]) implies that *absolute* λ-definability for fourth-order types is undecidable too; absolute λ-definability for third-order types is decidable [20].

The question about extra variables is generally asked for the *instance* of a rule, not for the rule itself. We have already seen that this difference matters in the presence of higher-order variables and indeed we do not need extra variables in the rules themselves to solve unification problems:

Theorem 4. *For any second-order unification problem $t \doteq u$ there is a finite (generalised) second-order rewrite system R (with all rules $l \to r$ satisfying $\mathrm{FV}_\beta(l) \supseteq \mathrm{FV}_\beta(r)$) and terms C, D such that $t \doteq u \iff C \to_R ; \to_R D$.*

Proof. We choose fresh symbols D, F, G such that the result type of F and G is first-order, and a fresh second-order variable y. The rules of R are $G(y(F\,t\,u)) \to (F\,t\,u)$ and as before $F\,x\,x \to D$. We can only apply the first rule to $C \equiv G\,D$ by taking *any* substitution σ such that $\sigma(y) = \lambda z.\, D$ and have the same situation as in the proof of theorem 3. □

This observation is based on *generalised* higher-order rewrite systems [14]. Originally, HRSs were defined with an additional condition for left-hand sides [16] which we shall not consider here; suffice it to say that the subterm $(y\,(F\,t\,u))$ in the above proof does not satisfy this condition.

Theorem 4 shows that $\mathrm{FV}_\beta(l) \supseteq \mathrm{FV}_\beta(r)$ is clearly not the right condition for general higher-order rewriting if we want to ban extra variables and keep n-step rewriting decidable. We need something stronger, a property which is also a rewrite relation.

There is a general principle behind the last remark. A rewrite relation is a relation closed under substitution application and context application, i.e. $>$ is a rewrite relation iff $t > u$ implies $t^\theta > u^\theta$ for arbitrary substitutions θ, and $C[t] > C[u]$ for arbitrary contexts $C[\;]$. In a typed scenario, the "arbitrary" comes with a typing proviso.

The typical use of the term "rewrite relation" is to form the rewrite closure of a relation R, i.e. the smallest rewrite relation $\rightarrow_R$ which contains R. This is well-defined, because rewrite relations are closed under arbitrary intersections. As they are also closed under arbitrary unions, the dual concept is also well-defined: the *rewrite interior* of R is the largest rewrite relation $\rightarrow^R$ contained in R.

The notion of rewrite interior is useful for the following reason. Sometimes we want to show that all terms in a rewrite relation (given by a rewrite system R) satisfy a certain property, i.e. $t \rightarrow_R u$ implies tSu, more briefly $\rightarrow_R \subseteq S$. The proof will hardly ever work directly, because $\rightarrow_R$ is almost always an infinitary object, it relates infinitely many terms. The solution is to prove instead a property about R itself, since R is typically a finite relation.

Theorem 5. *Let R and S be relations on terms. Then $\rightarrow_R \subseteq S \iff R \subseteq \rightarrow^S$.*

Proof. Trivial by exploiting the following facts: (i) Rewrite interior and rewrite closure are monotonic w.r.t. to $\subseteq$; (ii) any rewrite relation is a fixpoint of both the closure and the interior operator, in particular this applies to $\rightarrow_R$ and $\rightarrow^S$; (iii) $R \subseteq \rightarrow_R$ and $\rightarrow^S \subseteq S$. □

In words: to show that a rewrite closure $\rightarrow_R$ satisfies an invariant S we can show that R satisfies $\rightarrow^S$. In our situation, S is the relation $tSu \iff \mathrm{FV}_\beta(t) \supseteq \mathrm{FV}_\beta(u)$ and *variable containment* is the interior of this relation.

2 Variable Containment in General

Definition 6. Given two terms $t, u \in \Lambda$, their *variable containment problem*, $t \succeq u$, is defined as the following property:

$$t \succeq u \overset{def}{\iff} \forall C[\;].\, \forall\theta.\, \mathrm{FV}_\beta(C[t^\theta]) \supseteq \mathrm{FV}_\beta(C[u^\theta]).$$

For the untyped λ-calculus this is obviously an undecidable problem as even the sets $\mathrm{FV}_\beta(t)$ are generally non-recursive. We can also ignore the "$\forall\theta$" quantifier as any substitution application can occur as the substitution derived from a β-reduction.

For typed λ-calculi the problem has to be slightly restated, restricting t and u to be well-typed preterms in some context[2] Γ and θ a substitution mapping variables in Γ to preterms that type-check (with the same type) in some context Δ. An analogous restriction applies to $C[\,]$. The exact formulation depends on the particular λ-calculus, though the general principle should be clear.

It is possible to formalise it uniformly for all type systems expressible in the formalism of Pure Type Systems (short: PTS; see [2, 3]), especially the "PTS with signature" as in [5] which support a proper treatment of constant symbols. However, this goes somewhat beyond the scope of this paper and therefore we concentrate on the simply typed λ-calculus $\lambda^{\rightarrow}$ and its fragments.

In order to formulate the appropriate notion of variable containment for typed λ-calculi we have to adapt the notion of substitution accordingly.

Definition 7. We write $\theta : \Gamma \rightarrow \Delta$ if θ is a function from variables to preterms and Γ and Δ are contexts such that:

$$\forall x \in \mathit{Dom}\,\Gamma.\, \Gamma \vdash_{\Sigma} x : \tau \Longrightarrow \Delta \vdash_{\Sigma} \theta(x) : \tau$$

For arbitrary type systems, we would have to formulate a similar though more awkward adaptations for contexts (in the sense: term with hole). However, for $\lambda^{\rightarrow}$ and its n-th order restrictions this is not really necessary due to the following observations. Suppose t and u have a function type, then $t \succeq u \iff t\,x \succeq u\,x$ for some fresh x. Thus we can reduce variable containment of arbitrary types to variable containment of base types. Moreover, if t and u have a base type then $\mathrm{FV}_{\beta}(t) \supseteq \mathrm{FV}_{\beta}(u)$ iff for all $C[\,]$ we have $\mathrm{FV}_{\beta}(C[t]) \supseteq \mathrm{FV}_{\beta}(C[u])$. This way we can avoid the quantification over contexts by restricting our attention to variable containment for base types. To be precise: this trick requires that substitution does not affect the types, i.e. it does not apply to $\lambda\!\rightarrow$ as presented in [3] where base types are variables — we need them to be constants.

Definition 8. The variable containment problem for $\lambda^{\rightarrow}$ with signature Σ is the following:

$$\begin{aligned}\Gamma \vdash_{\Sigma} t \succeq u : \tau \overset{def}{\Longleftrightarrow} & (\tau : *) \in \Sigma \ \wedge \Gamma \vdash_{\Sigma} t : \tau \ \wedge \ \Gamma \vdash_{\Sigma} u : \tau \ \wedge \\ & \forall \Delta.\, \forall \theta : \Gamma \rightarrow \Delta.\, \mathrm{FV}_{\beta}(t^{\theta}) \supseteq \mathrm{FV}_{\beta}(u^{\theta})\end{aligned}$$

The property $(\tau : *) \in \Sigma$ just means that τ is a base type in the signature.

To decide the variable containment problem we would generally need that $\mathrm{FV}_{\beta}(t)$ is recursive which is the case for all strongly normalising type systems. Then we have to find a semantic domain $(D, \geq)$ to interpret the judgements $\Gamma \vdash_{\Sigma} t : \tau$ such that the predicate $\geq$ and the interpretation function are total

[2] I use the word "context" for terms with holes $C[\,]$ and also for sets of pairs of variables and types Γ, since it is established terminology for both.

recursive functions and $[\![t]\!] \geq [\![u]\!]$ iff $t \succeq u$. We follow tradition by using double brackets $[\![\_]\!]$ for denoting the semantic interpretation of syntactic objects.

There is no other requirement we need for these domains, i.e. D is just a set and $\geq$ a binary relation on D. Since $\succeq$ is a preorder (easy to show), we would need that $\geq$ is a preorder as well if $[\![\_]\!]$ is surjective.

3 Variable Containment for $\lambda_2^{\rightarrow}$

We begin with the type theory $\lambda_2^{\rightarrow}$, the second-order fragment of the simply typed λ-calculus $\lambda^{\rightarrow}$. In $\lambda_2^{\rightarrow}$, free variables are restricted to at most second-order types, i.e. types of the form $\tau_1 \rightarrow \tau_2 \rightarrow \cdots \rightarrow \tau_n$ such that all τ_i are base types. It is possible to define $\lambda_2^{\rightarrow}$ as a PTS, but we shall not do that here, for it would distract too much from the major issues we want to tackle.

Combinatory reduction systems (CRSs) [10, 11] can be seen as a special class of rewrite systems in the type theory $\lambda_2^{\rightarrow}$ over extensions of the signature $\Sigma = \langle 0{:}\ *, \Lambda{:}\ (0 \rightarrow 0) \rightarrow 0\rangle$. To get an exact match, no further type constants (other that 0) or third-order constants (other than Λ) should be allowed. CRSs come equipped with an additional restriction for left-hand side of rules (each free variable is applied to a sequence of distinct bound variables) that makes the variable containment problem trivial — for CRS rules $l \rightarrow r$ the property $l \succeq r$ is equivalent to $\mathrm{FV}(l) \supseteq \mathrm{FV}(r)$.

However, we can drop the restriction for left-hand sides and generalise the definition of second-order rewrite rule.

Definition 9. Given a $\lambda_2^{\rightarrow}$-signature Σ, a Σ-rule is a tuple (Γ, l, r, τ), written $\Gamma \vdash_\Sigma l \rightarrow r : \tau$, such that (i) τ is a base type, (ii) $\Gamma \vdash_\Sigma l : \tau$, (iii) $\Gamma \vdash_\Sigma r : \tau$

An instance of a rule (Γ, l, r, τ) is given by a substitution $\theta : \Gamma \rightarrow \Delta$ and a context $C[\,]$ such that $E \vdash C[x] : \sigma$ for some type σ, some fresh variable x, and some context E such that $(x : \tau) \in E$ and $\Delta \vdash x : \tau$ is a premise of $E \vdash C[x] : \sigma$, i.e. Δ is the context in which the hole of $C[\,]$ is being type-checked. We omit the formal definition of the latter, but it can easily be formalised in the style of a type system. We have $t \rightarrow_R u$ for terms t, u with $E \vdash t : \tau$ (analogously for u) if there is a rule $\Gamma \vdash_\Sigma l \rightarrow r : \tau$, a substitution $\theta : \Gamma \rightarrow \Delta$ and a context $C[\,]$ (as described) such that $C[l^\theta] =_{\beta\eta} t$ and $C[r^\theta] =_{\beta\eta} u$.

Since second-order matching is decidable, we can decide whether we have an instance of a rule, i.e. the rewrite relation $\rightarrow_R$ is decidable for finitely many rules. As for first-order rewriting, the transitive closure of $\rightarrow_R$ is undecidable. As already explained, the two-step rewrite relation $\rightarrow_R ; \rightarrow_R$ is undecidable for general second-order rewrite systems because of extra variables.

Therefore, it makes sense to require $\Gamma \vdash_\Sigma l \succeq r : \tau$ for all rules $\Gamma \vdash_\Sigma l \rightarrow r : \tau$. Since we require that the type τ of a rule is a base type, the proposition $\Gamma \vdash_\Sigma l \rightarrow r : \tau$ is well-formed and according to our general observations for base types it is equivalent to variable containment for the rewrite relation generated from this rule. In the following we shall omit the subscript Σ for judgements.

How can we decide variable containment in $\lambda_2^{\rightarrow}$ for two terms t and u? Take for instance the terms $t = F\,(x\,(y\,z))\,(y\,w)$ and $u = G\,(y\,(x\,z))$ (where w, x, y, z are variables, x and y second-order), do we have $t \succeq u$, or $u \succeq t$, or both, or neither, and how can we find out? For $t \succeq u$ we have to check $\mathrm{FV}_\beta(t^\theta) \supseteq \mathrm{FV}_\beta(u^\theta)$ for all substitutions θ, but this is an infinitary condition. For second-order variable containment only two things matter for a substitution: (i) which variables are free in the substitute, and (ii) for second-order variables v with substitute $\lambda v_1, \cdots, vn.s$, which of the (first-order) variables v_i is free in s? The former limits which variables can be free in the substituted term, from the latter we can find out which subterms will be erased during normalisation. Consider the variables x and y from the example and their substitutes $\theta(x) = \lambda x'.p$ and $\theta(y) = \lambda y'.q$: if x' is free in p and y' is free in q then the free variables in u^θ are exactly $\mathrm{FV}_\beta(\theta(x)) \cup \mathrm{FV}_\beta(\theta(y)) \cup \mathrm{FV}_\beta(\theta(z))$ and $\mathrm{FV}_\beta(t^\theta)$ contains those and also $\mathrm{FV}_\beta(\theta(w))$. Thus $u \not\succeq t$. If y' is not free in q then $\mathrm{FV}_\beta(u^\theta) = \mathrm{FV}_\beta(\theta(y))$ and $\mathrm{FV}_\beta(t^\theta) = \mathrm{FV}_\beta(\theta(x)) \cup \mathrm{FV}_\beta(\theta(y))$. Finally, if $y \in \mathrm{FV}_\beta(q), x \notin \mathrm{FV}_\beta(p)$ then $\mathrm{FV}_\beta(t^\theta) = \mathrm{FV}_\beta(\theta(w)) \cup \mathrm{FV}_\beta(\theta(x)) \cup \mathrm{FV}_\beta(\theta(y))$ and $\mathrm{FV}_\beta(u^\theta) = \mathrm{FV}_\beta(\theta(x)) \cup \mathrm{FV}_\beta(\theta(y))$. So, in all cases we have $\mathrm{FV}_\beta(t^\theta) \supseteq \mathrm{FV}_\beta(u^\theta)$ and thus we have $t \succeq u$.

The general picture is that we have to consider all free variable occurrences in a term and see in which argument positions of which other variables these occurrences are. The following semantic interpretation of terms captures these observations.

Definition 10. For $\lambda_2^{\rightarrow}$, we interpret judgements $\Gamma \vdash t : \tau$ as pointwise ordered functions in $\mathit{Dom}\,\Gamma \rightarrow \wp(\wp((\mathit{Dom}\,\Gamma) \times \mathcal{N}))$ where the order on the codomain is given as:

$$A \geq B \overset{def}{\iff} \forall M \in B.\, \exists N \in A.\, N \subseteq M.$$

We assume in the following that t is in β-normal form, i.e. if it is not then we take $[\![\Gamma \vdash t : \tau]\!] = [\![\Gamma \vdash t\!\downarrow : \tau]\!]$ where $t\!\downarrow$ is the normal form of t.

If t has the form $x\,t_1 \cdots t_n$ $(n \geq 0)$ with $x \in \mathit{Dom}\,\Gamma$ then:

$$[\![\Gamma \vdash x\,t_1 \cdots t_n : \tau]\!](x) = \{\emptyset\}$$
$$[\![\Gamma \vdash x\,t_1 \cdots t_n : \tau]\!](y) = \{M \cup \{(x,i)\} \mid 1 \leq i \leq n, M \in [\![\Gamma \vdash t_i : \tau_i]\!](y)\} \quad \text{if } x \neq y$$

If t has the form $f\,t_1 \cdots t_n$ with $f \in \mathit{Dom}\,\Sigma$ then

$$[\![\Gamma \vdash f\,t_1 \cdots t_n : \tau]\!](y) = \bigcup_{1 \leq i \leq n} [\![\Gamma \vdash t_i : \tau_i]\!](y)$$

If t is an abstraction $\lambda x\colon \tau.\, u$ then

$$[\![\Gamma \vdash \lambda x\colon \tau.\, u : \tau \rightarrow \sigma]\!](x) = \emptyset$$
$$[\![\Gamma \vdash \lambda x\colon \tau.\, u : \tau \rightarrow \sigma]\!](y) = \{M \setminus (\{x\} \times \mathcal{N}) \mid M \in [\![\Gamma, x\colon \tau \vdash u : \sigma]\!](y)\} \quad \text{if } x \neq y$$

The subtraction of $\{x\} \times \mathcal{N}$ (for the abstraction) is only necessary if the type of x is second-order. This situation can only occur on outermost level, and it cannot in our second-order rules. One could argue whether these terms exist in $\lambda_2^{\rightarrow}$, but they do indeed in a PTS-like formalisation.

The interpretation can be explained as follows: if $[\![\Gamma \vdash t : \tau]\!](x) = M$ then M contains for each free occurrence of x in t the set of argument positions in variable applications that lie above that occurrence. In particular: if $M = \emptyset$ then x is not free in t and if $\emptyset \in M$ then there is a topmost occurrence of x in t and all variables free in $\theta(x)$ will be free in t^θ as well.

Definition 11. A substitution $\theta : \Gamma \rightarrow \Delta$ *preserves* a set $M \subseteq (\mathcal{V} \times \mathcal{N})$, written $\theta \models M$, iff

$$\forall (x,i) \in M.\, x \in \mathit{Dom}\,\Gamma \Rightarrow \exists y_1, \ldots, y_i, t.\quad \theta(x) =_{\beta\eta} \lambda y_1, \ldots, y_i.\, t \wedge y_i \in \mathrm{FV}_\beta(t)$$

We can read the property $\exists M \in [\![\Gamma \vdash t : \tau]\!](x).\, \theta \models M$ as follows: "there is a free occurrence of x in t which is not erased when we apply θ to t".

Definition 12. A substitution $\theta : \Gamma \rightarrow \Delta$ is called *first-order* iff for all $(x : \sigma) \in \Gamma$ the preterm $\theta(x)$ is *not* an abstraction.

Thus, if t is a normal form and θ is a first-order substitution with only normal forms in its codomain then t^θ is in normal form too. Obviously, a first-order substitution preserves any M. This means that $\exists M \in [\![\Gamma \vdash t : \tau]\!](x).\, \theta \models M$ is equivalent to $x \in \mathrm{FV}_\beta(t)$ for first-order θ.

Lemma 13. *Let $\Gamma \vdash t : \tau$, $\theta : \Gamma \rightarrow \Delta$ and $y \in \mathit{Dom}\,\Delta$ be arbitrary. We have*
$y \in \mathrm{FV}_\beta(t^\theta) \iff$
$\exists x \in \mathit{Dom}\,\Gamma.\, y \in \mathrm{FV}_\beta(\theta(x)) \wedge \exists M \in [\![\Gamma \vdash t : \tau]\!](x).\, \theta \models M$

Proof. We can w.l.o.g. assume that t is in normal form and that θ maps variables to terms in normal form.

First we prove the lemma for first-order substitutions. Using our assumptions about t and θ and the above observations about first-order substitutions, the lemma reduces to $y \in \mathrm{FV}(t^\theta) \iff \exists x \in \mathit{Dom}\,\Gamma.\, y \in \mathrm{FV}(\theta(x)) \wedge x \in \mathrm{FV}(t)$ which is an obvious property of substitutions.

Now let θ be arbitrary. We prove the lemma by induction on the term structure. We just show "$\Rightarrow$", "$\Leftarrow$" is similar. We only have to consider variable applications $z\, t_1 \cdots t_n$, constant applications $f\, t_1 \cdots t_n$ and abstractions $\lambda z.\, t'$.

Let t be a variable application $z\, t_1 \cdots t_n$. Let $\theta(z) = \lambda y_1 \cdots y_n.\, u$. Then $t^\theta\!\downarrow = u^v$ where $v : E \rightarrow \Delta$ is given by $v(y_i) = t_i^\theta\!\downarrow$ and $v(v) = v$ for $v \notin \{y_1, \ldots, y_n\}$. Observe that v is first-order, i.e. we can apply the lemma to it. We get: $y \in \mathrm{FV}_\beta(t^\theta) \iff y \in \mathrm{FV}_\beta(u^v) \iff \exists x' \in \mathit{Dom}\,E.(y \in \mathrm{FV}_\beta(v(x')) \wedge \exists M \in [\![E \vdash u : \tau]\!](x').\, v \models M) \iff \exists x' \in \mathit{Dom}\,E.\, y \in \mathrm{FV}_\beta(v(x')) \wedge x' \in \mathrm{FV}_\beta(u)$. For $x' \in \mathit{Dom}\,E$, we either have $x' = y_i$ for some y_i or $x' \in \mathit{Dom}\,\Delta$.

In the former case the condition reduces to $y_i \in \mathrm{FV}_\beta(u) \wedge y \in \mathrm{FV}_\beta(t_i^\theta)$ for some i. The first part means that θ preserves $M \cup \{(z,i)\}$ iff it preserves M. For the second we can apply the induction hypothesis and get a variable $x \in Dom\,\Gamma$ and a set $M_i \in [\![\Gamma \vdash t_i : \sigma_i]\!](x)$ with $\theta \models M_i$. Thus $\theta \models M'_i = M_i \cup \{(z,i)\}$ and clearly $M'_i \in [\![\Gamma \vdash z\,t_1 \cdots t_n : \tau]\!](x)$.

In the latter case, $x' \in Dom\,\Delta$, we have $y = x'$ and can choose $x = z$: since $\emptyset \in [\![\Gamma \vdash z\,t_1 \cdots t_n : \tau]\!](z)$ we only have to show $\theta \models \emptyset$, but this is trivially true.

For constant applications $f\,t_1 \cdots t_n$ we can directly apply the induction hypothesis: $y \in \mathrm{FV}_\beta(f\,t_1^\theta \cdots t_n^\theta) \iff \exists i.\, y \in \mathrm{FV}_\beta(t_i^\theta) \iff \exists x \in Dom\,\Gamma.\, y \in \mathrm{FV}_\beta(\theta(x)) \wedge \exists M \in [\![\Gamma \vdash t_i : \sigma_i]\!](x).\, \theta \models M \iff \exists x \in Dom\,\Gamma.\, y \in \mathrm{FV}_\beta(\theta(x)) \wedge \exists M \in [\![\Gamma \vdash f\,t_1 \cdots t_n : \tau]\!](x).\, \theta \models M$.

Finally, let t be an abstraction $\lambda z.u$. We define $\theta' = \theta[z \mapsto z]$. We get: $y \in \mathrm{FV}_\beta(t^\theta) \iff y \in \mathrm{FV}_\beta(u^{\theta'}) \wedge y \neq z \iff y \neq z \wedge \exists x \in Dom\,\Gamma \cup \{z\}.\, y \in \mathrm{FV}_\beta(\theta'(x)) \wedge \exists M \in [\![\Gamma, z:\sigma \vdash u : \sigma']\!](x).\, \theta' \models M \iff y \neq z \wedge \exists x \in Dom\,\Gamma.\, y \in \mathrm{FV}_\beta(\theta(x)) \wedge \exists M \in [\![\Gamma, z:\sigma \vdash u : \sigma']\!](x).\, \theta' \models M \iff y \neq z \wedge \exists x \in Dom\,\Gamma.\, y \in \mathrm{FV}_\beta(\theta(x)) \wedge \exists M \in [\![\Gamma, z:\sigma \vdash u : \sigma']\!](x).\, \theta \models M \setminus (\{z\} \times \mathcal{N}) \iff \exists x \in Dom\,\Gamma.\, y \in \mathrm{FV}_\beta(\theta(x)) \wedge \exists M \in [\![\Gamma \vdash \lambda x{:}\,\sigma.\, u : \sigma']\!](x).\, \theta \models M$. □

Lemma 14. $\Gamma \vdash t \succeq u : \tau \iff [\![\Gamma \vdash t : \tau]\!] \geq [\![\Gamma \vdash u : \tau]\!]$

Proof. This follows easily from a pointwise extension of lemma 13. Considering the "⇒" direction, notice that for each $N \in [\![\Gamma \vdash u : \tau]\!](x)$ we can construct a substitution θ such that $\theta \models M$ iff $M \subseteq N$ and $y \in \mathrm{FV}_\beta(\theta(x))$. □

Clearly, $[\![\_]\!]$ is a total computable function and so is the order $\geq$ when restricted to total computable functions. Therefore:

Theorem 15. *The variable containment problem for* $\lambda_2^{\rightarrow}$ *is decidable.*

4 Variable Containment for $\lambda^{\rightarrow}$

We are going to reduce the general variable containment problem for $\lambda^{\rightarrow}$ to a more specific situation, in which we only consider a particular signature and substitutions into a particular context. This reduction also links the problem to a problem in the semantics of PCF.

Definition 16. A *pseudo-constant* in a $\lambda^{\rightarrow}$-signature Σ is a term c with $\langle\rangle \vdash_\Sigma c : \sigma$ for some type σ and:

$$\forall \Gamma.\, \forall \sigma.\, \forall t_1, \ldots, t_n.\, \Gamma \vdash_\Sigma c\,t_1 \cdots t_n : \sigma \implies \forall x \in Dom\,\Gamma.\, (x \in \mathrm{FV}_\beta(c\,t_1 \cdots t_n) \iff \exists i.\, x \in \mathrm{FV}_\beta(t_i))$$

Any symbol in the signature is obviously a pseudo-constant. The identity function $\lambda x{:}\,\sigma.\, x$ is a pseudo-constant if and only if σ is a type constant. The idea behind pseudo-constants is that they behave like constants in many ways, in particular with respect to the variable containment problem. It is sometimes useful to assume a constant for any type (for freezing variables), but this would require an infinite signature. For our purposes it is sufficient to have pseudo-constants for any type.

Definition 17. A $\lambda^{\rightarrow}$-signature Σ is called *rich* if (i) it includes a base type 0, (ii) there are constants $A : 0$ and $B : 0 \rightarrow 0 \rightarrow 0$ in Σ and (iii) for any other base type $\alpha \in \Sigma$ there are constants $C_\alpha : 0 \rightarrow \alpha$ and $D_\alpha : \alpha \rightarrow 0$ in Σ.

We can extend any signature to a rich signature just by adding the missing constants. One could also view signatures as rich if they have *pseudo*-constants of the required types, but we shall not do that as it only complicates the technicalities without adding anything substantial. In the following, we assume for simplicity that there is only one base type 0 in Σ. The corresponding adjustments to the general case are straightforward.

Definition 18. Let Σ be rich. For any type σ we define a term con_σ as follows:

$$\begin{aligned}
\mathrm{con}_0 &= A \\
\mathrm{con}_{0\rightarrow 0} &= \lambda x\colon 0.\, x \\
\mathrm{con}_{0\rightarrow(\sigma\rightarrow\tau)} &= \lambda x\colon 0.\, \lambda y\colon \sigma.\, \mathrm{con}_{0\rightarrow\tau}(\mathrm{B}\,\mathrm{x}\,(\mathrm{con}_{\sigma\rightarrow 0}\,\mathrm{y})) \\
\mathrm{con}_{(\sigma\rightarrow\tau)\rightarrow v} &= \lambda f\colon \sigma\rightarrow\tau.\, \mathrm{con}_{\tau\rightarrow v}\,(\mathrm{f}\,\mathrm{con}_\sigma)
\end{aligned}$$

The function con is well-defined as the right-hand sides of the equations use "fewer arrows" in the types of con than the corresponding left-hand sides. Clearly, each con_σ has type σ in the empty context.

Remark: it is worth noting that the terms con_σ have a more general significance, e.g. they show up in [19] where A is 0 and B is addition. As explained in [9], the map $\mathrm{con}_{\sigma\rightarrow 0}$ is the inverse of $\mathrm{con}_{0\rightarrow\sigma}$ whenever A and B form a monoid; moreover, in the terminology of category theory [13], they are even morphisms of (some) actions of this monoid.

Proposition 19. *Each* con_σ *is a pseudo-constant.*

One consequence of having pseudo-constants for all types is that we can slightly simplify the variable containment problem.

Lemma 20. *The variable containment problem* $\Gamma \vdash_\Sigma t \succeq u : \tau$ *is equivalent to the following property for a rich extension* Σ' *of* Σ:

$$\forall\theta : \Gamma \rightarrow X.\, \mathrm{FV}_\beta(t^\theta) \supseteq \mathrm{FV}_\beta(u^\theta) \quad (*)$$

where X *is the fixed context* $\langle x : 0\rangle$.

Proof. We prove both implications by contradiction, first ($\Rightarrow$). The property ($*$) is an instance of the variable containment problem if Σ is already rich. Otherwise, let $\theta : \Gamma \rightarrow X$ be a Σ'-substitution such that $x \in \mathrm{FV}_\beta(u^\theta)$ and $x \notin \mathrm{FV}_\beta(t^\theta)$. We can create a counter-example for variable containment as follows: the context is $\Delta = \langle a : 0, b : 0 \rightarrow 0 \rightarrow 0, x' : 0\rangle$ and the substitution $\phi : \Gamma \rightarrow \Delta$ is given by $\phi(y) = \theta(y)[a/A, b/B, x'/x]$. Clearly, u^ϕ contains the variable x' essentially free whilst t^ϕ does not.

Now ($\Leftarrow$): suppose variable containment does not hold, i.e. for some context Δ, some variable $y \in \mathit{Dom}\,\Delta$, and some substitution $\theta : \Gamma \rightarrow \Delta$ we have that

$y \in \mathrm{FV}_\beta(u^\theta)$ but $y \notin \mathrm{FV}_\beta(t^\theta)$. We can define a substitution $\phi : \Delta \to X$ as follows:

$$\phi(z) = \quad \mathrm{con}_\sigma \text{ if } z \neq y,\ z{:}\,\sigma \in \Delta$$
$$\phi(y) = (\mathrm{con}_{0\to\sigma}\ \mathrm{x}) \text{ if } y{:}\,\sigma \in \Delta$$

and from this we get a contradiction of $(*)$ using the substitution $\phi \circ \theta : \Gamma \to X$: the pseudo-constant property of all con_σ makes sure that x is in $\mathrm{FV}_\beta(u^{\phi\circ\theta})$ while $\mathrm{FV}_\beta(t^{\phi\circ\theta}) = \emptyset$. □

Variable containment is unaffected by replacing constants by pseudoconstants. Based on this observation and lemma 20 we can design a semantic interpretation for types, terms, and judgements to model variable containment. Since X has only one variable x of type 0, $\mathrm{FV}(t)$ is just a boolean information for any t with $X \vdash_\Sigma t : 0$. For higher types, we also have to model how the freeness of x can be affected by β-reduction.

Thus we can interpret $0 : * \in \Sigma$ by the partially ordered set $\{\bot, \top\}$ (with $\bot \leq \top$) and each function space $\sigma \to \tau$ by the set of λ-definable monotonic functions from $[\![\sigma]\!]$ to $[\![\tau]\!]$, ordered pointwise. Here, we take the constants $\bot$ and $\top$ and the function $\wedge \in [\![0 \to 0 \to 0]\!]$ (greatest lower bound) as primitively λ-definable. Thus, our semantic domain is a fully abstract model for PCF_1, i.e. finitary PCF over the unit type. We come to that later in more detail.

The restriction of the function space to λ-definable functions is crucial: the terms $f\,x\,x$ and $B\,(f\,A\,x)\,(f\,x\,A)$ are equivalent w.r.t. to variable containment but are different in the full Poset model over $[\![0]\!]$.

Definition 21. Given a context Γ an *environment* ρ for Γ is a finite map from the domain of Γ to the union of all $[\![\sigma]\!]$ (with $\langle\rangle \vdash_\Sigma \sigma : *$) such that: $\forall x : \sigma \in \Gamma.\ \rho(x) \in [\![\sigma]\!]$.

Let Σ be rich (otherwise we can make it rich by a signature extension). Given any context Γ and environment ρ for Γ, we can interpret judgements $\Gamma \vdash_\Sigma t : \tau$ as follows.

$$[\![\Gamma \vdash_\Sigma \lambda x{:}\,\sigma.\,t : \sigma \to \tau]\!]_\rho = (v \mapsto [\![\Gamma, x{:}\,\sigma \vdash_\Sigma t : \tau]\!]_{\rho[x\mapsto v]})$$
$$[\![\Gamma \vdash_\Sigma t\,u : \tau]\!]_\rho = [\![\Gamma \vdash_\Sigma t : \sigma \to \tau]\!]_\rho([\![\Gamma \vdash_\Sigma u : \sigma]\!]_\rho)$$
$$[\![\Gamma \vdash_\Sigma x : \sigma]\!]_\rho = \rho(x)$$
$$[\![\Gamma \vdash_\Sigma c : 0]\!]_\rho = \top$$
$$[\![\Gamma \vdash_\Sigma c : 0 \to 0 \to 0]\!]_\rho = (x \mapsto (y \mapsto x \wedge y))$$
$$[\![\Gamma \vdash_\Sigma c : \sigma]\!]_\rho = [\![\langle\rangle \vdash_\Sigma \mathrm{con}_\sigma : \sigma]\!]_{[]} \text{ if } \sigma \notin \{0, 0 \to 0 \to 0\}$$

The reason for the special treatment of types 0 and $0 \to 0 \to 0$ is that con_σ terms can contain constants of only these two types, so this stops the recursion. The definition of the interpretation function $[\![\_]\!]$ is well-defined as the interpretation of each judgement $[\![\Gamma \vdash_\Sigma t : \sigma]\!]$ is in $[\![\sigma]\!]$. Moreover, for any given environment ρ, the function $[\![\_]\!]_\rho$ is clearly recursive.

The interpretation of judgements is in fact independent from the choice of signature, as all constants of the same type have equal interpretations. The idea behind this interpretation is the following: we use the fixed context $X = \langle x : 0 \rangle$ and take $\top$ for "x is not essentially free" and $\bot$ for "x is essentially free". Apparently, x does not occur free in any constant c of type 0, which we model by interpreting c as $\top$. Then, x is essentially free in $B\,t\,u$ iff it is essentially free in either t or u — this explains the interpretation of B (and any other constant of type $0 \to 0 \to 0$) as $\wedge$, the greatest lower bound. The rest of the definition is just book-keeping and reducing more complicated situations to simpler ones. In particular, $\beta\eta$-equivalent terms have equal interpretations as syntactic abstraction and application are modelled by semantic abstraction and application, and constants of any type can be replaced by pseudo-constants of the same type as they behave the same w.r.t. the variable containment problem.

As usual, we can compose substitutions and environments.

Definition 22. Let $\theta : \Gamma \to \Delta$ be a substitution and ρ be an environment for Δ. We define a function $\rho \circ \theta$ as follows:

$$(\rho \circ \theta)(x) = [\![\Delta \vdash \theta(x) : \tau]\!]_\rho \quad \text{if } \Gamma \vdash x : \tau$$

Lemma 23. *Let $\theta : \Gamma \to \Delta$ be a substitution and ρ be an environment for Δ.*

1. *$\rho \circ \theta$ is an environment for Γ.*
2. *For all $\Gamma \vdash t : \tau$ we have $[\![\Delta \vdash t^\theta : \tau]\!]_\rho = [\![\Gamma \vdash t : \tau]\!]_{\rho\circ\theta}$.*

Lemma 23 is standard for semantic interpretations of the λ-calculus, the proof is routine and needs hardly any adaptation from (for example) the proof of lemma 24 in [18].

Definition 24. We define an order $\leq$ on judgements of the same type and context as follows:

$$(\Gamma \vdash t : \tau) \leq (\Gamma \vdash u : \tau) \iff \forall \rho.\, [\![\Gamma \vdash t : \tau]\!]_\rho \leq_\tau [\![\Gamma \vdash u : \tau]\!]_\rho$$

Lemma 25. *Let J_1, J_2 be judgements $J_i = \Gamma \vdash_\Sigma t_i : \tau$.*
We have $J_1 \leq J_2 \iff \Gamma \vdash t_1 \succeq t_2 : \tau$.

Proof. By lemma 20 we can w.l.o.g. assume that Σ is rich and restrict our attention to variable containment w.r.t. the context X. Similarly we can require Σ to be the rich extension of the empty signature, because variable containment and $[\![\_]\!]$ are unaffected by replacing constants by arbitrary pseudo-constants. Since the interpretation of syntactic abstraction and application is by semantic abstraction and application, β-reduction does not affect the interpretation. From this it follows (by a straightforward induction on normal forms) that $[\![X \vdash u : 0]\!]_{x\mapsto\bot} = \bot \iff x \in \mathrm{FV}_\beta(u)$.

To prove ($\Leftarrow$) we need to be able to construct a substitution counterexample for $\Gamma \vdash t_1 \succeq t_2 : \tau$ whenever we have an environment ρ such that $\neg[\![J_1]\!]_\rho \leq [\![J_2]\!]_\rho$. Because we required each value in the model to be λ-definable relative to $\bot$, $\top$, and $\wedge$, we can find for each value v in $[\![\sigma]\!]$ a term t_v such that $X \vdash t_v : \sigma$ and $[\![X \vdash t_v : \sigma]\!]_{x\mapsto\bot} = v$. The substitution $\theta : \Gamma \to X$ with $\theta(y) = t_{\rho(y)}$ is then the substitution we were looking for. □

In other words, the variable containment problem is equivalent to deciding the inequality $\leq$ in a fully abstract model of PCF_1 (PCF over the unit type with constants $\bot$ and $\top$ and $\wedge$). To decide $\leq$, it would be sufficient to effectively construct such a model, because each type is interpreted by a finite poset. The connection is rather tight indeed: if we have a partial construction of the model for types up to order n then we can decide variable containment for $\lambda_n^{\rightarrow}$.

A recent result by Zaionc [21] means[3] that variable containment is decidable for $\lambda_3^{\rightarrow}$, as his technique of creating all λ-definable values by some grammar easily extends to the situation with predefined constants A and B. Sieber's PCF model of "logically sequential" elements [20] seems to be effective for finitary PCF and it is fully abstract up to order 4 and term-generated up to order 3; this also implies the decidability of variable containment of $\lambda_3^{\rightarrow}$, though the connection is less direct than in the case of Zaionc's result. This improves upon my theorem 15; but the decision procedures obtained that way are extremely inefficient and of solely theoretical interest, while the decision procedure outlined earlier for $\lambda_2^{\rightarrow}$ is of polynomial complexity.

For PCF_2 (PCF over the booleans with constants $\bot$, `tt`, `ff`, `if`), effectively constructing a fully abstract model was posed as an open problem by Jung and Stoughton in [8]; it is yet unclear whether this is equivalent to our problem.

We can also show that the "effective" construction of a model for PCF_1 is *necessary* to decide $\leq$ and even the indistinguishability relation $\approx$.

Theorem 26. *The problem of deciding the indistinguishability relation $\approx$ for* PCF_1 *is equivalent to effectively constructing a fully abstract model.*

Proof. As explained before, one implication is trivial. It remains to show that $\approx$ gives us a way of constructing a fully abstract model.

We can construct $[\![0]\!] = \{\bot, \top\}$ with $\bot \leq \top$. Suppose we have constructed the sets $[\![\sigma_i]\!]$ then we can construct $[\![\sigma_1 \to \cdots \to \sigma_n \to 0]\!]$ as follows. Each element in this set is a function mapping n-tuples to either $\bot$ or $\top$. There are only finitely many such functions (as all the $[\![\sigma_i]\!]$ are finite).

To decide whether a particular function F is λ-definable we consider the term $\chi_F = \lambda f.\, f\, a_{11} \cdots a_{1n} \wedge \cdots \wedge f\, a_{k1} \cdots a_{kn}$ where the tuples $a_{i1} \cdots a_{in}$ are tuples of terms representing exactly those tuples (of values) mapped by F to $\top$. Since the construction of each $[\![\sigma_j]\!]$ is assumed to be complete we can effectively find a term a for each value v in these sets.

Now take $\succeq$ to be the pointwise extension of the $\geq_j$ such that it is defined on *all* monotonic functions, not just the λ-definable ones. Now consider any other function $G \succ F$ and its characteristic function χ_G. Suppose F is defined by a term t then $[\![\chi_G\, t]\!] = \bot$ and $[\![\chi_F\, t]\!] = \top$. Consequently, χ_F and χ_G are distinguishable if F is λ-definable, and thus if $\chi_G \approx \chi_F$ for any $G \succ F$ then F cannot be λ-definable. Now suppose that χ_F is distinguishable from χ_G for each $G \succ F$. This means that there has to be a term t_G for each G such that $[\![\chi_F t_G]\!] = \top$ and $[\![\chi_G t_G]\!] = \bot$. We can define a term $t = \lambda x_1 \cdots x_n . t_{G_1}\, x_1 \cdots x_n \wedge$

[3] The title of Zaionc's paper is a little misleading — he gives the base type order 0 instead of 1, following a deplorable custom in programming language semantics.

$\cdots \wedge t_{G_r}\, x_1 \cdots x_n$ where $G_1 \cdots G_r$ are all functions greater than F. We obviously have $[\![\chi_F\, t]\!] = \top$ and $[\![\chi_{G_i}\, t]\!] = \bot$ for all G_i. But this exactly means $[\![t]\!] = F$, i.e. F is λ-definable.

Hence we can construct $[\![\sigma_1 \to \cdots \to \sigma_n \to 0]\!]$ as the set of all monotonic functions that pass the outlined test, i.e. whose characteristic functions are distinguishable. □

Unsurprisingly a similar result holds for PCF_2, though the proof is a bit messier, involving pairs of characteristic functions (one for `tt`, one for `ff`). We do not go into that.

5 Conclusion and Open Problems

We have explained why the usual condition $\mathrm{FV}(l) \supseteq \mathrm{FV}(r)$ for higher-order rewrite rules $l \to r$ is inadequate and why it should be replaced by the "variable containment" property $l \succeq r$, the rewrite interior of $\mathrm{FV}_\beta(l) \supseteq \mathrm{FV}_\beta(r)$.

We have shown that variable containment is decidable for the third-order fragment of $\lambda^\to$, also giving a constructive solution for the second-order fragment. The general problem for $\lambda^\to$ is equivalent to effectively constructing a fully abstract model for finitary PCF over the unit type.

Open problems are:

- Is the problem for $\lambda^\to$ decidable? I have seen a preliminary version of an unpublished paper which claims that it is indeed. The proof in the paper is rather complicated and without thorough revision I would not say that the problem is settled.
- Is variable containment equivalent to providing a fully abstract model for PCF_2? This is very delicate. I had a promising proof idea which I pursued for a few weeks without getting it to work. One of the referees conjectured that the PCF_2 model is not recursive.
- For which type systems is variable containment undecidable?
- Finally: what about other type systems of the λ-cube, is there a similar correspondence between full abstraction and variable containment for those systems? Probably yes, but to make any sense of this, one first has to generalise the definition of full abstraction to these type systems.

Acknowledgments

Many thanks to Alex Simpson with whom I had a number of fruitful discussions on the subject, especially on the PCF part and theorem 26. Also thanks to the HOA referees.

References

1. Hendrik P. Barendregt. *The Lambda-Calculus, its Syntax and Semantics.* North-Holland, 1984.

2. Hendrik P. Barendregt. Introduction to generalised type systems. *Journal of Functional Programming*, 1(2):124–154, 1991.
3. Hendrik P. Barendregt. Lambda calculi with types. In *Handbook of Logic in Computer Science, Vol.2*, pages 117–309. Oxford Science Publications, 1992.
4. Gilles Dowek. Third order matching is decidable. In *Proceedings of the 7th Symposium on Logic in Computer Science*, pages 2–10, 1992.
5. Philippa Gardner. *Representing Logics in Type Theory.* PhD thesis, University of Edinburgh, 1992.
6. W. D. Goldfarb. The undecidability of the second-order unification problem. *Theoretical Computer Science*, 13:225–230, 1981.
7. Gérard Huet and Bernard Lang. Proving and applying program transformations expressed with second-order patterns. *Acta Informatica*, 11:31–55, 1978.
8. Achim Jung and Allen Stoughton. Studying the fully abstract model of PCF within its continuous function model. In *Typed Lambda Calculi and Applications*, 1993. LNCS 664.
9. Stefan Kahrs. Towards a domain theory for termination proofs. In *Rewriting Techniques and Applications*, pages 241–255, 1995. LNCS 914.
10. Jan Willem Klop. *Combinatory Reduction Systems.* PhD thesis, Centrum voor Wiskunde en Informatica, 1980.
11. Jan Willem Klop, Vincent van Oostrom, and Femke van Raamsdonk. Combinatory reduction systems: Introduction and survey. *Theoretical Computer Science*, 121:279–308, 1993.
12. Ralph Loader. The undecidability of λ-definability, 1994.
13. Saunders MacLane. *Categories for the Working Mathematician.* Springer, 1971.
14. Richard Mayr and Tobias Nipkow. Higher-order rewrite systems and their confluence. Technical Report TUM-I94333, Technische Universität München, 1994.
15. Aart Middeldorp. Modular aspects of properties of term rewriting systems related to normal forms. In *Rewriting Techniques and Applications*, pages 263–277, 1989. LNCS 355.
16. Tobias Nipkow. Higher order critical pairs. In *Proceedings of the 6th Symposium on Logic in Computer Science*, pages 342–349, 1991.
17. Vincent Padovani. On equivalence classes of interpolation equations. In *Typed Lambda Calculi and Applications*, pages 335–249, 1995. LNCS 902.
18. Jaco van de Pol. Termination proofs for higher-order rewrite systems. In *Higher-Order Algebra, Logic, and Term Rewriting*, pages 305–325, 1993. LNCS 816.
19. Jaco van de Pol and Helmut Schwichtenberg. Strict functionals for termination proofs. In *Typed Lambda Calculi and Applications*, pages 350–364, 1995. LNCS 902.
20. Kurt Sieber. Reasoning about sequential functions via logical relations. In M.P. Fourman, P.T. Johnstone, and A.M. Pitts, editors, *Applications of Categories in Computer Science*, pages 258–269. Cambridge University Press, 1992.
21. Marek Zaionc. Lambda definability is decidable for second order types and for regular third order types. Unpublished Manuscript, University of New York at Buffalo, 1995.

Higher-Order Equational Logic for Specification, Simulation and Testing

Karl Meinke

Department of Computer Science,
University of Wales, Swansea,
Swansea SA2 8PP,
Great Britain.
K.Meinke@swansea.ac.uk

Abstract We consider higher-order equational logic as a formalism for the specification, simulation and testing of systems. We survey recent theoretical results on the expressiveness and proof theory of higher-order equations. These results are then interpreted within the context of specification language design to show that higher-order equational logic, used as a specification language, provides a useful compromise between the conflicting requirements of logical expressiveness and computational tractability.

1. Introduction. In designing a practical specification language for computer software or hardware which is to be based on some logical formalism, two essentially conflicting requirements can be observed. On the one hand it is natural to require that a specification language be as *expressive* as possible. This requirement can be formulated in a rather weak (perhaps even vague) form as a condition on the syntax of the language: that it provides a rich set of notations for describing designs. It can also be formulated in a stronger form, as a condition on the semantics of a specification language, that it allows the designer to express complex descriptions, where complexity is measured in some precise mathematical way (such as recursion theory). Generally speaking, foundational studies of specification languages can only illuminate this latter semantical approach to expressiveness.

On the other hand, the design of a logically based specification language which is to have some practical utility, requires that the underlying logic chosen should be *computationally tractable*. The primary motivation for adopting a formal, logically based approach to systems design comes from the fact that substantial machine support can be made available to assist the design process using such languages. In this way the principles of computer aided design can be applied to the discipline of software engineering itself. In practise, machine assistance ranges from the provision of low level support such as tools for editing, parsing and pretty printing of specifications, to more sophisticated tools for the simulation and testing of specifications which are able to answer logical queries about the specification and thus the system under design. The latter tools come under the loose heading of "rapid prototyping tools" in software engineering. There seems to be widespread agreement among the software engineering community that rapid prototyping is a desirable facility for software specification and development (see for example the survey described in Gordon and Bieman

[1995] and the empirical observations contained therein). Indeed disagreement only seems to arise over whether the facility can be made at all practicable within industrial strength systems. For a specification language based on logic, rapid prototyping and the ability to compute responses to logical queries about a specification require the implementation of a computational logic for the language in question. The utility of this implementation then depends heavily upon the space and time complexity of the computational logic.

Experience with the theoretical foundations of specification languages and with computational logic suggests that these two requirements of logical expressiveness and computational tractability seem to be essentially in conflict. The design of a practical specification language therefore requires some compromise to be made between these two requirements. In this paper we will attempt to substantiate a claim, using recent theoretical results of the author and others that, used as a specification language, higher-order equational logic represents one such useful compromise between expressiveness and tractability.

The structure of the paper is as follows. In Section 2 we will survey the fundamentals of higher-order algebraic specifications including the syntax of higher-order equational specifications, their higher-order initial algebra semantics, the associated proof systems and their completeness properties. In Section 3 we will consider recent expressiveness results for higher-order equational specifications under higher-order initial semantics. These results characterise the recursion theoretic complexity of the algebras which can be specified (up to isomorphism) by means of the arithmetical and analytical hierarchies. The fundamental observation here is the axiomatisability of the so called *search* or *quantifier functional*, using a finite set of second-order equations, which is primarily responsible for giving higher-order equational logic its expressive power. In particular the quantifier functional allows us to express first-order existential quantification within a quantifier free logical system (equations). Thus within higher-order equational logic we may formalise non-constructive specifications quite different in structure and complexity from the systems of recursion equations which are the mainstay of first-order equational specifications. Finally in Section 4 we consider the computational logic of higher-order equations and its algorithmic tractability. In particular we consider the problem of automating higher-order equational inference for ground first-order equations (data) by means of term rewriting. We survey the results of a topological approach to the proof theory of higher-order equations developed in Meinke [1996b]. These results establish that discontinuity of any higher-order operations in the initial model with respect to a topology of finite information is the only obstacle to automation by this approach. (The quantifier functional is the classic example of a discontinuous functional with respect to this topology.) Thus we observe that the price to be paid for supporting rather powerful non-constructive specifications is the absence of an efficient algorithm for answering all logical queries about a higher-order equational specification. This finding is not surprising in retrospect. However, we will conclude that given a higher-order equational specification, the logical queries about data (tests and simulations) which *can* be answered by means of automated deduction, can be answered efficiently by means of term rewriting.

2. Higher-Order Equational Logic. In this section we introduce the fundamentals of higher-order equational specifications, including higher-order signatures, algebras, equations and their initial semantics, and the logic of higher-order equations and its completeness properties. We consider two basic forms of higher-order equational specification, namely:

(i) *specifications with hidden sorts and operations*, and
(ii) *specifications with constructors and hidden sorts and operations.*

We define the higher-order initial semantics of both these types of specifications, and consider the proof systems required to concretely construct the higher-order initial model in each case. Constructors provide an extension of the basic method which is useful for overcoming certain problems associated with specifying partial functions. Further discussion of this technical device (in the first-order case) can be found in Wirsing [1990].

We begin by making precise our notation for many-sorted universal algebra which is adapted from Meinke and Tucker [1993]. We let **N** denote the set of natural numbers, $[\mathbf{N} \to \mathbf{N}]$ denotes the set of all total functions from **N** to **N**. For any set S, we let S^* denote the set of all *words* or *strings* over S, including the *empty word* λ. Then S^+ denotes the set of all non-empty words, $S^+ = S^* - \{ \lambda \}$.

2.1. Definition. *An S-sorted* **signature** *Σ is a pair*

$$\Sigma = (S, \langle \Sigma_{w,s} \mid w \in S^*, s \in S \rangle)$$

consisting of a non-empty set S, each element $s \in S$ is termed a **sort**, *and an $S^* \times S$-indexed family $\langle \Sigma_{w,s} \mid w \in S^*, s \in S \rangle$ of sets of constant and operation symbols. For any sort $s \in S$, each element $c \in \Sigma_{\lambda, s}$ is termed a* **constant symbol** *of sort s; for each non-empty word $w = s(1) \ldots s(n) \in S^+$ and any sort $s \in S$, each element $f \in \Sigma_{w,s}$ is termed an* **operation symbol** *of* **domain type** *w,* **codomain type** *s and* **arity** *n.*

If (S_0, Σ^0) and (S_1, Σ^1) are signatures, we say that (S_0, Σ^0) is a **subsignature** *of (S_1, Σ^1) if, and only if, $S_0 \subseteq S_1$ and for each $w \in S_0^*$ and $s \in S_0$, we have $\Sigma^0_{w,s} \subseteq \Sigma^1_{w,s}$.*

Let Σ be an S-sorted signature. An S-sorted **Σ algebra** *A is a pair*

$$A = (\langle A_s \mid s \in S \rangle, \langle \Sigma^A_{w,s} \mid w \in S^*, s \in S \rangle)$$

consisting of an S-indexed family $\langle A_s \mid s \in S \rangle$ of sets, the set A_s being termed the **carrier set** *of sort s for A, and an $S^* \times S$ indexed family $\langle \Sigma^A_{w,s} \mid w \in S^*, s \in S \rangle$ of sets of constants and operations. For each sort $s \in S$,*

$$\Sigma^A_{\lambda,s} = \langle c_A \mid c \in \Sigma_{\lambda,s} \rangle$$

where $c_A \in A_s$ is a **constant** *that interprets c in the algebra A. For each $w = s(1) \ldots s(n) \in S^+$ and each $s \in S$,*

$$\Sigma^A_{w,s} = \langle f_A \mid f \in \Sigma_{w,s} \rangle$$

where $f_A : A^w \to A_s$ *is an* **operation** *with* **domain** $A^w = A_{s(1)} \times \ldots \times A_{s(n)}$, **codomain** A_s *and* **arity** n *which interprets* f *in* A.

If $(S_0,\ \Sigma^0)$ *and* $(S_1,\ \Sigma^1)$ *are signatures, and* $(S_0,\ \Sigma^0)$ *is a subsignature of* (S_1, Σ^1) *then for any* Σ^1 *algebra* A *there is a unique* Σ^0 *algebra* B, *termed the* Σ^0 **reduct** *of* A, *such that for each* $s \in S_0$, $B_s = A_s$ *and for each* $w \in S_0^*$ *and* $s \in S_0$ *and each* $f \in \Sigma^0_{w,\ s}$, $f_B = f_A$. *We let* $A|_{\Sigma^0}$ *denote the* Σ^0 *reduct of* A. •

Where no ambiguity arises, we allow A to denote both a Σ algebra and its S-indexed family of carrier sets. A Σ algebra A is said to be *minimal* if, and only if, A has no proper subalgebra. We let $Min_\Sigma(A)$ denote the unique minimal Σ subalgebra of A.

Let Σ be an S-sorted signature and $X = \langle\ X_s \mid s \in S\ \rangle$ be an S-indexed family of sets of variable symbols, then $T(\Sigma,\ X)$ denotes the *free* or *term algebra* over Σ generated by X. We let $T(\Sigma)$ denote the algebra of all *ground terms* or *words* over Σ. If A is a Σ-algebra and $\alpha = \langle\ \alpha_s : X_s \to A_s \mid s \in S\ \rangle$ is an S-indexed family of mappings then $\overline{\alpha} = \langle\ \overline{\alpha}_s : T(\Sigma,\ X)_s \to A_s \mid s \in S\ \rangle$ denotes the unique homomorphic extension of α, also termed the *valuation mapping* on terms (under the *assignment* α).

The theory of higher-order universal algebra is developed within the framework of many-sorted universal algebra. We recall the basic definitions of Meinke [1992] beginning with notations for higher-order types.

2.2. Definition. *Let* B *be any non-empty set, the members of which will be termed* **basic types**, *the set* B *being termed a* **type basis**. *The* **type hierarchy** $H(B)$ *generated by* B *is the set* $H(B) = \cup_{n\in\omega} H_n(B)$ *of formal expressions defined inductively by,*

$$H_0(B) = B$$

and

$$H_{n+1}(B) = H_n(B) \cup \{\ (\sigma \times \tau),\ (\sigma \to \tau) \mid \sigma,\ \tau \in H_n(B)\ \}.$$

Each element $(\sigma \times \tau) \in H(B)$ *is termed a* **product type** *and each element* $(\sigma \to \tau) \in H(B)$ *is termed a* **function type** *or* **arrow type**.

We can assign an **order** *to each type* $\sigma \in H(B)$ *as follows. Each basic type* $\sigma \in B$ *has order* 0. *If* $\sigma,\ \tau \in H(B)$ *have order* m *and* n *respectively then* $(\sigma \times \tau)$ *has order* $sup\{\ m,\ n\ \}$ *and* $(\sigma \to \tau)$ *has order* $sup\{\ m+1,\ n\ \}$.

A **type structure** S *over a type basis* B *is a subset* $S \subseteq H(B)$ *which is closed under subtypes in the sense that for any* $\sigma,\ \tau \in H(B)$, *if* $(\sigma \times \tau) \in S$ *or* $(\sigma \to \tau) \in S$ *then both* $\sigma \in S$ *and* $\tau \in S$. *A type structure* S *over a basis* B *is said to be an* n**-th-order type structure** *if, and only if, the order of each type* $\tau \in S$ *is strictly less than* n. *We say that* S *is an* ω**-order type structure** *if, and only if, there is no* $n \in \mathbf{N}$ *which bounds the order of every type* $\tau \in S$. •

Given a type structure S, a higher-order signature Σ is an S-sorted signature with distinguished operation symbols for *projection* and *evaluation* on higher types.

2.3. Definition. *Let* S *be a type structure over a type basis* B. *An* S**-typed signature** Σ *is an* S*-sorted signature such that for each product type* $(\sigma \times \tau) \in S$

we have two unary **projection operation symbols**

$$proj^{(\sigma\times\tau),\,\sigma} \in \Sigma_{(\sigma\times\tau),\,\sigma}, \quad proj^{(\sigma\times\tau),\,\tau} \in \Sigma_{(\sigma\times\tau),\,\tau}.$$

Also for each function type $(\sigma \to \tau) \in S$ *we have a binary* **evaluation operation symbol**

$$eval^{(\sigma\to\tau)} \in \Sigma_{(\sigma\to\tau)\,\sigma,\,\tau}.$$

•

An S-typed signature Σ is also termed an *n-th-order signature* when S is an n-th-order type structure. Commonly, when the types σ and τ are clear, we let $proj^1$ and $proj^2$ denote the projection operation symbols $proj^{(\sigma\times\tau),\,\sigma}$ and $proj^{(\sigma\times\tau),\,\tau}$ respectively and we let *eval* denote the evaluation operation symbol $eval^{(\sigma\to\tau)}$.

Next we introduce the intended models of a higher-order signature Σ.

2.4. Definition. *Let* S *be a type structure over a type basis* B, *let* Σ *be an* S*–typed signature and let* A *be an* S*–sorted* Σ *algebra. We say that* A *is an* S**–typed** Σ **algebra** *if, and only if, for each product type* $(\sigma \times \tau) \in S$ *we have* $A_{(\sigma\times\tau)} \subseteq A_\sigma \times A_\tau$, *and for each function type* $(\sigma \to \tau) \in S$ *we have* $A_{(\sigma\to\tau)} \subseteq [A_\sigma \to A_\tau]$, *i.e.* $A_{(\sigma\to\tau)}$ *is a subset of the set of all (total) functions from* A_σ *to* A_τ. *Furthermore, for each product type* $(\sigma \times \tau) \in S$ *the operations*

$$proj_A^{(\sigma\times\tau),\,\sigma} : A_{(\sigma\times\tau)} \to A_\sigma, \quad proj_A^{(\sigma\times\tau),\,\tau} : A_{(\sigma\times\tau)} \to A_\tau$$

are the **first** *and* **second projection operations** *defined on each* $a = (a_1, a_2) \in A_{(\sigma\times\tau)}$ *by*

$$proj_A^{(\sigma\times\tau),\,\sigma}(a) = a_1, \quad proj_A^{(\sigma\times\tau),\,\tau}(a) = a_2;$$

also, for each function type $(\sigma \to \tau) \in S$, $eval_A^{(\sigma\to\tau)} : A_{(\sigma\to\tau)} \times A_\sigma \to A_\tau$ *is the* **evaluation operation** *on the function space* $A_{(\sigma\to\tau)}$ *defined by*

$$eval_A^{(\sigma\to\tau)}(a, n) = a(n)$$

for each $a \in A_{(\sigma\to\tau)}$ *and* $n \in A_\sigma$. •

An S-typed Σ algebra A is also termed an *n-th-order* Σ *algebra* when Σ is an n-th-order signature.

The first step towards establishing the expressiveness of higher-order equational specification methods (which will be investigated in depth in section 3) is to characterise the structure of a higher-order algebra up to isomorphism. We will show that the S-typed Σ algebras are, up to isomorphism, precisely the *extensional* Σ *algebras*, i.e. those which satisfy the following set *Ext* of extensionality axioms over Σ. By measuring the quantifier complexity of the extensionality formulas themselves we gain some insight into the structural complexity of extensional algebras, and their closure properties.

2.5. Definition. *Let* $S \subseteq H(B)$ *be a type structure over a type basis* B, *let* Σ *be an* S*–typed signature and let* X *be an* S*–indexed family of infinite sets of* *variables. The set $Ext = Ext_\Sigma$ of* **extensionality sentences** *over Σ is the set of all Σ sentences of the form*

$$\forall x \forall y \Big(\forall z \big(eval^{(\sigma \to \tau)}(x, z) = eval^{(\sigma \to \tau)}(y, z) \big) \Rightarrow x = y \Big),$$

for each function type $(\sigma \to \tau) \in S$, where $x, y \in X_{(\sigma \to \tau)}$, $z \in X_\sigma$, and

$$\forall x \forall y \Big(proj^1(x) = proj^1(y) \ \wedge \ proj^2(x) = proj^2(y) \Rightarrow x = y \Big),$$

for each product type $(\sigma \times \tau) \in S$, where $x, y \in X_{(\sigma \times \tau)}$. A Σ algebra A is **extensional** *if, and only if, $A \models Ext$. We let $Alg_{Ext}(\Sigma)$ denote the class of all extensional Σ algebras, i.e.*

$$Alg_{Ext}(\Sigma) = \{ \ A \in Alg(\Sigma) \mid A \models Ext \ \}.$$

•

It is easily shown (see Meinke [1993]) that a Σ algebra A is isomorphic to an S–typed Σ algebra if, and only if, A is extensional. Observe that the logical complexity of the extensionality sentence for a function type $(\sigma \to \tau)$ is Π^0_2 Horn. This observation has immediate consequences for the closure properties of classes of extensional Σ algebras, and thus indirectly for the structure of the higher-order initial model. (See Meinke [1993] for a more detailed discussion.)

For the fundamental principles of equational specification using first-order initial and first-order final algebra semantics we refer the reader to Ehrig and Mahr [1985] and the survey Wirsing [1990]. We will henceforth assume that the reader is familiar with construction of the initial model $I(\Sigma, E)$ and final model $Z(\Sigma, E)$ of a simple first-order equational specification (Σ, E), and with its basic properties such as the completeness of ground first-order equational deduction over this model. We now review the fundamentals of higher-order equational specification under *higher-order initial semantics.* We begin by making precise the notion of a higher-order equation.

2.6. Definition. *Let S be a type structure over a type basis B. Let Σ be an S-typed signature and X be an S–indexed family of sets of variables. By a* **higher-order equation** *(over Σ and X) we mean a formula of the form*

$$t = t',$$

where for some sort $s \in S$, t, $t' \in T(\Sigma, X)_s$ are terms of sort s over Σ and X.
•

Given any Σ algebra A, we have the usual notion of *truth* for an equation e under an assignment $\alpha : X \to A$, and the usual *validity relation* $\models$ for an equation e or set E of equations with respect to a Σ algebra A or a class K of Σ algebras. If E is any set of higher-order equations over Σ then we let $Alg_{Ext}(\Sigma, E)$ denote the class of all extensional Σ algebras which are models of E, i.e.

$$Alg_{Ext}(\Sigma, E) = \{ \ A \in Alg_{Ext}(\Sigma) \mid A \models E \ \}.$$

Let us observe at this point that all the definitions and results obtained for higher-order equations in this paper can be generalised to *higher-order conditional equations* of the form

$$t_1 = t_1' \,\&\, \ldots \,\&\, t_n = t_n' \;\Rightarrow\; t = t',$$

and with some minor qualifications, to arbitrary (quantifier free) *higher–order Horn formulas.* Next we introduce two basic forms of higher-order equational specification.

2.7. Definition. *Let B be a type basis. By a* **higher-order equational specification with constructors and hidden sorts and operations** *we mean a 4-tuple*

$$Spec = (\Sigma^{vis},\, \Sigma^{cons},\, \Sigma^{extrn},\, E),$$

which satisfies the following properties.
(a) The signatures Σ^{vis}, Σ^{cons} and Σ^{extrn} are respectively S^{vis}, S^{cons} and S^{extrn}–typed signatures over a common type basis B.
(b) The signature Σ^{vis} is a subsignature of Σ^{extrn} termed the **visible signature.** *The signature Σ^{extrn} is termed the* **external signature.**
(c) The signature Σ^{cons} is a subsignature of Σ^{extrn} termed the **constructor signature.**
(d) E is a set of higher-order equations over Σ^{extrn}.
(e) For any function type $(\sigma \to \tau) \in S^{extrn}$ we have $\sigma, \tau \in S^{cons}$.

We say that Spec is a **higher-order equational specification with hidden sorts and operations** *(i.e. without constructors) if, and only if, $\Sigma^{cons} = \Sigma^{extrn}$. In this case every operation symbol is viewed as a constructor symbol.*

We say that Spec is **recursive** *(respectively* **recursively enumerable***) if, and only if, Σ^{extrn} is countable and recursive, and Σ^{vis} and Σ^{cons} are recursive subsignatures of Σ^{extrn}, i.e. given $f \in \Sigma^{extrn}_{w,s}$ it is decidable whether $f \in \Sigma^{vis}_{w,s}$ and whether $f \in \Sigma^{cons}_{w,s}$. Furthermore E is a recursive (respectively recursively enumerable) set of equations.* •

A full discussion of the role of constructor subsignatures for higher–order signatures is given in Meinke [1996a]. Note condition (e) above ensures that the ground term model constructed using an extensionality rule with respect to constructors (rule 2.8.(viii) below) is extensional.

We now turn our attention to proof systems for higher-order equational specifications. The standard finitary proof system, in which extensionality axioms are encoded as proof rules, is sound and complete for all extensional models of a specification. By adding an appropriate infinitary version of the extensionality rule which is relativised to the constructor subsignature we obtain a second calculus which can be used to construct the higher-order initial model.

2.8. Definition. *Let B be a type basis. Let*

$$Spec = (\Sigma^{vis},\, \Sigma^{cons},\, \Sigma^{extrn},\, E),$$

be a higher-order equational specification with constructors and hidden sorts and operations. We introduce the following finitary and infinitary proof rules for Spec.

(i) For any equation $t = t' \in E$,

$$\overline{t = t'}$$

is an **axiom introduction** *rule.*

(ii) For any type $\tau \in S^{extrn}$ and any term $t \in T(\Sigma^{extrn}, X)_\tau$,

$$\overline{t = t}$$

is a **reflexivity** *rule.*

(iii) For any type $\tau \in S^{extrn}$ and any terms $t_0, t_1 \in T(\Sigma^{extrn}, X)_\tau$,

$$\frac{t_0 = t_1}{t_1 = t_0}$$

is a **symmetry** *rule.*

(iv) For any type $\tau \in S^{extrn}$ and any terms $t_0, t_1, t_2 \in T(\Sigma^{extrn}, X)_\tau$,

$$\frac{t_0 = t_1, \quad t_1 = t_2}{t_0 = t_2}$$

is a **transitivity** *rule.*

(v) For each type $\sigma \in S^{extrn}$, any terms $t, t' \in T(\Sigma^{extrn}, X)_\sigma$, any type $\tau \in S^{extrn}$, any variable symbol $x \in X_\tau$ and any terms $t_0, t_1 \in T(\Sigma^{extrn}, X)_\tau$,

$$\frac{t = t', \quad t_0 = t_1}{t[x/t_0] = t'[x/t_1]}$$

is a **substitution** *rule.*

(vi) For each product type $(\sigma \times \tau) \in S^{extrn}$ and any terms $t_0, t_1 \in T(\Sigma^{extrn}, X)_{(\sigma\times\tau)}$,

$$\frac{proj^1(t_0) = proj^1(t_1), \quad proj^2(t_0) = proj^2(t_1)}{t_0 = t_1}$$

is a **projection** *rule.*

(vii) For each function type $(\sigma \to \tau) \in S^{extrn}$ and any terms $t_0, t_1 \in T(\Sigma^{extrn}, X)_{(\sigma\to\tau)}$, and any variable $x \in X_\sigma$ not occurring in t_0 or t_1,

$$\frac{eval^{(\sigma\to\tau)}(t_0, x) = eval^{(\sigma\to\tau)}(t_1, x)}{t_0 = t_1}$$

is a (finitary) **extensionality rule**.

(viii) For each function type $(\sigma \to \tau) \in S^{extrn}$ and any terms $t_0, t_1 \in T(\Sigma^{extrn}, X)_{(\sigma\to\tau)}$,

$$\frac{\langle eval^{(\sigma\to\tau)}(t_0, t) = eval^{(\sigma\to\tau)}(t_1, t) \mid t \in T(\Sigma^{cons})_\sigma\rangle}{t_0 = t_1}$$

is an (infinitary) ω*–***extensionality rule with respect to constructors.** •

The finitary projection and extensionality rules (vi) and (vii) above encode the extensionality sentences of Definition 2.5 as proof rules. Taking $Spec \vdash_{Ext}$ to be the derivability predicate on higher-order equations using rules (i) ,..., (vii) above we have the the following completeness result. (For the usual technical reasons, see for example Goguen and Meseguer [1982] or Meinke and Tucker [1993], we impose the assumption of *non–voidness* on Σ^{extrn}, i.e. for each type $\tau \in S^{extrn}$ we assume that there exists a ground term $t \in T(\Sigma^{extrn})_\tau$.)

2.9. Completeness Theorem. *Let* $Spec = (\Sigma^{vis}, \Sigma^{cons}, \Sigma^{extrn}, E)$ *be a higher-order equational specification with constructors and hidden sorts and operations. Suppose that* Σ^{extrn} *is non-void. For any equation e over* Σ^{extrn} *and* X,

$$Spec \vdash_{Ext} e \Leftrightarrow Alg_{Ext}(\Sigma^{extrn}, E) \models e.$$

Proof. See Meinke [1993]. •

Let us turn to the problem of constructing a higher-order initial semantics for a higher-order equational specification *Spec*. In general, if we take the congruence $\equiv^{Spec}$ of provable equivalence on ground $T(\Sigma^{extrn})$ terms induced by the provability relation $Spec \vdash_{Ext}$ then this congruence will not be fine enough to yield an extensional quotient term model $T(\Sigma^{extrn})/\equiv^{Spec}$. However taking the infinitary extensionality rule 2.8.(viii) instead of 2.8.(vii) (either in the presence or absence of constructors), letting $Spec \vdash_{\omega-Ext}$ denote the corresponding derivability relation on equations, and taking the induced congruence of provable equivalence

$$\equiv^{E,\omega} = \langle \equiv_\tau^{Spec,\,\omega} \mid \tau \in S^{extrn} \rangle$$

on the term algebra $T(\Sigma^{extrn})$ defined by

$$t \equiv_\tau^{Spec,\,\omega} t' \Leftrightarrow Spec \vdash_{\omega-Ext} t = t'$$

for each type $\tau \in S^{extrn}$ and any terms $t, t' \in T(\Sigma^{extrn})_\tau$, we can concretely construct the higher-order initial semantics of *Spec* as a quotient term model. Notice that rule (viii) allows us to identify two higher-order terms t_0, t_1 of type $(\sigma \rightarrow \tau)$ when they are provably identical on all *constructor terms* of type σ. Obviously this rule implies that on any non-constructor term t of type σ, t_0 and t_1 are identified.

2.10. Definition. *Let* $Spec = (\Sigma^{vis}, \Sigma^{cons}, \Sigma^{extrn}, E)$ *be a higher-order equational specification with constructors and hidden sorts and operations.*
(i) The **higher-order initial algebra semantics** *of Spec is given by the algebra*

$$I_{Ext}(Spec) = Min_{\Sigma^{vis}}(T(\Sigma^{extrn})/\equiv^{Spec,\,\omega}|_{\Sigma^{vis}}).$$

(ii) Let A be a Σ^{vis} *algebra. We say that Spec* **specifies** *A* **under higher-order initial algebra semantics** *if, and only if,*

$$I_{Ext}(Spec) \cong A.$$ •

Thus according to 2.10.(i) above the higher-order initial model, for a specification involving constructors, is obtained by factoring the algebra $T(\Sigma^{extrn})$ of ground external terms by provable equivalence (modulo extensionality with respect to constructors), taking the Σ^{vis} reduct and then the minimal Σ^{vis} subalgebra to remove all elements not denotable by ground visible terms.

3. The Expressiveness of Higher-Order Equations. In this section we consider higher-order equations, as a formal specification language, from the viewpoint of expressive power. As we have already observed in section 1, this is not the only criterion for judging their effectiveness in this role. Therefore in section 4 we will contrast the expressiveness results described in this section (which are quite strong) with results on the complexity of theorem proving for higher-order equational logic.

From a theoretical viewpoint, the expressive power of algebraic specification methods such as first and higher-order equations can be studied using recursion theory to measure the complexity of the algebras (and hence systems) which can be specified. This approach has been extensively developed to characterise first-order algebraic specification methods in for example Bergstra and Tucker [1982, 1983], Goguen and Meseguer [1985] and Bergstra and Tucker [1987] (a general survey can be found in Wirsing [1990]). This recursion theoretic approach was extended to higher-order equational specifications in Meinke [1994], Kosiuczenko and Meinke [1995] and Meinke [1995]. In this section we will survey some basic definitions and fundamental results on the expressiveness of higher-order specifications.

We begin by recalling the definition of the recursion theoretic complexity of a Σ algebra A in terms of the so called *arithmetical* and *analytical hierarchies.* Recall that for any signature Σ, any Σ algebra A can be obtained (up to isomorphism) by factoring an appropriate term algebra $T(\Sigma, X)$ by a congruence. This follows from the so called *freeness property* of term algebras. In particular, any minimal Σ algebra is (up to isomorphism) a quotient of $T(\Sigma)$. The basic idea behind measuring the complexity of (countable) algebras recursion theoretically, is first to formalise the notion of a "structurally simple" Σ algebra as a *recursive Σ number algebra*, which has carrier sets that are recursive subsets of **N**. A useful fact in the light of the above remarks is that for countable Σ the ground term algebra $T(\Sigma)$ is (up to isomorphism) a recursive number algebra, simply by Gödel numbering of terms.

A measure of the complexity of A can be obtained as the complexity of factoring a recursive Σ number algebra (conventionally $T(\Sigma)$ for the reasons above) to obtain A up to isomorphism. Importantly, this measure is isomorphism invariant. The factor structure consisting of a recursive Σ number algebra R and an epimorphism onto A is termed an *effective coordinatization.*

3.1. Definition. *Let S be a sort set and let Σ be an S-sorted signature. A Σ algebra R is said to be a* **recursive number algebra** *if, and only if, for each sort $s \in S$ the carrier set R_s is a recursive subset of* **N**, *and for each*

$w = s(1)\dots s(n) \in S^+$, each sort $s \in S$ and each operation symbol $f \in \Sigma_{w,s}$, the operation $f_R : R^w \to R_s$ is a recursive function. An **effective coordinatization** *of a Σ algebra A is a pair (R, θ) consisting of a Σ recursive number algebra R and an epimorphism $\theta : R \to A$.* •

To measure the complexity of a countable algebra A we now measure the complexity of the congruence $\equiv \subseteq \mathbf{N} \times \mathbf{N}$ required to obtain A, up to isomorphism, as a quotient $R/\equiv$ of a recursive number algebra R. The complexity of $\equiv$ is defined to be the first level at which it occurs in the arithmetical or analytical hierarchy. We recall these hierarchies following the account of Hinman [1978].

3.2. Definition.

(i) The class of **arithmetical relations** *is the smallest class of relations which contains all recursive relations $r \subseteq \mathbf{N}^j \times [\mathbf{N} \to \mathbf{N}]^k$, for all $j, k \in \mathbf{N}$, and is closed under existential ($\exists^0$) and universal ($\forall^0$) number quantification. (The definition of a recursive relation $r \subseteq \mathbf{N}^j \times [\mathbf{N} \to \mathbf{N}]^k$ can be given using oracle computations. We refer the reader to Hinman [1978].)*

The **arithmetical hierarchy** *is the set of classes Σ^0_n, Π^0_n and Δ^0_n defined by induction on n.*

(a) $\Sigma^0_0 = \Pi^0_0$ = the class of all recursive relations $r \subseteq \mathbf{N}^j \times [\mathbf{N} \to \mathbf{N}]^k$ for all $j, k \in \mathbf{N}$.

(b) $\Sigma^0_{n+1} = \{\ \exists^0 r \mid r \in \Pi^0_n\ \}$.

(c) $\Pi^0_{n+1} = \{\ \forall^0 r \mid r \in \Sigma^0_n\ \}$.

(d) $\Delta^0_n = \Sigma^0_n \cap \Pi^0_n$.

We say that a relation $r \subseteq \mathbf{N}^k \times [\mathbf{N} \to \mathbf{N}]^j$ is Σ^0_n (respectively Π^0_n, Δ^0_n) if, and only if, $r \in \Sigma^0_n$ (respectively $r \in \Pi^0_n$, $r \in \Delta^0_n$).

(ii) The class of **analytical relations** *is the smallest class of relations which contains all arithmetical relations $r \subseteq \mathbf{N}^j \times [\mathbf{N} \to \mathbf{N}]^k$ for all $j, k \in \mathbf{N}$, and is closed under existential ($\exists^1$) and universal ($\forall^1$) function quantification.*

The **analytical hierarchy** *is the set of classes Σ^1_n, Π^1_n and Δ^1_n defined by induction on n.*

(a) $\Sigma^1_0 = \Pi^1_0$ = the class of all arithmetical relations $r \subseteq \mathbf{N}^j \times [\mathbf{N} \to \mathbf{N}]^k$, for all $j, k \in \mathbf{N}$.

(b) $\Sigma^1_{n+1} = \{\ \exists^1 r \mid r \in \Pi^1_n\ \}$.

(c) $\Pi^1_{n+1} = \{\ \forall^1 r \mid r \in \Sigma^1_n\ \}$.

(d) $\Delta^1_n = \Sigma^1_n \cap \Pi^1_n$.

We say that a relation $r \subseteq \mathbf{N}^k \times [\mathbf{N} \to \mathbf{N}]^j$ is Σ^1_n (respectively Π^1_n, Δ^1_n) if, and only if, $r \in \Sigma^1_n$ (respectively $r \in \Pi^1_n$, $r \in \Delta^1_n$). •

These hierarchies are applied to measure the complexity of a countable algebra A as follows.

3.3. Definition. *Let Σ be an S-sorted signature and let A be a countable Σ algebra. We say that A* **has complexity** *Σ_n^i (respectively Π_n^i, Δ_n^i), for $i = 0, 1$ and $n \in \mathbf{N}$, if, and only if, there exists an effective coordinatization (R, θ) of A such that for each sort $s \in S$, the kernel $\equiv_s^\theta$ is Σ_n^i (respectively Π_n^i, Δ_n^i).*

We say that A **has arithmetical complexity** *if, and only if, there exists $n \in \mathbf{N}$ such that A has complexity Σ_n^0 or Π_n^0. We say that A* **has analytical complexity** *if, and only if, there exists $n \in \mathbf{N}$ such that A has complexity Σ_n^1 or Π_n^1.* •

At this point it is appropriate to recall the expressive power of first-order equational specification methods. Of the many results which are known, the most important from our point of view are the following upper bound results on complexity.

3.4. Fact. *Let S be a countable sort set, Σ be a countable S-sorted signature and E be a recursively enumerable set of Σ equations:*

(i) the initial model $I(\Sigma, E)$ has complexity Σ_1^0;

(ii) the final model $Z(\Sigma, E)$, if it exists, has complexity Π_1^0.

Proof. A simple analysis of the closure ordinal associated with the inductive definition of provable equivalence on terms for each of these term model constructions. See for example Wirsing [1990]. •

An algebra of complexity Σ_1^0 is more commonly known as a *semicomputable algebra*, while an algebra of complexity Π_1^0 is also known as a *cosemicomputable algebra.*

Now let us compare Fact 3.4 with upper bounds on the recursion theoretic complexity of the higher-order initial model of a higher-order equational specification.

3.5. Theorem. *Let $Spec = (\Sigma^{vis}, \Sigma^{cons}, \Sigma^{extrn}, E)$ be a recursively enumerable higher-order equational specification with constructors and hidden sorts and operations. The higher-order initial model*

$$I_{Ext}(Spec)$$

has complexity Π_1^1.

Proof. A similar analysis of the closure ordinal associated with the inductive definition of provable equivalence on terms associated with the higher-order initial model. For details see Meinke [1996a]. •

Theorem 3.5 suggests that the complexity of a higher-order initial model can greatly exceed the complexity of any model obtained using first-order equational specifications (and either first-order initial or first-order final semantics). To confirm this hypothesis we need to consider the "converse" of Theorem 3.5: *can every countable algebra of complexity Π_1^1 be specified by a recursively enumerable, or even recursive, set of higher-order equations under higher-order initial semantics?* The answer turns out to be "yes"!

First let us consider algebras having just arithmetical complexity, that is Σ_n^0 or Π_n^0 for some $n \in \mathbf{N}$. For such algebras, second-order equations are sufficient. (Equations of third and higher-order may provide nice "syntactic sugar" to the formalism, but they add nothing to its expressive power.) Furthermore, for the large class of cases involving just a finite signature Σ a finite number of additional hidden sorts and operations, bounded by the size of Σ, suffices.

The proof of the following theorem involves considerable technical detail so it will be omitted here. Nevertheless the idea is straightforward. If an algebra A has arithmetical complexity then the congruence $\equiv \subseteq \mathbf{N} \times \mathbf{N}$ used to obtain A as a quotient of $T(\Sigma)$ is definable by a first-order formula of Peano arithmetic over the standard model $\mathbf{N}$ of arithmetic. The crucial fact is that we can capture the truth definition for all first-order formulas of Peano arithmetic over $\mathbf{N}$ (which have arbitrary first-order quantifier complexity) by means of a recursive set of (quantifier free) second-order equations. In essence we can encode the first-order theory of arithmetic inside a second-order quantifier free equational theory of arithmetic. This fact substantiates the power of second-order equations, since the expressive power of first-order formulas comes from quantifier alternations (defining the classes Σ_n^0 and Π_n^0) and these quantifier alternations are wholly absent from second-order equational formulas. So where does this additional expressive power come from, and how do we capture the truth definition for quantified formulas in a quantifier free language? Here the crucial fact is that we can *finitely* axiomatise the so called *quantifier functional* $E : [\mathbf{N} \to \mathbf{N}] \to \mathbf{N}$ defined by

$$E(\alpha) = \begin{cases} 0, & \text{if } \exists m.\alpha(m) = 0; \\ 1, & \text{otherwise,} \end{cases}$$

for any $\alpha : \mathbf{N} \to \mathbf{N}$. It is an instructive exercise to write down a finite second-order equational axiomatisation of E. Clearly this second-order functional captures the semantics of existential quantification. The rest of the proof is then a matter of suitably encoding first-order formulas, so let us state the result.

3.6. Theorem. *For any countable S-sorted signature Σ and any minimal Σ algebra A, if A has arithmetical complexity then A has a recursive second-order equational specification with hidden sorts and hidden functions. If $|S| = n$ then two hidden sorts and $27 + 2n$ hidden functions suffice.*

Proof. See Meinke and Kosiuczenko [1996]. •

Theorem 3.6 is not quite strong enough to act as a converse to Theorem 3.5. For this we must look at a different class of encodings. An important characterisation of the class of Π_1^1 relations is that this is precisely the class of relations which are *semicomputable in the quantifier functional* E, i.e. E is available as an oracle during computations. A proof of this fairly difficult result can be found in Hinman [1978]. Now the semicomputable relations on $\mathbf{N}$ can be recursively axiomatised without difficulty, provided that constructors are used to identify the class of numerals to which terminating computations normalise. Furthermore, by the remarks above, the quantifier functional E can be axiomatised using a finite set of second-order equations. These two facts are the essential ideas behind the following theorem. Again we omit the proof which involves considerable encoding details.

3.7. Theorem. *Let Σ be a many-sorted signature and let A be a Σ algebra. Then A is specifiable by means of a recursively enumerable higher-order equational specification with constructors and hidden sorts and operations under higher-order initial algebra semantics if, and only if, A has complexity Π_1^1.*

Proof. See Meinke [1996a]. •

It is an interesting open question whether the use of constructors in Theorem 3.7 is actually necessary to close the gap between Theorem 3.6 and Theorem 3.5.

At this point it is natural to ask whether this very considerable increase in specification power (Π_1^1 versus Σ_1^0 or Π_1^0) actually serves any useful purpose. We claim that it does for exactly the reason that the quantifier functional (and related functionals) allow us to encode existential statements of logic and thus to give *non-constructive specifications* of systems. An example of this approach may be found in Steggles [1995] where a non-constructive specification of the Hamming Stream problem due to Dijkstra [1976] is given. Typically such non-constructive specifications will arise at early stages in the design process for computing systems, when a problem is logically clarified but no algorithmic solution is yet in sight (or perhaps even known). Thus non-constructive specifications are an integral part of the design process by stepwise refinement. Perhaps for this reason richer logical formalisms which include quantification, such as first-order logic (as represented by for example VDM, Jones [1986]) and set theory (as represented by for example Z, Spivey [1992]), have found increasing favour among the practical formal methods community, while the simpler (first-order) algebraic methods are usually considered as special purpose tools.

We see from the above discussion that by using second-order equational logic we can enjoy the expressive benefits of a richer logical language with quantifiers, without losing the advantages of initial model semantics! If this situation seems to be too good to be true we will quickly point out that there are also some disadvantages incurred in relation to first-order equational specifications. As is commonly observed among logical languages, expressive power comes at the cost of some tractability of the associated computational logic. In the final section of this survey we will investigate this tradeoff.

4. Term Rewriting, Simulation and Testing. In this last section we consider the tractability of higher-order equational logic from a computational point of view. This question is clearly an important issue, since on the one hand we can expect that the dramatic increase in expressiveness of higher-order equations incurs a penalty in this area. On the other hand, following the main theme of this paper, simulation and testing of formal specifications demand a computational logic which can answer logical queries about a specification in an efficient manner.

Recall that for first-order equational specifications the methods of *term rewriting* (see for example Klop [1993]) can be applied to automate first-order equational logic in a highly efficient way. Thus successful practical software systems exist, such as OBJ (Goguen et al [1983]) and ASF (Bergstra et al [1989]), which can be used to parse, typecheck and execute first-order algebraic specifications. In fact a well known basic result establishes that term rewriting and first-order equational deduction are exactly equivalent. This result is of significance here since looking back on the (complete) finitary proof system for higher-order equations (Definition 2.8) we observe that this system clearly extends the rules of first-order equational deduction, and thus term rewriting alone can no longer suffice as a computational logic, at least for all possible queries. Furthermore any naive approach to automating the additional extensionality rules 2.8.(vi) and 2.8.(vii) is obviously computationally too expensive.

However on reflection it becomes clear that for the purposes of simulating and testing a specification we can restrict the kind of logical queries allowed to *ground first-order equations* (equations on data). Such queries arise when the system specifier attempts to simulate the system being designed using its specification (rapid prototyping), and perform tests and observations using this simulation for comparison with user expectations (evaluation of a formal specification against informal requirements). At a later stage in the software lifecycle such queries can also arise through comparisons between with the system specification and an actual system implementation (testing), possibly prior to attempting a full formal verification. Returning to the proof system of Definition 2.8, it is not at all clear at first sight whether the additional proof rules (vi) and (vii) are actually conservative over first-order equational logic for ground first-order equations. If they are, then we can apply ordinary term rewriting as an efficient computational logic for our restricted type of query. So it is clearly important to investigate this conservativity property. Let us define the property more precisely. Let *Spec* $\vdash$ denote the inference relation using the ordinary rules of many-sorted first-order equational deduction, i.e. rules (i),...,(v) of Definition 2.8.

4.1. Definition. Let $Spec = (\Sigma^{vis}, \Sigma^{cons}, \Sigma^{extrn}, E)$ be a higher-order equational specification with constructors and hidden sorts and operations. We say that *Spec* **admits ground first-order term rewriting** if, and only if, for any ground first-order terms t and t' over Σ^{extrn}, if $Spec \vdash_{\omega-Ext} t = t'$ (or equivalently $I_{Ext}(Spec) \models t = t'$) then $Spec \vdash t = t'$. •

It is not too difficult to find both examples of higher-order equational theories which *do* admit ground first-order term rewriting (such as systems of higher-order recursion equations) and examples which *do not* admit this property (hint: consider ω-incomplete first-order equational theories). So some characterisation of this property is required which gives an indication of the size of the class of all specifications which enjoy the property.

Important insight into the characterisation problem for ground first-order term rewriting can be gained by using topology. If E does not admit ground first-order term rewriting then the proof of at least one equation in ground first-order terms makes essential use of the infinitary ω-extensionality rule. Thus in an intuitive sense some operation named in the signature, and occurring in such a proof, requires an infinite amount of information about one of its (higher-order) arguments. This intuition can be made precise by means of a ***topology of finite information*** on the data sets of a higher-order algebra, for which the continuous operations on data are precisely the operations using finite information about their arguments to yield a value. The basic definitions and techniques of topology used in the sequel can all be found in a standard textbook such as Dugundji [1989] or Kelley [1955].

For simplicity we will restrict ourselves to the case of a fixed but arbitrarily chosen second-order type structure S (without product types) over a type basis B, an S-sorted second-order signature Σ, and a Σ algebra A (not necessarily extensional!). The results which we state below extend to higher orders although the finite information topology becomes more difficult to define (but more interesting to study!) in higher types.

The algebra A has a natural topology by taking the discrete topology on A_τ for each basic type τ and (a subspace of) the product or Tychonoff topology on $A_{(\sigma\rightarrow\tau)}$ for each function type $(\sigma \rightarrow \tau) \in S$. Continuity of an operation f_A of A with respect to this topology then captures a natural computational intuition that determining the value of f_A on any arguments $a_1, \ldots, a_n$ requires only a finite amount of information about each argument a_i.

We turn A into a topological algebra as follows.

4.2. Definition. Define the **finite information topology** on A to be the S-indexed family of topologies

$$T = \langle\, T_\tau \mid \tau \in S \,\rangle$$

on the carrier sets of A as follows.

(i) For each basic type $\tau \in S$ define $T_\tau = \wp(A_\tau)$, i.e. T_τ is the discrete topology on A_τ.

(ii) For each function type $(\sigma \rightarrow \tau) \in S$ define $T_{(\sigma\rightarrow\tau)}$ to have subbasic open sets of the form

$$O_{a,\,b} = \{\ a' \in A_{(\sigma\rightarrow\tau)} \mid eval_A(a',\, b) = eval_A(a,\, b)\ \}$$

for each $a \in A_{(\sigma\rightarrow\tau)}$ and $b \in A_\sigma$.

A function $f : A^w \rightarrow A_\tau$ is **continuous** if, and only if, for each open set $U \in T_\tau$, $f^{-1}(U) = \{\ \overline{a} \in A^w \mid f(\overline{a}) \in U\ \}$ is open in the induced product topology T^w on A^w.

We say that A **is continuous** if, and only if, for each $w \in S^+$ and each $\tau \in S$, each operation $f_A : A^w \rightarrow A_\tau$ is continuous. •

It is easily shown that the evaluation mapping $eval : A_{(\sigma\rightarrow\tau)} \times A_\sigma \rightarrow A_\tau$ is continuous in this topology. Thus the continuity of A depends only on the continuity of the other operations.

It is important to point out that this topology is well defined whether A is extensional or not. In the case that A is extensional then although the finite information topology agrees with the Tychonoff topology up to second-order types, it cannot be described in this way above such types, which is why we have restricted our attention to this simple case. Furthermore, above second-order types the finite information topology is not homeomorphic with the well known topology on the Kleene Kreisel total continuous functionals (as described in for example Normann [1980]). For example the finite information topology is second countable in every type (assuming A_τ is countable for every basic type $\tau \in B$). Also if A is extensional then T_τ is *metrisable* for *every* higher type τ. Further information about this topology can be found in Meinke [1996b] where all the results of this section are presented in detail.

To establish whether a given Σ algebra A is continuous it is helpful to characterise the continuous mappings $f : A^w \to A_\tau$ for $w \in S^+$ and $\tau \in S$. First consider the case that τ is a basic type.

4.3. Proposition. *Let $w \in S^+$, let $\tau \in B$ and let $f : A^w \to A_\tau$ be any mapping. Then f is continuous if, and only if, for any $\overline{a} \in A^w$ there is a basic open set $U \in T^w$ with $\overline{a} \in U$ and f constant valued on U.*

Proof. Exercise. •

If $w = \tau(1), \ldots, \tau(n)$ and every $\tau(i)$ is a basic type then $f : A^w \to A_\tau$ is trivially continuous. However, if one or more of the domain types $\tau(i)$ is a function type, then f is continuous if, and only if, f is *finitely determined* in the sense that for any $(a_1, \ldots, a_n) \in A^w$, $f(a_1, \ldots, a_n)$ is determined by just a finite part of each function argument a_i.

Let us characterise continuity in the case that τ is a function type.

4.4. Proposition. *Let $w \in S^+$, let $(\sigma \to \tau) \in S$ and let $f : A^w \to A_{(\sigma\to\tau)}$ be any mapping. Then f is continuous if, and only if, its uncurried form*

$$uc(f) : A^w \times A_\sigma \to A_\tau$$

given by $uc(f)(a, b) = f(a)(b)$ is continuous.

Proof. Exercise. •

We now have some idea of what it means for an arbitrary Σ algebra A to be continuous in the finite information topology. So let us return to the problem of ground first-order term rewriting. We can obtain a sufficient (but not necessary) condition for a specification to admit ground first-order term rewriting in terms of continuity. The proof of the following theorem is again lengthy and will be omitted in this survey.

4.5. Theorem. *Let $Spec = (\Sigma^{vis}, \Sigma^{cons}, \Sigma^{extrn}, E)$ be a higher-order equational specification with constructors and hidden sorts and operations. If the initial model $I(\Sigma^{extrn}, E)$ is continuous in the finite information topology then Spec admits ground first-order term rewriting.*

Proof. See Meinke [1996b]. •

At first sight it is surprising that the ordinary first-order initial model (which is usually not extensional) should play a role. However when this model is continuous then the higher-order initial model (our intended semantics) is also continuous and the two are structurally very similar (one is a direct extensional collapse of the other by virtue of the conservativity property).

Theorem 4.5 does not give necessary conditions for admitting ground first-order term rewriting. These are somewhat more generous, in fact continuity is not necessary in general, but a sufficient and necessary condition is rather more complicated to describe and the reader is referred to Meinke [1996b]. Nevertheless continuity provides a simple and clear benchmark for specifications. For example all systems of higher-order recursion equations, by virtue of their effectiveness involving only finite information for computation, lead to continuous models.

Let us return to the initial question of this section: *to what extent is higher-order equational logic computationally tractable?* Theorem 4.5 indicates that for a large class of examples it is highly tractable by means of term rewriting. Where term rewriting no longer suffices (which is possible) this is due to some discontinuous operation in the specification. However we observed in section 3 that it is precisely such kinds of discontinuous operation which give higher-order equations their expressive power, and which allow non-constructive specifications by implicit quantification. Not surprisingly, what is not constructively described cannot be effectively computed, and this is both the advantage and disadvantage of such specifications. The model theory of higher-order equational logic makes this issue clearer than is normally the case, since computable and semi-computable algebras lie in the lowest classes of our complexity hierarchy for algebras.

It is important not to interpret these results too negatively. Even if a specification does not admit ground first-order term rewriting, this does not mean that no interesting queries can be answered in an efficient way. In fact during the process of refining a non-constructive specification to a constructive one (which is the process of algorithm design) we will observe that more and more useful queries become answerable in an efficient way.

5. Conclusions. In this paper we have given a brief survey of higher-order algebraic specification and we have attempted to compare the conflicting requirements of expressiveness and computational tractability as these apply to the formalism. Our results are intended to support the claim that higher-order equations provide a useful compromise between these two requirements. Much further work remains to be done on classifying non-constructive specifications and our results represent only a first approach to the issues discussed here.

We would like to thank M. Dezani, J. Heering, B. Möller, L.J. Steggles and M. Wirsing for stimulating discussions about this work. The research presented here was partially supported by funds from EU Esprit Working Groups 7232 (Gentzen) and 8533 (Nada).

References

Bergstra, J.A., Heering, J., Klint, P.: Algebraic specification, Addison Wesley, New York, 1989

Bergstra, J.A., Tucker, J.V.: The completeness of the algebraic specification methods for computable data types, Inform. and Control **54**, 186–200 (1982)

Bergstra, J.A., Tucker, J.V.: Initial and final algebra semantics for data type specifications; two characterisation theorems, SIAM J. Comput. **12**, 366–387 (1983)

Bergstra, J.A., Tucker, J.V.: Algebraic specifications of computable and semi-computable data types. Theoret. Comput. Sci. **50**, 137–181 (1987)

Dijkstra, E.W.: A Discipline of Programming, Prentice Hall, Englewood Cliffs, 1976

Dugundji, J.: Topology, William C. Brown, Dubuque, 1989.

Ehrig, H., Mahr, B.: Fundamentals of algebraic specification I: equations and initial semantics, Berlin, Heidelberg, New York: Springer Verlag 1985

Goguen, J.A., Meseguer, J.: Completeness of many-sorted equational logic. Assoc. for Computing Machinery SIGPLAN notices **17**, 9–17 (1982)

Goguen, J.A., Meseguer, J.: Initiality, induction and computability. In: Nivat, M., Reynolds, J.C. (eds.) Algebraic methods in semantics, pp. 459–541. Cambridge: Cambridge University Press 1985

Goguen, J.A., Meseguer, J., Plaisted, D.: Programming with parameterized abstract objects in OBJ, 163-193 in: D. Ferrari, M. Bolognani and J.A. Goguen (eds), Theory and practice of software technology, North Holland, 1983.

Gordon, V.S., Bieman, J.: Rapid prototyping: lessons learned, IEEE Software **12**, 85-95, (1995)

Heering, J., Meinke, K., Möller, B., Nipkow, T., (eds.) HOA '93: an international workshop on higher-order algebra, logic and term rewriting, (Lect. Notes Comput. Sci., vol. 816) Berlin, Heidelberg, New York: Springer Verlag 1994

Hinman, P.G.: Recursion-theoretic hierarchies, Berlin, Heidelberg, New York: Springer Verlag 1978

Jones, C.B.: Systematic software development using VDM, Prentice Hall, Englewood Cliffs, 1986

Kelley, J.L.: General Topology, Springer Verlag, Berlin, 1955

Klop, J.W.: Term rewriting, in: Abramsky, S., Gabbay, D., Maibaum, T.S.E., (eds.) Handbook of logic in computer science, Vol II, pp. 1–111 . Oxford: Oxford University Press 1993

Kosiuczenko, P., Meinke, K.: On the power of higher-order algebraic specification methods, Information and Computation, **124**, 85-101, (1996).

Meinke, K.: Universal algebra in higher types, Theoretical Computer Science, **100**, 385-417, (1992)

Meinke, K.: A recursive second-order initial algebra specification of primitive recursion, Acta Informatica, **31**, 329-340, (1994)

Meinke, K.: A completeness theorem for the expressive power of higher-order algebraic specifications, Journal of Computer and System Sciences, to appear, 1996.a

Meinke, K.: Proof theory of higher-order equational logic: normal forms, continuity and term rewriting, technical report, Department of Computer Science, University College of Swansea, 1996.b to appear

Meinke, K., Tucker, J.V.: Universal algebra, in: Abramsky, S., Gabbay, D., Maibaum, T.S.E., (eds.) Handbook of logic in computer science, pp. 189–411 Oxford: Oxford University Press 1993

Möller, B.: Higher-order algebraic specifications. Facultät für Mathematik und Informatik, Technische Universität München, Habilitationsschrift, 1987

Spivey, J.M.: The Z notation, Prentice Hall, Englewood Cliffs, 1992

Steggles, L.J.: Extensions of Higher-Order Algebra, Case Studies and Fundamental Theory, Ph.D. Thesis, Dept. of Computer Science, University of Wales, Swansea, 1995

Normann, D.: Recursion on the countable functionals, Lecture Notes in Mathematics 811, Springer Verlag, Berlin, 1980

Wirsing, M.: Algebraic specification. In: van Leeuwen, J. (ed.) Handbook of theoretical computer science, pp. 675–788. Amsterdam: North Holland 1990

The Correctness of a Higher–Order Lazy Functional Language Implementation: An Exercise in Mechanical Theorem Proving

Sava Mintchev and David Lester

Department of Computer Science, Manchester University,
Oxford Road, Manchester M13 9PL, UK.
{smintchev,dlester}@cs.man.ac.uk

Abstract. Higher–order lazy functional languages have had a reputation for inefficiency, and recent work has both addressed and corrected these defects. An unfortunate side–effect of the improved efficiency is the increasing complexity of the resultant implementations. In this paper we present the correctness proof of a new abstract machine. The abstract machine deals with the issues of currying, nonstrictness and sharing. Like most modern abstract machines, it employs spineless reduction to tackle partial function applications, and graph rewriting to deal with both nonstrictness and sharing. The abstract machine has been defined as a state monad and is directly implementable in an imperative language.
The machine is proved correct with respect to the direct denotational semantics of the functional language. The proof is carried out in two stages, with the introduction of an intermediate operational semantics. The correctness of the intermediate semantics is stated in terms of soundness and adequacy. Because of the higher–order nature of the language, inclusive predicates over reflexive domains are used in the adequacy proof. A large part of the proof has been constructed with the aid of a theorem prover.

1 Introduction

Modern functional languages possess a number of desirable features — higher-order functions, partial function application, lazy evaluation — which are hard to implement efficiently. Fast implementations are becoming increasingly more complex, and are often designed as sophisticated abstract machines [2, 16]. Justifying the correctness of those designs is a problem of growing importance.

The lambda calculus, although considered to be the basis for lazy functional languages, does not reflect their operational semantics accurately — it does not take account of the sharing of expressions. Various calculi which bear closer resemblance to functional language implementations have been proposed, most recently the Call–by–need lambda calculus [1]. We are interested, however, in the correctness of a particular implementation, and have decided to prove it directly with respect to the denotational semantics of the language.

In the classical approach to correctness, an implementation (operational semantics, abstract machine) is proved congruent to the standard denotational semantics of the language. Unfortunately, the higher-order nature of functional languages is a major source of difficulty — it requires the use of *inclusive predicates* to describe the congruence [3, 9, 19]. When the language to be implemented is untyped or polymorphically typed, the existence of such inclusive predicates is a problem in its own right [13, 17].

Proofs of various functional language implementations have been published [6, 14], but few of the proofs have been machine-checked. We believe that the complexity of proofs needed for modern abstract machines requires the use of machine checking, before we can claim confidence in the implementation's correctness.

This paper describes our work on the correctness of a lazy functional abstract machine. The proof of correctness has been constructed with the help of a theorem proving tool for functional programs [11]. The abstract machine is a simplified version of the implementation semantics used in [10] to compile core Haskell into C. The work in the present paper is a development of [7], where a continuation semantics was used rather than an abstract machine, and the proof was done in Isabelle [15].

The meta-notation used in this paper is reminiscent of the pure functional programming language Haskell. The similarity is no coincidence — in fact all definitions and propositions have been pretty-printed by the theorem prover, which uses a subset of Haskell as its meta-language.

A theorem prover for functional programs. A detailed description of the theorem prover can be found in [11]; here we give just a brief overview.

The theorem proving tool accepts two kinds of input files: core Haskell program files, and theory files. A theory is a sequence of axioms and theorems. The syntax of a theorem (or axiom) is:

$$name : logic\_formula$$

The logic is a first-order predicate calculus, with predicates for semantic equivalence ($\equiv$) and approximation ($\sqsupseteq$, $\sqsubseteq$) of core Haskell expressions. Complex formulae are built up of conjunction, disjunction, implication and universal quantification.

Every logical formula has a well defined operational interpretation. For example, equivalence ($e_1 \equiv e_2$) is handled by alternating syntactic comparisons and reductions (to weak head normal form), until the two sides become syntactically equivalent or reach different head normal forms. An implication ($[P_1,\ P_2,\ \cdots P_n] \Rightarrow Q$) is proved by checking Q on the assumption that $P_1,\ P_2,\ \cdots P_n$ hold. A universally quantified formula ($\forall\ x.\ P$) is proved by proving P in which a fresh constant identifier has been substituted for x. There is a variant of the universal quantifier ($\forall$) — the *induction quantifier* ($\underline{\forall}$). A formula with an induction quantifier ($\underline{\forall}\ x.\ P$), where x is inferred to be of type *datatype*, is proved with the help of the internally generated structural induction axiom for *datatype*. Since induction is distinguished syntactically from universal quantification, the prover does not have to search heuristically for a suitable induction schema.

Syntax and semantics. The language for which an abstract machine is to be built is identical to the untyped lambda calculus: it has only variables (*AVar Id*), lambda abstractions (*ALam* (*List Id*) *SyntExpr*), basic constants (*ANum Int*) and applications (*AAp SyntExpr SyntExpr*). This minimal language can illustrate the important implementation problems without loss of generality. The language is given a direct nonstrict denotational semantics, defined in terms of the reflexive domain of semantic values — a separated sum of a domain of basic constants and a domain of functions:

$$\begin{aligned} SemValue \ ::= \ & LNum\ Int \mid \\ & LLam\ (SemValue \rightarrow SemValue) \end{aligned}$$

The main semantic function, *evalExpr*, converts syntactic expressions into values. The function *deFl* is a projection function. The data constructor *P* is used for making pairs (lifted products). The function *upd* takes an identifier–value pair ($P\ v\ x$) and an environment *env*, and returns the environment augmented with a binding of v to x.

Definition 1.

```
evalExpr = λ env e →
   case e of
   AVar v  → env v
   ALam vs e → case vs of
                  []      → evalExpr env e
                  v : vs → LLam (λ x →
                              evalExpr (upd (P v x) env) (ALam vs e))
   ANum n  → LNum n
   AAp e1 e2 → deFl (evalExpr env e1) (evalExpr env e2)
deFl = λ g x → case g of LLam f → f x
```

2 The abstract machine

The abstract machine for the lambda language is defined as a function from an initial state of the computation to a final state.

We define the abstract machine in the same functional metalanguage that we used for the denotational semantics. However, since we want our abstract machine to be implementable in an imperative language, we must ensure that the state is single–threaded. There are various mechanisms for guaranteeing single–threadedness in the functional metalanguage, for example:

- linear types
- continuations
- state monads

Monads have recently found increasing use in functional languages [20], most notably for implementing structured interpreters [8] inspired by the idea of Moggi to employ monads in structuring denotational semantic definitions [12]. State monads, in particular, have gained wide popularity for representing state–based computation in functional languages [4]. We have used monads in order to produce a customizable definition of the abstract machine, which paid off in the proof of correctness.

A state monad is a function which takes a state and returns a pair of a result and a new state. There are two main operations for making monads:

$$(\#) :: \text{Monad } a \rightarrow (a \rightarrow \text{Monad } b) \rightarrow \text{Monad } b \qquad \mathit{unit} :: a \rightarrow \text{Monad } a$$
$$(\#) = \lambda\, f\, g \rightarrow \mathit{unc}\, g\, .\, f \qquad \mathit{unit} = \lambda\, a\, s \rightarrow P\, a\, s$$

The *unit* operation turns an arbitrary expression into a monad. The combinator (#) (also known as '*then*' or '*bind*') corresponds to the sequencing operator in imperative languages. The function *unc* is the 'uncurry' function:

$$\mathit{unc} = \lambda\, f\, p \rightarrow \texttt{case}\ p\ \texttt{of}\ P\, a\, s \rightarrow f\, a\, s$$

The state of the abstract machine is simply a *heap*. The heap is needed to store *closures*, which contain unevaluated expressions, partially applied functions, or data structures. The heap has an essential role to play in supporting laziness (reevaluation of an unevaluated shared expression is avoided by providing multiple pointers to a closure containing the expression; on evaluation the closure is updated with the value).

The notion of a closure is formalized in the type definitions below:

$$\begin{aligned} &\mathit{Closure} ::= \mathit{CAp}\ \mathit{Op}\ (\mathit{List}\ \mathit{Addr}) \\ &\mathit{Op} ::= \mathit{SNum}\ \mathit{Int}\ \mid \\ &\qquad\quad \mathit{SAddr}\ \mathit{Addr}\ \mid \\ &\qquad\quad \mathit{SFun}\ (\mathit{List}\ \mathit{Id})\ (\mathit{Id} \rightarrow \mathit{Addr})\ \mathit{SyntExpr} \end{aligned}$$

Each closure contains an operation *Op* and a list of argument closure addresses (*List Addr*). The operation can be a basic constant (*SNum Int*), a closure address (*SAddr Addr*) or the body of a lambda abstraction (*SFun* (*List Id*) (*Id* $\rightarrow$ *Addr*) *SyntExpr*), where (*List Id*) is a list of formal parameters of the abstraction, and (*Id* $\rightarrow$ *Addr*) is an environment binding any other names that occur free in the body. Note that an (*SAddr a*) closure with a nonempty argument list *as* represents the application of an *unknown* function to the arguments *as*.

In fact we shall use a more general definition of closures, which enables us to replace addresses in a closure by the closures (at those addresses) themselves. Let us define a type of *items*, which will be either addresses or closures:

$$\mathit{Item}\ a\ c ::= A\ a \mid C\ c$$

Items are built up by constructors A and C and taken apart by projections *projA* and *projC*. Using *Item*, a closure can be defined as:

Definition 2.

$$\begin{aligned}
\mathit{Closure} &::= \mathit{SAp}\ \mathit{Op}\ (\mathit{List}\ (\mathit{Item}\ \mathit{Addr}\ \mathit{Closure}))\\
\mathit{Op} &::= \mathit{SNum}\ \mathit{Int}\ \mid\\
&\quad\ \ \mathit{SAddr}\ (\mathit{Item}\ \mathit{Addr}\ \mathit{Closure})\ \mid\\
&\quad\ \ \mathit{SFun}\ (\mathit{List}\ \mathit{Id})\ (\mathit{Id} \rightarrow \mathit{Item}\ \mathit{Addr}\ \mathit{Closure})\ \mathit{SyntExpr}
\end{aligned}$$

This slightly complicated definition has some advantages in subsequent proofs.

The heap is accessed by three functions (left-hand side of Figure 1): *hAlloc*

State monad and heap access functions	Identity monad
$\# = \lambda\ f\ g\ \rightarrow\ unc\ g\ .\ f$	$iThen\ =\ \lambda\ f\ g\ \rightarrow\ g\ f$
$unit\ =\ \lambda\ a\ s\ \rightarrow\ P\ a\ s,$	$iUnit\ =\ \lambda\ a\ \rightarrow\ a$
$unc\ =\ \lambda\ f\ p\ \rightarrow$ **case** p **of** $P\ a\ s\ \rightarrow\ f\ a\ s$	
$hAlloc\ =\ \lambda\ c\ heap\ \rightarrow\ P\ (A\ (getFreeAddr\ heap))$ $(upd\ heap\ (P\ (getFreeAddr\ heap)\ c))$	$iAlloc\ =\ \lambda\ c\ \rightarrow\ C\ c$
$hLookup\ =\ \lambda\ a\ heap\ \rightarrow$ $P\ (C\ (lookup\ heap\ a))\ heap$	$iLookup\ =\ \lambda\ a\ \rightarrow\ a$
$hUpdate\ =\ \lambda\ a\ c\ heap\ \rightarrow$ $P\ c\ (upd\ heap\ (P\ (projA\ a)\ (lookup\ heap\ c)))$	$iUpdate\ =\ \lambda\ a\ b\ \rightarrow\ b$

Fig. 1. State monad and identity monad

allocates a new cell in the heap and returns a pair of the address of that cell and the modified heap; *hLookup* looks up the contents of a heap address; *hUpdate* takes two addresses and a heap, updates the first address with the contents of the second address, and returns a pair of the second address and the modified heap.

The abstract machine has two main functions. The first one, *constr*, constructs in the heap a graphical representation of a syntactic expression. The function takes an environment *env* that binds identifiers to heap addresses, a syntactic expression *e* and a list of heap addresses *as*. It returns a state monad which, given a heap, constructs a graph representing *e* (applied to the arguments *as*) in that heap.

$$constr\ =\ \lambda\ env\ e\ as\ \rightarrow\ (constrGen\ (\#)\ unit\ hAlloc\ constr)\ env\ e\ as$$

The function *constr* is defined as a specialized version of a generic higher-order function, *constrGen*[1] (Figure 2). The generic function is parameterized on the state–monad manipulation operations (#, *unit*, *hAlloc*), and can be specialized to use different monads. The advantages of introducing a generic definition like *constrGen* will become apparent in subsequent sections.

```
constrGen (#) unit hAlloc constr  =  λ env e as →
  case e of
  AAp e1 e2  →  constr env e2 [] # (λ a2  →  constr env e1 (a2 : as))
  AVar v  →  hAlloc (SAp (SAddr (env v)) as)
  ANum n  →  hAlloc (SAp (SNum n) as)
  ALam vs e  →  hAlloc (SAp (SFun vs env e) as)

forceFGen (#) hLookup forceS force
          =  λ a  →  hLookup a # forceS force a
forceSGen (#) unit hAlloc hUpdate constr addArgs force1 force2
  =  λ a c  →  case projC c of
  SAp f as  →
        case f of
        SNum n  →  case as of
                   []  →  unit a
        SAddr b  →  ((force2 b #
                   addArgs as) #
                   force1) #
                   hUpdate a
        SFun vs env e  →
              case length as < length vs of
              True  →  unit a
              False  →  ((constr (foldl upd env (zip2 vs as)) e [] #
                        addArgs (drop (length vs) as)) #
                        force1) #
                        hUpdate a
addArgsGen (#) hLookup hAlloc  =  λ as a  →  hLookup a # (λ c  →
  case projC c of
  SAp f xs  →  hAlloc (SAp f (xs ++ as)))
```

Fig. 2. Abstract machine: generic definitions

[1] Some standard functions used in the definition of *constrGen* are listed in Appendix A.

The second function in the definition of the abstract machine is *force*. Just as the name suggests, it forces the evaluation of closures. It is defined as the fixed point of the functional *forceF*. When given an address (a) and a heap, it evaluates the closure at a, and returns the address of the result paired with the heap (possibly modified during the evaluation).

Definition 3.

$$
\begin{aligned}
&force \;=\; fix\ ww\ forceF\\
&forceF \;=\; \lambda\ force\ a \;\rightarrow\; (forceFGen\ (\#)\ hLookup\ forceS\ force)\ a\\
&forceS \;=\; \lambda\ force\ a\ c \rightarrow\\
&\quad (forceSGen\ (\#)\ unit\ hAlloc\ hUpdate\ constr\ addArgs\ force\ force)\ a\ c\\
&addArgs \;=\; \lambda\ as\ a \;\rightarrow\; (addArgsGen\ (\#)\ hLookup\ hAlloc)\ as\ a
\end{aligned}
$$

In the theorem prover we use for the correctness proof (Section 4) fixed point induction is not supported directly, and must be implemented by numerical induction. That is why the fixed point operator, *fix*, takes the number of iterations as its first argument. The number *ww* allows unlimited iteration (because the successor constructor *Succ* is nonstrict):

$$
\begin{aligned}
&fix \;=\; \lambda\ k\ f \;\rightarrow\; \mathbf{case}\ k\ \mathbf{of}\ Succ\ k \;\rightarrow\; f\ (fix\ k\ f)\\
&ww = \; Succ\ ww
\end{aligned}
$$

Just like the heap constructing function, the forcing function is defined in terms of generic higher-order functions, *forceFGen* and *forceSGen* (Figure 2). The function *forceSGen* analyses the type of closure to be forced, and takes appropriate action. If the closure contains a constant (*SNum*), or a function (*SFun*) with fewer arguments *as* than formal parameters *vs*, no action is taken, since the closure is irreducible. If the closure contains an application of an unknown function from another closure at address (*SAddr b*), then the closure at b is forced first, and the result is forced again after adding the arguments *as*. Forcing the closure at b first is necessary in order to avoid a possible loss of laziness (in case b points to an unevaluated shared closure).

The forcing of a closure containing a known function embodies the concept of *spineless reduction* [2, 16]: function arguments are collected together and used when their number reaches the arity of the function.

Note that in both *constr* and *force*, the heap is not given explicitly as an argument, thanks to the use of the state monadic combinators.

3 Intermediate semantic definitions

We start work on the correctness proof by looking at the direct semantics from Definition 1 and transforming it into an equivalent function which is closer to the operational model. One of the differences between the direct semantic function and the abstract machine lies in the way functions are applied to their arguments: in the direct semantics, a function is applied to one argument at a time, while in the abstract machine arguments are collected together and used when their number reaches the arity of the function.

Definition 4 shows the function *eval*, which was derived from the semantic function *evalExpr* from Definition 1, and which behaves more like the abstract machine:

Definition 4.

$$
\begin{array}{l}
eval \;=\; \lambda\; env\; e\; as \;\rightarrow\; \mathbf{case}\; e\; \mathbf{of} \\
\qquad\qquad\qquad\qquad AAp\; e1\; e2 \;\rightarrow\; eval\; env\; e1\; (eval\; env\; e2\; []\; :\; as) \\
\qquad\qquad\qquad\qquad \_ \;\rightarrow\; foldl\; deFl\; (evall\; env\; e)\; as \\
evall \;=\; \lambda\; env\; e \;\rightarrow\; \mathbf{case}\; e\; \mathbf{of} \\
\qquad\qquad\qquad AVar\; v \;\rightarrow\; env\; v \\
\qquad\qquad\qquad ANum\; n \;\rightarrow\; LNum\; n \\
\qquad\qquad\qquad ALam\; vs\; e \;\rightarrow\; mkLLam\; env\; vs\; e \\
mkLLam \;=\; \lambda\; env\; vs\; e \;\rightarrow \\
\quad \mathbf{case}\; vs\; \mathbf{of} \\
\quad [] \;\rightarrow\; evall\; env\; e\; [] \\
\quad v\; :\; vs \;\rightarrow\; LLam\; (\lambda\; x \;\rightarrow\; mkLLam\; (upd\; env\; (P\; v\; x))\; vs\; e)
\end{array}
$$

In order to show that *eval* has the same meaning as the original direct semantic function *evalExpr*, we prove that *eval* satisfies the following proposition:

$$
\begin{array}{l}
\forall e.\; \forall vs.\; \forall envx.\; mkLLam\; envx\; vs\; e \;\equiv\; evalExpr\; envx\; (ALam\; vs\; e) \;\wedge \\
\quad \forall as.\; \forall env.\; eval\; env\; e\; as \;\equiv\; foldl\; deFl\; (evalExpr\; env\; e)\; as
\end{array}
$$

The proof is by induction on *e* and *vs*.

We now look at the abstract machine from Definition 3 and Figure 2. The function *forceS* uses heap updates sparingly — just like any sensible implementation, it does not update expressions already in weak head normal form. On the other hand, we would like to make sure that *forceS* does all the updates necessary, i.e. does not lose laziness. We can get closer to ensuring this by defining a function (*uforce*) which updates all expressions, and then by showing *uforce* equivalent to *force*.

Definition 5.

$$
\begin{array}{l}
uforce \;=\; fix\; ww\; uforceF \\
uforceF \;=\; \lambda\; force\; a \;\rightarrow \\
\quad forceFGen\; (\#)\; hLookup\; uforceS\; force\; a\; \#\; hUpdate\; a \\
uforceS \;=\; \lambda\; force\; a\; c \rightarrow \\
\quad forceSGen\; (\#)\; unit\; hAlloc\; uUpdate\; constr\; addArgs\; force\; force\; a\; c \\
uUpdate \;=\; \lambda\; a\; b\; hp \;\rightarrow\; P\; b\; hp
\end{array}
$$

The major part of the proof of correctness of our abstract machine lies in establishing the relationship between the denotational semantics from Definition 4 and the operational semantics from Definition 5. There are two main factors which complicate the proof. The first one is the state, present in the abstract machine, but absent from the direct semantics of the language. The second one is the higher–orderness of the language.

To avoid having to deal with the complexities imposed by both factors simultaneously, we can split up the proof into two parts. To that end we can introduce an intermediate operational semantics. The intermediate semantics, in contrast with the abstract machine, has no heap, and thus cannot model sharing. We shall call the closures it uses 'quasi-closures' or q-closures. There is no need to give a new definition for quasi-closures; Definition 2 is general enough to represent both the usual heap closures (when all the items in a closure are addresses) and q-closures (when items are closures).

It is easy to define the intermediate operational semantics using the generic functions from Figure 2. All we need to do is replace the state–monad functions (#, *unit*, *hAlloc*, *hLookup*, *hUpdate*) with the corresponding identity–monad functions from Figure 1 on page 5:

Definition 6.

$$
\begin{array}{l}
qconstr \;=\; \lambda\; env\; e\; as \;\rightarrow\; constrGen\; iThen\; iUnit\; iAlloc\; qconstr\; env\; e\; as \\
qforce \;=\; fix\; ww\; qforceF \\
qforceF \;=\; \lambda\; qforce\; a \;\rightarrow\; forceFGen\; iThen\; iLookup\; qforceS\; qforce\; a \\
qforceS \;=\; \lambda\; qforce\; a\; c \;\rightarrow \\
\quad forceSGen\; iThen\; iUnit\; iAlloc\; iUpdate\; qconstr \\
\qquad\qquad\qquad\qquad qaddArgs\; qforce\; iUnit\; a\; c \\
qaddArgs \;=\; \lambda\; as\; a \;\rightarrow\; addArgsGen\; iThen\; iLookup\; iAlloc\; as\; a
\end{array}
$$

We define a function, *remHp* (remove heap), to translate a pair of a heap address and a heap into a q-closure. It just replaces all the heap addresses in a closure by the closures at those addresses.

```
remHp = λ p → case p of
                  P a hp → remHpS hp (lookup hp a)
remHpArg = λ hp a → remHpS hp (lookup hp a)
remHpS = λ hp c → C (
    case c of
    SAp op as → SAp (remHps hp op) (map (remHpArg hp) as))
remHps = λ hp op →
    case op of
    SNum n → SNum n
    SAddr a → SAddr (remHpArg hp a)
    SFun vs env e → SFun vs (remHpArg hp . env) e
```

Finally, in order to be able to state the correctness proposition for the abstract machine, we need two functions which abstract semantic values from heap closures / q-closures. The functions do not have to be written in a monadic style, since they are not meant to be implemented in an imperative language. Each of the two functions (*abstr* and *qabstr*) takes a pair of an address and a heap, and returns a semantic value:

```
abstr = λ p → case p of
                P a hp → abstrSGen abstrArg hp (lookup hp a)
abstrArg = λ hp a → abstrSGen abstrArg hp (lookup hp a)
qabstr = λ a → abstrSGen qabstrArg ⊥ (projC a)
qabstrArg = λ hp a → abstrSGen qabstrArg ⊥ (projC a)
```

Both functions are defined in terms of the generic function *abstrSGen*:

```
abstrSGen = λ abstrArg hp c →
   case c of
   SAp f as →
        foldl deFl (abstrsGen abstrArg hp f) (map (abstrArg hp) as)
abstrsGen = λ abstrArg hp f →
   case f of
   SNum n → LNum n
   SAddr a → abstrArg hp a
   SFun vs envr e → mkLLam (abstrArg hp . envr) vs e
```

The syntactic and semantic domains, as well as the functions between them are illustrated in Figure 3.

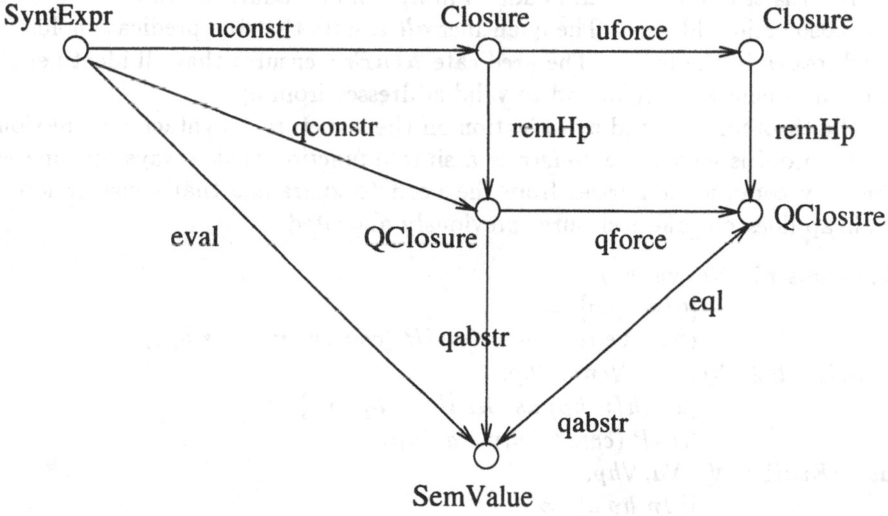

Fig. 3. Domains and functions for the abstract machine

4 Correctness of the abstract machine

With these definitions in place, we can state relationships between the abstract machine and the intermediate semantics, on the one hand, and between the intermediate semantics and the denotational semantics, on the other hand. The denotational semantics is represented by the function *eval*, which is derived from the direct denotational semantics. Here we shall just state the main theorems that relate the different semantic representations. The reader may find Figure 3 useful for illustrating the relationships. We begin with establishing the correctness of the graph construction function.

4.1 Correctness of graph construction

The correctness of graph construction performed by the function *constr* is stated in Theorem 7.

Theorem 7.

$$\begin{array}{l} abstrConstr : \\ \quad \underline{\forall} e.\ \forall as.\ \forall env.\ \forall hp. \\ \qquad [all\ (hIn\ hp)\ as,\ hInEnv\ hp\ env] \Rightarrow \\ \qquad (abstr\ (constr\ env\ e\ as\ hp) \equiv \\ \qquad eval\ (abstrArg\ hp\ .\ env)\ e\ (map\ (abstrArg\ hp)\ as)) \end{array}$$

Here *hIn* is a predicate which takes a heap *hp* and an address *a* as arguments, and returns true if *a* is a valid address in *hp*, and all addresses in the closure at *a* are also valid addresses. The quantifier *all* asserts that the predicate holds for all addresses in the list *as*. The predicate *hInEnv* ensures that all identifiers in an environment *env* are bound to valid addresses from *hp*.

The theorem is proved by induction on the structure of syntactic expressions *e*. The proof is easy since *constr* is a simple function that always terminates. The only complication arises from the need to guarantee that construction in the heap does not affect closures previously allocated.

$$\begin{array}{ll} hInConstr1 : & \underline{\forall} e.\ \forall a.\ \forall hp. \\ & \quad [hIn\ hp\ a] \Rightarrow \\ & \quad (\forall as.\ \forall env. \quad hIn\ (sndP\ (constr\ env\ e\ as\ hp))\ a) \\ hInConstr2 : & \underline{\forall} e.\ \forall as.\ \forall env.\ \forall hp. \\ & \quad [all\ (hIn\ hp)\ as,\ hInEnv\ hp\ env] \Rightarrow \\ & \quad hInP\ (constr\ env\ e\ as\ hp) \\ abstrEval1 : & \underline{\forall} e.\ \forall a.\ \forall hp. \\ & \quad [hIn\ hp\ a] \Rightarrow \\ & \quad (\forall as.\ \forall env. \\ & \qquad abstrArg(sndP(constr\ env\ e\ as\ hp))a \equiv abstrArg\ hp\ a \\ & \quad) \end{array}$$

Lemma *hInConstr*1 states that closures already in the heap are still in the heap after more construction; furthermore, *abstrEval*1 guarantees that the semantics of those closures remains unchanged. The predicate *hInP* in *hInConstr*2 takes an address–heap pair as its argument, and is true when the address is valid in the heap.

The next theorem regards the construction of q–closures, and is also proved by induction on the structure of expressions.

Theorem 8.

$$\begin{array}{l} constr : \underline{\forall} e.\ \forall as.\ \forall env.\ \forall hp. \\ \quad [all\ (hIn\ hp)\ as,\ hInEnv\ hp\ env] \Rightarrow \\ \quad (remHp\ (constr\ env\ e\ as\ hp)\ \equiv \\ \quad qconstr\ (remHpArg\ hp\ .\ env)\ e\ (map\ (remHpArg\ hp)\ as)) \end{array}$$

4.2 Soundness of forcing closures

The correctness of the forcing function is proved in two parts — soundness and adequacy. Soundness guarantees that forcing a closure does not change its denotation (so long as forcing terminates).

Theorem 9.

$$sound : (\forall k.\ soundForce\ k)$$

where the predicate *soundForce* is defined as:

$$\begin{array}{l} soundForce\ =\ \lambda\ k\ \rightarrow\ \forall a.\ \forall p.\ soundForceP\ k\ a\ p \\ soundForceP\ =\ \lambda\ k\ a\ p\ \rightarrow\ (unc\ (fix\ k\ uforceF)\ p\ \equiv \bot) \vee \\ \quad ([hInP\ p]\ \Rightarrow \\ \quad ((abstr\ (unc\ (fix\ k\ uforceF)\ p)\ \equiv\ abstr\ p) \wedge \\ \quad ([hIn\ (sndP\ p)\ a]\ \Rightarrow \\ \quad (abstrArg\ (sndP\ (unc\ (fix\ k\ uforceF)\ p))\ a\ \equiv\ abstrArg\ (sndP\ p)\ a)))) \end{array}$$

The argument k to *soundForce* is the maximum number of force iterations, i.e. the number of times *uforceF* may be called while forcing a closure. Forcing is sound if for any address a and address–heap pair p (provided that a and $(fstP\ p)$ are valid addresses in the heap $(sndP\ p)$), either

1. the forcing of p diverges, or
2. (a) the denotation of p is preserved by forcing, and
   (b) the denotation of the closure at any address a in the heap $(sndP\ p)$ is preserved after forcing p.

The theorem is proved by induction on k, in other words by fixed point induction on *uforceF*. The proof uses the properties of *hUpdate*, namely that the denotation of a closure is not changed if it is updated with a closure having the same denotation.

$$\begin{array}{l} soundUpd : \forall b.\ \forall p. \\ \qquad [abstrArg\ (sndP\ p)\ b\ \equiv\ abstr\ p]\ \Rightarrow \\ \qquad ((\forall a.\ abstrArg\ (sndP\ (unc\ (hUpdate\ b)\ p))\ a\ \equiv \\ \qquad\quad abstrArg\ (sndP\ p)\ a) \wedge \\ \qquad (abstr\ (unc\ (hUpdate\ b)\ p)\ \equiv\ abstr\ p) \\ \qquad) \end{array}$$

The proof of the soundness theorem proceeds by case analysis of the closure at address $(fstP\ p)$ in the heap $(sndP\ p)$. The interesting cases are when the closure is reducible — either an application of an unknown function (i.e. the address of another closure *SAddr*) or an application of a known function (*SFun*).

In the case of an unknown function a number of preliminary results are required. For example, it must be established that any valid address in a heap remains valid after:

- a new graph is constructed in the heap (Lemma *hInConstr*1 from Section 4.1);
- valid addresses are added as arguments in a closure;
- a closure in the heap is forced, and the forcing terminates.

In the case when a closure contains a known function applied to an insufficient number of arguments, soundness follows immediately. For dealing with the case when a closure is an application of a known function, and the number of arguments in the closure is higher than the arity of the function, the *abstrConstr* lemma from Section 4.1 and the following result are necessary:

$$
\begin{array}{l}
applyFalse:\\
\quad(\underline{\forall} as.\ \forall vs.\\
\qquad[(length\ as < length\ vs) \equiv False] \Rightarrow\\
\qquad(\forall e.\ \forall env.\ foldl\ deFl\ (mkLLam\ env\ vs\ e)\ as \equiv\\
\qquad\qquad\qquad eval\ (foldl\ upd\ env\ (zip2\ vs\ as))\ e\ (drop\ (length\ vs)\ as))\\
\quad)
\end{array}
$$

This lemma is proved by induction on the list *as* and case analysis on *vs*.

4.3 Correspondence between heapful and heapless abstract machines

The proof of computational adequacy is the second part of the proof of correctness of *uforce*. Adequacy guarantees that the forcing of a closure terminates as long as the closure represents a non–bottom semantic value.

The proof of adequacy presented here consists of two parts. In the first part, the abstract machine from Definition 5 is related to the simplified 'heapless' abstract machine from Definition 6. In the second part, which will be considered in the next section, the adequacy of the heapless machine with respect to the denotational semantics is established.

In this section we look at the relationship between the heapful and heapless machines. The relationship is one of approximation — the heapless machine approximates (i.e. diverges or produces the same result as) the original machine. Theorem 10 formulates the relationship precisely:

Theorem 10.

$$cong: (\underline{\forall} k.\ congForce\ k)$$

where the predicate $congForce$ is defined as:

$$congForce = \lambda\ k \rightarrow \forall p.\ [hInP\ p] \Rightarrow$$
$$(abstr\ (unc\ (fix\ k\ uforceF)\ p) \sqsupseteq qabstr\ (fix\ k\ qforceF\ (remHp\ p)))$$

Just as in the definition of $soundForce$ (Section 4.2), the argument k limits the number of forcing iterations. The predicate holds if the denotation of any forced closure is approximated by the denotation of that closure after removing the heap and q-forcing.

The theorem is proved by induction on k, i.e. by fixed point induction on $uforceF$ and $qforceF$. The preliminary results needed for proving the theorems in Section 4.2 are used here as well. Case analysis of the closure from the address–heap pair p is carried out. Just as with soundness, the interesting cases are closures representing applications of unknown or known functions. In the former case the proof relies on the fact that the forcing of the closure at address b (the unknown function) can be omitted in $uforceS$ (notice that it is omitted by Definition 6 in $qforceS$):

$$abstrSAddr:$$
$$(\forall a.\ \forall as.\ \forall b.\ \forall hp.\ \forall k.$$
$$abstr\ (uforceS\ (fix\ k\ uforceF)\ a\ (C\ (SAp\ (SAddr\ b)\ as))\ hp) \sqsupseteq$$
$$abstr\ (unc\ (fix\ k\ uforceF)\ (unc\ (addArgs\ as)\ (P\ b\ hp)))$$
$$)$$

The second interesting case in the proof of $cong$ — that of an application of a known function — is justified by the facts that construction in the heap and heap removal 'commute' (Theorem 8), that likewise addition of arguments and heap removal 'commute', and that abstracting the value of a closure is the same as removing the heap and q-abstracting the result.

4.4 Adequacy of the heapless machine

For the proof of computational adequacy, we need a relation between denotational values and q–closures which captures the fact that the forcing of a q-closure terminates whenever the q–closure denotes a constant. We say nothing about the termination of forcing of closures which denote function values (but of course the forcing of a closure denoting a constant may involve the forcing of other closures denoting functions). It is sufficient that the relation satisfies the following equation:

$$eql = \lambda\ l\ q \rightarrow$$
$$\textbf{case}\ l\ \textbf{of}$$
$$Bot \rightarrow TT$$
$$LNum\ n \rightarrow fix\ ww\ qforceF\ q \equiv C\ (SAp\ (SNum\ n)\ [])$$
$$LLam\ f \rightarrow \forall l2.\ \forall s2.\ [eql\ l2\ s2] \Rightarrow$$
$$eql\ (f\ l2)\ (qaddArgs\ [s2]\ (fix\ ww\ qforceF\ q))$$

A constant semantic value (*LNum n*) is related to a q-closure if forcing the closure produces the same constant. A function value (*LLam f*) is related to a closure q if for any pair of related value l_2 and closure q_2, the application (f $e2$) and the closure q (with an added argument q_2) are related.

The existence of a solution of the equation for *eql* is not guaranteed by monotonicity, since *eql* appears on both sides of the nonmonotonic ($\Rightarrow$) operator. Fortunately, the existence of a solution for this equation has been established by direct construction and other methods; a recent elegant proof appears in [17]. Thanks to the introduction of the heapless operational semantics, the relation *eql* is easy to construct.

The adequacy theorem uses the predicate *eql* to relate q-closures to semantic values.

Theorem 11.

$$\begin{aligned} adeq : \underline{\forall} e.\ \forall envl.\ \forall envs.\ \forall ls.\ \forall ss.\ & \\ [eqL\ eql\ ls\ ss,\ eqlEnv\ envl\ envs] & \Rightarrow \\ eql\ (eval\ envl\ e\ ls)\ (qconstr\ envs\ e\ ss) & \end{aligned}$$

where predicate *eqlEnv* relates two environments, and the predicate *eqL* relates with *eql* the corresponding elements of the two lists *ls* and *ss*. The proof of Theorem 11 is by induction on the structure of expressions *e*. In the case of variables, the theorem follows from the lemmas below:

$$\begin{aligned} foldEql :\ & \underline{\forall} ls.\ \forall ss.\ \forall l.\ \forall s.\ [eqL\ eql\ ls\ ss] \Rightarrow [eql\ l\ s] \Rightarrow \\ & \qquad eql\ (foldl\ deFl\ l\ ls)\ (qaddArgs\ ss\ s) \\ eqlSAddr :\ & \forall as.\ \forall b.\ \forall l. \\ & \qquad eql\ l\ (C\ (SAp\ (SAddr\ b)\ as)) \equiv eql\ l\ (qaddArgs\ as\ b) \end{aligned}$$

Lemma *foldEql* is proved by induction on the lists *ls* and *ss*. The proof uses two properties of *qforce*: namely, that forcing is idempotent, i.e. repeated forcing of a closure produces the same result as a single forcing; and that the forcing of a function before forcing an application of that function gives the same result as forcing the application.

The cases of constants (*ANum*) and applications (*AAp*) in the proof of Theorem 11 are easy. The interesting case is that of lambda abstractions:

$$\begin{aligned} & eqlLam : \\ & \quad \forall e.\ \forall envl.\ \forall envs.\ \forall ls.\ \forall ss. \\ & \qquad [eqL\ eql\ ls\ ss,\ eqlEnv\ envl\ envs,\ \forall envll.\ \forall envss.\ \forall lss.\ \forall sss. \\ & \qquad\quad [eqL\ eql\ lss\ sss,\ eqlEnv\ envll\ envss] \Rightarrow \\ & \qquad\quad eql\ (eval\ envll\ e\ lss)\ (qconstr\ envss\ e\ sss) \\ & \qquad] \Rightarrow \\ & \qquad (\forall vs.\ eql\ (eval\ envl\ (ALam\ vs\ e)\ ls)\ (qconstr\ envs\ (ALam\ vs\ e)\ ss)) \end{aligned}$$

This is proved by case analysis on (*length ss* $<$ *length vs*), where *ss* is the list of q-closure arguments to the function, and *vs* is the list of formal parameters of the function[2]. Let us assume first that the number of arguments is smaller than

[2] Note that the length of the list of arguments *ls* is the same as the length of *ss*.

the arity of the function. In that case *eval* turns the syntactic expression *e* into a function value, as confirmed by the following lemma:

$$
\begin{array}{l}
applyTrue : \\
\quad \forall vs.\ \forall as. \\
\qquad [(length\ as < length\ vs)\ \equiv\ True]\ \Rightarrow \\
\qquad (\forall e.\ \forall env. \\
\qquad\quad foldl\ deFl\ (mkLLam\ env\ vs\ e)\ as\ \equiv \\
\qquad\quad mkLLam\ (foldl\ upd\ env\ (zip2\ vs\ as))\ (drop\ (length\ as)\ vs)\ e \\
\qquad)
\end{array}
$$

On the other hand, we have that:

$$
\begin{array}{l}
eqlLamT2 : \\
\quad \forall e.\ [\forall envll.\ \forall envrr.\ \forall sss.\ \forall xss. \\
\qquad [(length\ sss == length\ xss)\ \equiv\ True, \\
\qquad eqlEnv\ envll\ (foldl\ upd\ envrr\ (zip2\ xss\ sss))]\ \Rightarrow \\
\qquad eql\ (eval\ envll\ e\ [])\ (qconstr\ (foldl\ upd\ envrr\ (zip2\ xss\ sss))\ e\ []) \\
\quad]\ \Rightarrow \\
\quad (\forall envl.\ \forall envr.\ \forall ss.\ \forall vs. \\
\qquad [(length\ ss < length\ vs)\ \equiv\ True, \\
\qquad eqlEnv\ envl\ (foldl\ upd\ envr\ (zip2\ vs\ ss))]\ \Rightarrow \\
\qquad eql\ (mkLLam\ envl\ (drop(length\ ss)vs)\ e)\ (C(SAp(SFun\ vs\ envr\ e)ss)) \\
\quad)
\end{array}
$$

Theorem *eqlLam* (in the case when the number of arguments is smaller than the function arity) follows from *applyTrue*, *eqlLamT2*, and properties of the functions *foldl*, *zip2*, *length*, (==), (<).

The second case in the proof of *eqlLam*, when the number of arguments exceeds the function arity, is easier to establish. The proof uses *applyFalse* (from Section 4.2) and properties of *length*, *foldl* and *zip2*.

This concludes our brief summary of the proof of correctness of the abstract machine. The machine–executable proof is contained in over a hundred theorems in seven theories. Apart from the function definitions, axioms and theorems, there is no other input to the theorem prover (no scripts of tactics of any sort). The absence of a language of tactics and tacticals for controlling proof construction probably results in a larger number of theorems than would otherwise be the case; but on the other hand the proof is easier to read just from the theory files. The prover produces detailed traces of proofs, which are essential for constructing a new proof.

5 Related work

Recent work on the correctness of a lazy functional abstract machine is reported in [18]. Sestoft derives an abstract machine, starting from Launchbury's natural semantics [5], proves the machine correct, and extends it to handle data structures.

Although our work covers similar ground, it is different from Sestoft's in a number of ways. First of all, Sestoft considers the correctness of the abstract machine with respect to the natural *operational* semantics, while we consider it with respect to the direct *denotational* semantics of the language. Since an operational semantics for a language is more concrete than the denotational semantics, the denotational approach to correctness leads to easier proofs than the operational approach whenever the abstract machine differs significantly from the operational semantics. An added difficulty of an operational approach is the need to establish the (natural) operational semantics correct with respect to the denotational semantics (as in [5], where the natural semantics is shown to agree with a *resourced*, or a sequentialised denotational semantics).

Further differences appear in the design of abstract machines. In some respects, our abstract machine is simpler than Sestoft's. As our machine was designed as the basis for a high–level translation of the functional language into an imperative language, some of the operational detail was considered unnecessary. In particular, the machine has no explicit stack (relying on the stack of the target imperative language compiler), and variable names are not replaced by de Bruijn indices (since the names can be handled by the imperative compiler). On the other hand, the treatment of function applications in this abstract machine is more complicated, so that the function call mechanism of a conventional imperative implementation can be used. As far as the source functional language is concerned, Sestoft's work is wider in scope, as it considers let-expressions and data structures.

6 Conclusion and further work

The proof outlined in this report concerns the correctness of the lazy implementation of a minimal untyped language. The use of a theorem prover has helped us to avoid some pitfalls of informality, and has given us additional confidence in the abstract machine.

The language lacks two features which are normally present in intermediate languages used in compilers: data structures and let–expressions. While these features can be encoded in the untyped lambda calculus, they ought to be directly supported for efficiency and thus need to be included in the proof. The next step would be the proof of various optimizations which are essential for fast implementations: call–by–value for strict functions, use of 'unboxed' function arguments and results.

How easy is it to implement the abstract machine in an imperative language? The functions comprising the abstract machine (*constr* and *force*) can easily be rephrased in an imperative language because of their state monadic definitions. The abstract machine deals with the problems of partial function applications and lazy evaluation of source programs. If a source program is lambda–lifted, *i.e.* all lambda–abstractions are lifted to the top level, then it can be compiled into an imperative language.

References

1. Z.M. Ariola, M. Felleisen, J. Maraist, M. Odersky, and P. Wadler. The call-by-need lambda calculus. In *22'nd Annual ACM SIGACT-SIGPLAN Symposium on Principles of Programming Languages*, San Francisco, California, 1995. ACM Press.
2. G.L. Burn. *Lazy Functional Languages: Abstract Interpretation and Compilation.* Research Monographs in Parallel and Distributed Computing. Pitman in association with MIT Press, 1991.
3. M.J.C. Gordon. Models of pure LISP (a worked example in semantics). Experimental Programming Reports 31, Department of Machine Intelligence, University of Edinburgh, 1973.
4. S.P. Jones and P. Wadler. Imperative functional programming. In *20'th Symposium on Principles of Programming Languages*, Charlotte, North Carolina, 1993. ACM Press.
5. J. Launchbury. A natural semantics for lazy evaluation. In *Twentieth ACM Symposium on Principles of Programming Languages, Charleston, South Carolina*, pages 144–154. ACM, 1993.
6. D.R. Lester. *Combinator Graph Reduction: A Congruence and its Applications.* Dphil thesis, Oxford University, 1988. *Also* published as Technical Monograph PRG-73.
7. D.R. Lester and S. Mintchev. Towards machine–checked compiler correctness for higher–order pure functional languages. In L. Pacholski and J. Tiuryn, editors, *Proceedings of the 1994 Annual Conference of the European Association for Computer Science Logic*, pages 369–381. Springer-Verlag LNCS 933, 1995.
8. S. Liang, P. Hudak, and M. Jones. Monad transformers and modular interpreters. In *22'nd Annual ACM SIGACT-SIGPLAN Symposium on Principles of Programming Languages*, San Francisco, California, 1995. ACM Press.
9. R.E. Milne. *The Formal Semantics of Computer Languages and Their Implementation.* PhD thesis, University of Cambridge, 1974.
10. S. Mintchev. A denotational approach to compiling core Haskell. In Peter Fritzson, editor, *Proceedings of the poster session of the International Conference on Compiler Construction CC'94*, Linköping University research report LiTH-IDA-R-94-11, pages 123–133, 1994.
11. S. Mintchev. Mechanized reasoning about functional programs. In K. Hammond, D.N. Turner, and P. Sansom, editors, *Functional Programming, Glasgow 1994*, pages 151–167. Springer-Verlag Workshops in Computing, 1995.
12. E. Moggi. Computational lambda-calculus and monads. In *Proceedings of the Fourth Annual Symposium on Logic in Computer Science (LICS)*, Pacific Grove, California, 1989.
13. K. Mulmuley. *Full Abstraction and Semantic Equivalence.* MIT Press, Cambridge, Massachusetts, 1987. ACM Doctoral Dissertation Award 1986.
14. F. Nielson and H.R. Nielson. *Two-level Functional Languages.* Number 34 in Cambridge Tracts in Theoretical Computer Science. Cambridge University Press, 1992.
15. L.C. Paulson. *Isabelle: A Generic Theorem Prover.* Springer-Verlag LNCS 828, 1994.
16. S.L. Peyton Jones. Implementing lazy functional languages on stock hardware: the spineless tagless G-Machine. *Journal of Functional Programming*, 2(2):127–202, April 1992.

17. A.M. Pitts. Computational adequacy via 'mixed' inductive definitions. In *Mathematical Foundations of Programming Semantics, Proc. 9th Int. Conf., New Orleans, LA, USA, April 1993*, volume 802 of *Lecture Notes in Computer Science*, pages 72–82. Springer-Verlag, Berlin, 1994.
18. P. Sestoft. Deriving a lazy abstract machine. Technical Report ID-TR 1994–146, Department of Computer Science, Technical University of Denmark, 1994.
19. J.E. Stoy. The congruence of two programming language definitions. *Theoretical Computer Science*, 13(2):151–174, February 1981.
20. P. Wadler. Comprehending monads. *Mathematical Structures in Computer Science*, 1992. Special issue of selected papers from 6'th Conference on Lisp and Functional Programming.

Appendix A: Some standard function definitions

```
map   = λ f xs → case xs of
                   [] → []
                   y : ys → f y : map f ys
foldl = λ f acc xs → case xs of
                       [] → acc
                       y : ys → foldl f (f acc y) ys
zip2  = λ xs ys → case xs of
                    x : xs → case ys of
                               y : ys → P x y : zip2 xs ys
                               [] → []
                    [] → []
length = λ xs → case xs of
                  [] → 0
                  y : ys → 1 + length ys
take  = λ n xs → case n of
                   0 → []
                   Succ n → case xs of
                              [] → []
                              y : ys → y : take n ys
drop  = λ n xs → case n of
                   0 → xs
                   Succ n → case xs of
                              [] → []
                              y : ys → drop n ys
```

Assertions and Recursions

Bernhard Möller

Institut für Mathematik, Universität Augsburg, D-86135 Augsburg, Germany,
e-mail: moeller@uni-augsburg.de

Abstract. We provide an algebraic description of subtypes and the way they propagate through recursive functions. By abstracting from the concrete domain of functions or relations we obtain a framework which is independent of strict or non-strict, deterministic or non-deterministic semantics. Applications include efficiency increasing simplification of recursions as well as proofs about recursions by noetherian induction, such as termination proofs.

1 Introduction

The paper [12] presented a way of introducing and manipulating assertions in an applicative language, mainly for use in transformational program development. The treatment was based one particular functional language with strict call-time choice semantics, and the essential rule for strengthening and weakening assertions was proved only for the case of linear recursion. Although this was already a generalization of corresponding techniques from the refinement calculus (see e.g. [15]), one would have liked a rule that works for arbitrary recursions. Related investigations are reported in [17] for the case of a non-strict functional semantics.

In the present paper we abstract from the particular semantic framework; the central requirements on the objects involved lead to the notion of a composition algebra. Concrete instances of this are usual functions over domains, binary relations, but also formal languages and sets of paths in graphs. We come up with an algebraic characterization of invariants and a general rule for strengthening and weakening them in recursions over the respective composition algebra. The rule is a simple instantiation of the fusion rule for least fixpoints.

We extend the technique to cover proofs by noetherian induction, in particular, termination proofs, for general recursions.

2 Assertions

2.1 Representation

We want to model that at a certain occurrence of an expression E an assertion A is satisfied. To this end we replace the occurrence of E under consideration by another expression $A \rhd E$ where $A \rhd E$ is equivalent to E if A holds at that occurrence and to error otherwise, where error represents the semantical pseudo-value $\perp$ or "undefined". This construct corresponds to the assert-statement in CIP-L

[3, 2] and the refinement calculus (see e.g. [15]); its properties were investigated in [12] for a strict semantics.

As assertions we allow arbitrary logical formulas over expressions of the language as terms and relations such as semantic equivalence $\equiv$ or approximation $\sqsubseteq$ as predicates. The semantics of the assertion expression is then more precisely defined, using an evaluation function

$$[\![\_]\!] : \mathit{Expr} \rightarrow \mathit{Env} \rightarrow \mathit{Val} \ ,$$

by the equation

$$[\![A \rhd E]\!]\rho \stackrel{\text{def}}{=} \begin{cases} [\![E]\!] & \text{if } \rho \models A \ , \\ \bot & \text{otherwise} \ . \end{cases}$$

Note that this may lead to non-monotonic expressions. However, as long as for a recursively defined function the assertions do not contain recursive calls, the associated functional will still be monotonic.

It is convenient to admit also Boolean expressions B of the programming language as assertions; the expression $B \rhd E$ is then defined to be an abbreviation for $(B \equiv \mathit{true}) \rhd E$. This yields monotonic expressions; such an expression is equivalent to

$$\textsf{if } B \textsf{ then } E \textsf{ else error} \ .$$

The use of Boolean assertions admits assertions as first-class objects [17], in particular, recursively defined assertions.

While the above semantics draws on a conventional domain-theoretic approach, there is another possibility if one considers only strict operations. In this case $\bot$ is not needed and one can use simple argument-value-pairs as semantics for (partial) functions. This approach was used e.g. in [10]. In this case assertions can simply be represented as subsets of the domain of discourse, namely as the set of all values satisfying the assertions. The theory in this case becomes much smoother if one embeds the partial functions into the set of arbitrary binary relations between domain and co-domain. This approach is followed in [14, 1, 5]; it also allows a convenient treatment of non-determinacy. Expressions are then set-valued, giving the set of all possible results; the empty set $\emptyset$ plays the role of $\bot$.

The assertion expression can then be defined analogously. We want to avoid the associated technical details (see [12, 14]), since they are not needed in the sequel. Rather, we present a treatment that is independent of the particular representation of assertions and hence works for both of the above approaches. In any of them the following algebraic properties hold:

$$\mathit{TRUE} \rhd E = E \tag{1}$$

$$A \rhd (B \rhd E) = (A \wedge B) \rhd E \ , \tag{2}$$

$$A \rhd f(E) = A \rhd f(A \rhd E) \ , \tag{3}$$

where f is an arbitrary function. Moreover,

$$\begin{aligned} \mathit{true} \rhd E &= E \ , \\ \mathit{false} \rhd E &= \textsf{error} \ , \\ \textsf{error} \rhd E &= \textsf{error} \ , \end{aligned}$$

and, for determinate Boolean expression B,

$$\begin{aligned} & \textsf{if } B \textsf{ then } E \textsf{ else } F \\ = \; & \textsf{if } B \textsf{ then } B \rhd E \textsf{ else } F \\ = \; & \textsf{if } B \textsf{ then } E \textsf{ else} \neg B \rhd F \, . \end{aligned} \tag{4}$$

Of course there are useful combinations of (3) and (4).

2.2 Assertions as Restrictions

We mainly use assertions as parameter restrictions for functions (cf.[3]).

Definition 2.1 Let A be an assertion possibly involving the identifier x. The *partial identity* belonging to A is

$$id_A \stackrel{\text{def}}{=} \lambda x.\, A \rhd x \, .$$

A function with its parameter *restricted* by A is then

$$\lambda x : A.\, E \stackrel{\text{def}}{=} (\lambda x.\, E) \circ id_A \, .$$

Here, $\circ$ is the usual function composition given by $(f \circ g)(x) = f(g(x))$. The definition using the composition operator encapsulates the particular underlying semantics. In case of a strict semantics we have for $f \stackrel{\text{def}}{=} \lambda x : A.\, E$ that

$$f = \lambda x.\, A \rhd E \, . \tag{5}$$

This means that f is undefined for all arguments x that violate the restriction A, or, in other words, that A is a precondition for definedness of f. In case of a non-strict semantics, the assertion A is checked lazily, i.e., only if parameter x is actually used critically in evaluating E.

Characteristic properties of the restricting partial identities are, first, that they are weaker than the full identity and that they are idempotent under composition by (2). Moreover, the elements that satisfy the assertion are precisely the fixpoints of the corresponding restriction function.

Following [17], we shall call such functions ***subtypes*** and prove our laws about assertions for general subtypes in an algebraic style in Section 3.

2.3 Invariants

How can we now describe that an assertion holds for all recursive calls of a function and hence is "invariant" for that function? The key idea is to compare the behaviour of functions with and without the assertion. If it makes no difference whether we plug in one or the other in the place of recursive calls, then for each of them the assertion must be satisfied or the result of the call is irrelevant for the overall computation anyhow.

We introduce a higher-order function that provides a function with an assertion. We shall assume that the assertion is expressed by a unary predicate C.

Definition 2.2 The *partial identity* belonging to C is

$$id_C \stackrel{\text{def}}{=} \lambda x.\ C(x) \rhd x \ .$$

while the *restriction functional* for C is

$$F_C \stackrel{\text{def}}{=} \lambda f.\ f \circ id_C \ .$$

Consider now a declaration of the form

$$f \stackrel{\text{def}}{=} \mu G \ ,$$

where μ is the least fixpoint operator and G is the functional corresponding to the body of the recursively defined routine. Semantically, G "records" all the places where recursive calls occur. Hence we can describe a modified recursion where a check for assertion C is inserted for the argument of every recursive call by the functional $G \circ F_C$.

Checking assertion C for the argument of the original call of a function f amounts to evaluating $F_C(f)$. Informally, C is an invariant of a recursion when C holds for all recursive calls provided it holds for the initial call. Using functional notation, we therefore say

Definition 2.3 Assertion C is an *invariant* of functional G if

$$F_C \circ G = F_C \circ G \circ F_C \ .$$

This will be taken up at a more abstract level in the following section.

3 An Algebraic View of Subtypes and Assertions

3.1 Composition Algebras

We have shown how an assertion can be modeled by a particular object that represents the set of elements for which the assertion holds. Enforcing an assertion is done by composing with that object. The assertion holds for the range of a function or relation, if composing with that object on the range side does not change the function or relation; a dual remark applies to the domain side.

Since composition plays such an essential role, we define the notion of a composition algebra. We briefly repeat the necessary order-theoretic concepts. Let $(M, \leq)$ be a partial order. A subset $D \subseteq M$ is *directed* if it is non-empty and any two elements of D have a common upper bound in D. The order $(M, \leq)$ is called *complete* ("complete partial order", briefly *cpo*), if M has a least element and every directed subset of M has a supremum in M. A *complete lattice* is a partial order in which *every* subset has a supremum (and hence also an infimum). A complete lattice also is a cpo.

Let $(M, \leq)$ and $(N, \leq)$ be partial orders. A (total) map $f : M \to N$ is *monotonic*, if for all $x, y \in M$ with $x \leq y$ also $f(x) \leq f(y)$. Assume now that

$(M, \leq)$ and $(N, \leq)$ are cpos. Then f is *strict* if it preserves the least element, and *continuous* if it is monotonic and preserves the suprema of all directed subsets of M. If $(M, \leq)$ and $(N, \leq)$ are complete lattices then map $F : M \to N$ is *universally disjunctive* if it preserves the suprema of all subsets of M. A universally disjunctive map is both strict and continuous. Now we are ready for

Definition 3.1 A *composition algebra* is a quintuple $C = (M, \cdot, 1, \leq, 0)$ such that
- $(M, \cdot, 1)$ is a monoid; the operation $\cdot$ is called *composition*;
- $(M, \leq)$ is a cpo with least element 0;
- the operation $\cdot$ is monotonic in both arguments.

The algebra C is called *left-continuous* if $\cdot$ is continuous in its left argument. *Right-continuity* is defined analogously. If C is both left-continuous and right-continuous it is called *continuous*.

A prominent example of a composition algebra is the set of all monotonic functions between two cpos under function composition and the usual pointwise extension of the approximation ordering. This algebra is left-continuous but not right-continuous. A continuous algebra is obtained by considering continuous functions only. However, a number of specification constructs, such as the quantifiers in CIP-L [3], lead to functions that are monotonic but not continuous, and so the more general notion has its use.

Prompted by this example, we view in all composition algebras the range of an element as being "on its left", its domain as being "on its right".

Note that, by monotonicity, multiplication with a subidentity is strict:

$$x \leq 1 \Rightarrow x \cdot 0 = 0 = 0 \cdot x . \tag{6}$$

A particular and important class of composition algebras is introduced by (see e.g. [6])

Definition 3.2 A composition algebra $C = (M, \cdot, 1, \leq, 0)$ is called a *Kleene algebra* if $(M, \leq)$ is even a complete lattice and $\cdot$ is universally disjunctive. In Kleene algebras the general supremum operator is usually denoted by Σ, the binary one by $+$.

All Kleene algebras are continuous. Important examples are

1. the power set of a set under intersection or union and the inclusion ordering;
2. the set of all binary relations on a set under relational composition and the inclusion ordering;
3. the set of all formal languages over some alphabet under concatenation and the inclusion ordering;
4. the Kleene algebra of all sets of paths in a directed graph under the pointwise extension of path concatenation and the inclusion ordering;
5. all sequential algebras [9].

It turns out that the following relations are of interest for many properties; they reflect a kind of subsumption. In the power set Kleene algebra they coincide with inclusion.

Definition 3.3 The relations $\preceq_d, \preceq_r \subseteq M \times M$ are defined by

$$x \preceq_d y \overset{\text{def}}{\Leftrightarrow} x \cdot y = x ,$$
$$x \preceq_r y \overset{\text{def}}{\Leftrightarrow} y \cdot x = x .$$

Hence $x \preceq_d y$ means that the domain of x is stable under y; it implies that y is idempotent on the domain of x. Analogous remarks hold for $\preceq_r$. The following properties are immediate:

Lemma 3.4 1. For all $x \in M$ we have $x \preceq_d 1$ and $x \preceq_r 1$.
2. $\preceq_d$ and $\preceq_r$ are transitive.
3. $x \preceq_d x$ iff $x \preceq_r x$ iff x is idempotent, i.e., $x \cdot x = x$.
4. Suppose x and y commute, i.e., $x \cdot y = y \cdot x$. Then $\preceq_d$ and $\preceq_r$ are antisymmetric on x, y, i.e., $x \preceq_d y \wedge y \preceq_d x \Rightarrow x = y$ and $x \preceq_r y \wedge y \preceq_r x \Rightarrow x = y$.
5. If $x \preceq_d y$ then also $z \cdot x \preceq_d y$ for all $z \in M$. If $x \preceq_r y$ then also $x \cdot z \preceq_r y$ for all $z \in M$.

Next, we introduce ternary relations that express weak commutativity properties:

Definition 3.5 For $a, b, c \in M$ we set

$$x : y \rightharpoonup z \overset{\text{def}}{\Leftrightarrow} x \cdot y \leq z \cdot x ,$$
$$x : y \leftharpoondown z \overset{\text{def}}{\Leftrightarrow} z \cdot x \leq x \cdot y ,$$
$$x : y \rightleftharpoons z \overset{\text{def}}{\Leftrightarrow} x : y \rightharpoonup z \wedge x : y \leftharpoondown z$$
$$\Leftrightarrow x \cdot y = z \cdot x .$$

Roughly, the first relation says that the range of the restriction of x to y is contained in z. Analogous remarks apply to the other relations. A similar definition occurs in [7]. Connections between these relations and $\preceq_d, \preceq_r$ will be established in a later section. Their use is illustrated in the following fusion rules for least fixpoints. Consider monotonic functions $f, g, h : M \to M$, where $(M, \leq)$ is a cpo, under function composition $\circ$. As usual, we denote the least fixpoint of a function k by μk. We have

$$\text{(Right Subfusion)} \quad \frac{g : h \leftharpoondown f}{\mu f \leq g(\mu h)}$$

$$\text{(Left Subfusion)} \quad \frac{\begin{array}{l} f \text{ continuous and strict} \\ f : g \rightharpoonup h \end{array}}{f(\mu g) \leq \mu h}$$

$$\text{(Fusion)} \quad \frac{\begin{array}{l} f \text{ continuous and strict} \\ f : g \rightleftharpoons h \end{array}}{f(\mu g) = \mu h}$$

The right subfusion rule is immediate from the fact that μf is also the least element x with $f(x) \leq x$. Contrarily, the proof of the left subfusion rule needs fixpoint induction. For the case of $(M, \leq)$ being a complete lattice, a different derivation using the theory of Galois connections is given in [11]; then f has to be universally disjunctive. The fusion rule, finally, is immediate from the two subfusion rules.

3.2 Pretypes

We now return to the topic of assertions. As already stated, it is characteristic of restriction functions representing assertions that they are idempotent subidentities (see also [17]). Idempotence is also characteristic of so-called retractions and used centrally in Scott's classical paper [19] for the definition of data types: each such type is viewed as the range or, equivalently, the set of fixpoints of a retraction. We shall call idempotent subidentities pretypes, since under additional assumptions they will play the role of subtypes.

Definition 3.6 The set of *pretypes* of composition algebra C is

$$PT(C) \stackrel{\text{def}}{=} \{x \in M : x \leq 1 \wedge x \leq x \cdot x\} .$$

We list the pretypes for some of the Kleene algebras mentioned above:

1. for binary relations the pretypes are exactly the subrelations of the identity, called monotypes in [1, 7, 8];
2. for formal languages we only have ε, the language consisting of the empty word, and $\emptyset$, i.e., just 1 and 0;
3. for sets of paths in a graph with vertex set V the pretypes are exactly the subsets of $\varepsilon \cup V$, i.e., sets of paths with at most one vertex.

The following properties are immediate from the definition, (6), monotonicity of composition, neutrality of 1 and antisymmetry of $\leq$:

$$0, 1 \in PT(C) , \tag{7}$$

$$x \in PT(C) \wedge y \in M \Rightarrow x \cdot y \leq y \wedge y \cdot x \leq y , \tag{8}$$

$$x \in PT(C) \Rightarrow x \cdot x = x . \tag{9}$$

In the relation or function model, a pretype is characterized by its set of fixpoints which coincides with its range. We call multiplication by a pretype on the right *restriction* and multiplication on the left *corestriction* by that pretype. Hence for pretype y the relation $x \preceq_r y$ means that corestriction of x by y does not change x, so that the range of x is contained in the characteristic set of y. A dual remark holds for the $\preceq_d$ relation.

It turns out that for pretypes the subsumption relations coincide with the ordering $\leq$:

Lemma 3.7 For $x, y \in PT(C)$ we have

$$x \preceq_d y \Leftrightarrow x \leq y \Leftrightarrow x \preceq_r y \,.$$

Proof: We show the second equivalence, the first one being dual.

$$\begin{array}{ll} & x \leq y \\ \Rightarrow & \{\!\![\ \text{monotonicity of composition}\ ]\!\!\} \\ & x \cdot x \leq y \cdot x \\ \Rightarrow & \{\!\![\ x \in PT(C) \text{ and transitivity}\ ]\!\!\} \\ & x \leq y \cdot x \\ \Rightarrow & \{\!\![\ \text{by (8) and antisymmetry}\ ]\!\!\} \\ & x = y \cdot x \\ \Rightarrow & \{\!\![\ \text{reflexivity}\ ]\!\!\} \\ & x \leq y \cdot x \\ \Rightarrow & \{\!\![\ \text{by (8) and transitivity}\ ]\!\!\} \\ & x \leq y \,. \end{array}$$

■

The composition of two pretypes is their infimum, provided it is a pretype:

Lemma 3.8 Suppose $X \subseteq PT(C)$ and $x, y \in X$ such that $x \cdot y \in X$. Then $x \cdot y = x \sqcap_X y$.

Proof: From (8) we get $x \cdot y \leq x, y$. Assume now $z \leq x, y$ for some $z \in X$. Then $z = z \cdot z \leq x \cdot y$ by (9) and monotonicity of composition. ■

The infimum, i.e., the composition, of two pretypes plays the role of the conjunction of the corresponding assertions. We have the property

Corollary 3.9 The composition of two pretypes is a pretype iff they commute under composition.

Proof: (⇒) follows from the previous lemma and the commutativity of the binary infimum operator.
(⇐) Assume that $x, y \in PT(C)$ commute. By monotonicity of $\cdot$ and idempotence of 1 we get $x \cdot y \leq 1$. Moreover, since x and y commute, we have $x \cdot y \cdot x \cdot y = x \cdot x \cdot y \cdot y = x \cdot y \geq x \cdot y$. ■

The property that the composition of two pretypes is their infimum if they commute can already be found in [19], p. 541.

3.3 Pretype Propagation

We investigate now the relations $\rightharpoonup$, $\leftharpoondown$ and $\rightleftharpoons$ in the particular case of subtypes. For $x, y \in PT(C)$ and $a \in M$ the relation $a : x \rightharpoonup y$ means that the range of the restriction of a to x is included in y. A dual remark applies to $\leftharpoondown$. This is also shown by

Lemma 3.10 For $x, y \in PT(C)$ and $a \in M$ we have

$$a : x \leftharpoondown y \Leftrightarrow y \cdot a \preceq_d x \, ,$$
$$a : x \rightharpoonup y \Leftrightarrow a \cdot x \preceq_r y \, .$$

Proof: We show the second equivalence, the first one being dual.

$$a \cdot x \leq y \cdot a$$
$$\Rightarrow \quad \{\!\!\{ \text{ monotonicity } \}\!\!\}$$
$$a \cdot x \cdot x \leq y \cdot a \cdot x$$
$$\Rightarrow \quad \{\!\!\{ \text{ idempotence of } x \}\!\!\}$$
$$a \cdot x \leq y \cdot a \cdot x$$
$$\Rightarrow \quad \{\!\!\{ \text{ by (8) } \}\!\!\}$$
$$y \cdot a \cdot x = a \cdot x$$
$$\Rightarrow \quad \{\!\!\{ \text{ by (8) } \}\!\!\}$$
$$a \cdot x \leq y \cdot a \, .$$

■

Assume now $x, y, u, v, x \cdot u, y \cdot v \in PT(C)$. Then we have the following immediate inference rules for forming conjunctions:

$$\text{(Pretype Conjunction I)} \quad \frac{a : x \rightharpoonup y \qquad a : u \rightharpoonup v}{a : x \cdot u \rightharpoonup y \cdot v}$$

$$\text{(Pretype Conjunction II)} \quad \frac{a : x \leftharpoondown y \qquad a : u \leftharpoondown v}{a : x \cdot u \leftharpoondown y \cdot v}$$

With the help of the $\rightharpoonup$ and $\leftharpoondown$ relations it is straightforward to characterize invariants:

Definition 3.11 A pretype $x \in PT(C)$ is an *invariant* of $a \in M$ if $a : x \rightharpoonup x$; it is a *co-invariant* of a if $a : x \leftharpoondown x$.

Corollary 3.12 A type $x \in PT(C)$ is an invariant of $a \in M$ iff $a \cdot x = x \cdot a \cdot x$ iff $x : a \cdot x \rightleftharpoons a$. Similarly, x is a co-invariant of a iff $x \cdot a = x \cdot a \cdot x$ iff $x : a \rightleftharpoons x \cdot a$.

Proof: immediate from Lemma 3.10. ∎

3.4 Subtypes

The above inferences have shown us the need for sets of pretypes that are closed under composition, i.e., under conjunction.

Definition 3.13 We call $T \subseteq PT(C)$ a *set of subtypes* if $1 \in T$ and T is closed under composition.

Hence a set of subtypes is a submonoid of $(M, \cdot, 1)$, and $\{1\}$ is the smallest set of subtypes. Moreover, by Corollary 3.9, all elements of a set of subtypes commute.

How can we distinguish sets of subtypes? A sufficient criterion is the following, which in the case of the function algebra was given in [17] at the point level.

Definition 3.14 A pretype $x \in PT(C)$ is called *downward closed* if for all $a \in M$ we have $a \leq x \Rightarrow a \preceq_r x$.

In the function algebra this means that for every fixpoint of x also all smaller elements are fixpoints of x. Note that 1 is downward closed. Since the characteristic set of a pretype is considered to be its range, we do not bother to define a dual notion involving domains.

Lemma 3.15 Let x, y be downward closed pretypes. Then $z \stackrel{\text{def}}{=} x \cdot y$ is a downward closed pretype as well.

Proof: By (8) we have $z \leq x, y$ and hence also $z \leq 1$. Moreover, by downward closedness of x, y we have $z \preceq_r x, y$. Now

$$
\begin{array}{ll}
 & z \cdot z \\
= & \{\!\!\{ \text{ definition } \}\!\!\} \\
 & x \cdot y \cdot z \\
= & \{\!\!\{ \text{ since } z \preceq_r y \}\!\!\} \\
 & x \cdot z \\
= & \{\!\!\{ \text{ definition } \}\!\!\}
\end{array}
$$

$x \cdot x \cdot y$

$=$ ⦃ idempotence of x ⦄

$x \cdot y$

$=$ ⦃ definition ⦄

z .

So z is a pretype. Assume now $a \leq z$. Then also $a \leq x, y$ and

$z \cdot a$

$=$ ⦃ definition ⦄

$x \cdot y \cdot a$

$=$ ⦃ y downward closed ⦄

$x \cdot a$

$=$ ⦃ x downward closed ⦄

a .

∎

Corollary 3.16 The set of all downward closed pretypes is a set of subtypes.

From now on we assume a fixed set $T \subseteq PT(C)$ of subtypes.

3.5 Algebras of Functionals

We now lift our notions to the level of functions over composition algebras. Since for the functional and relational case these functions are of second order, we speak of algebras of functionals.

Definition 3.17 Given a composition algebra $C = (M, \cdot, 1, \leq, 0)$, we define the *algebra of functionals* over C as $F(C) = ([M \to M], \circ, id_M, \leq, \Omega)$ where $[M \to M]$ is the set of all monotonic functions from M into itself, $\circ$ is the usual function composition, id_M is the identity on M, $\leq$ is the pointwise function ordering and Ω is the constant 0-valued function.

It is clear that this defines a left-continuous composition algebra. We now prepare the lifting of subtypes to the level of functionals:

Definition 3.18 For $a \in M$ we define $F_a \in [M \to M]$ as the restriction functional $F_a(b) \stackrel{\text{def}}{=} b \cdot a$.

In the discussion about assertions in Section 2 only restriction was used. For that reason we only deal with that operation here; a dual treatment for co-restriction is obvious.

Lemma 3.19 For all $a, b \in M$ we have $F_{a \cdot b} = F_b \circ F_a$ and $F_a \leq F_b \Leftrightarrow a \leq b$. If C is left-continuous then all F_a are continuous. If $x \in PT(C)$ then $F_x \in PT(F(C))$ and F_x is strict. Finally, for a set T of subtypes over C the set $F(T) \stackrel{\text{def}}{=} \{F_x : x \in T\}$ is a set of subtypes over $F(C)$.

Proof: The first two claims are obvious. For $x \in PT(C)$ the first claim implies $F_x \in PT(F(C))$, whereas strictness follows from (6). The last claim follows from the first one and the definitions. ■

Invariants are now also lifted to the functional level by generalising Definition 2.3:

Definition 3.20 We call a subtype $x \in T$ an *invariant* of $G \in [M \rightarrow M]$ if

$$F_x \circ G = F_x \circ G \circ F_x .$$

A more concise statement of the invariance property is given in

Corollary 3.21 x is an invariant of G iff $F_x : G \rightleftharpoons F_x \circ G$ iff $G : F_x \leftharpoondown F_x$, i.e., if F_x is a co-invariant of G.

Proof: Immediate from Corollary 3.12. ■

This odd reversal seems unavoidable.

3.6 Invariants and Recursion

The semantics of a recursive definition over a composition algebra is, as usual, given as the least fixpoint of the associated functional in the respective algebra of functionals. We investigate in this section how invariants influence least fixpoints.

Frequently, one has to strengthen an invariant to achieve a certain simplification. This is done by showing that the conjunction of the original invariant with another assertion is an invariant, too. One then passes to a recursion with the stronger invariant and performs the simplification exploiting it. However, after that, one may want to get rid of the assertion again, in particular, when assertions are implemented as actually executed checks. Over a left-continuous composition algebra, a tool for this is the following inference rule which is an instance of the fusion rule of Section 3.1:

$$(\text{Invariant}) \ \frac{x \in PT(C) \text{ invariant for } G}{F_x(\mu G) = \mu(F_x \circ G)}$$

The first premise of the fusion rule is satisfied by Corollary 3.21 and strictness and continuity of F_x (see Lemma 3.19). Let us interpret this rule in the algebra of functionals over monotonic functions: on the left hand side of the conclusion, x is only checked for the argument of the first call to μG, whereas on the right it is also checked for all recursive calls. So passing from left to right means introduction of an invariant, the reverse direction its removal.

Conjunctions of assertions can be introduced by iterated application of this rule using the property

$$F_x \circ F_y = F_{x \sqcap_T y}$$

for subtypes x, y, which follows from Lemma 3.8 and Lemma 3.19. This allows then weakening and strengthening assertions.

As an application we show a property of Kleene algebras. To this end we recall that for $a, b \in M$ we have

$$a \cdot b^* = \mu G_{ab} , \tag{10}$$

where $G_{ab}(x) \stackrel{\text{def}}{=} a + x \cdot b$. In particular, $b^* \stackrel{\text{def}}{=} \mu G_{1b}$.

Lemma 3.22 Let C be a Kleene algebra and assume that $a \in M$ is an invariant of $b \in M$. Then

$$b^* \cdot a = a \cdot (b \cdot a)^* .$$

Proof: We first show that a is an invariant of G_{1b}:

$$(F_a \circ G_{1b} \circ F_a)(x)$$
$$= \quad \{\!\!\{ \text{ definitions } \}\!\!\}$$
$$(1 + x \cdot a \cdot b) \cdot a$$
$$= \quad \{\!\!\{ \cdot \text{ universally disjunctive, 1 neutral } \}\!\!\}$$
$$a + x \cdot a \cdot b \cdot a$$
$$= \quad \{\!\!\{ a \text{ invariant of } b \}\!\!\}$$
$$a + x \cdot b \cdot a$$
$$= \quad \{\!\!\{ \cdot \text{ universally disjunctive, 1 neutral } \}\!\!\}$$
$$(1 + x \cdot b) \cdot a$$
$$= \quad \{\!\!\{ \text{ definitions } \}\!\!\}$$
$$(F_a \circ G_{1b})(x) .$$

The last three steps also show $F_a \circ G_{1b} = G_{a,(b \cdot a)}$. Now the claim is immediate from the previous lemma and (10). ■

Corollary 3.23 Under the above assumptions,

$$b^* \cdot a = a \cdot b^* \cdot a .$$

Proof:

$$b^* \cdot a$$
$$= \quad \{\!\!\{ \text{ above lemma } \}\!\!\}$$
$$a \cdot (b \cdot a)^*$$
$$= \quad \{\!\!\{ a \text{ idempotent } \}\!\!\}$$

$$
\begin{array}{ll}
 & a \cdot a \cdot (b \cdot a)^* \\
= & \{\!\{ \text{ above lemma } \}\!\} \\
 & a \cdot b^* \cdot a \,.
\end{array}
$$

■

In the path Kleene algebra this means for a set R of edges and a set S of vertices, viewed as singleton paths, the following: If all edges of R that start in S also end in S, then all paths using only edges in R that start in S will pass only through vertices in S and end in S. This is used in [18] for the simplification of a general scheme for layer-oriented graph traversal.

4 Assertions Revisited

We now apply the general results of the previous section to the particular case of subtypes generated by assertions. In particular, we are interested in invariants. For a very simple case we can give a sufficient criterion:

Lemma 4.1 Suppose G is a functional in f such that, for fixed binary g and unary h, k, we have $G(f) = g \circ [k, f \circ h]$, where $[f_1, f_2] \stackrel{\text{def}}{=} \lambda x.\,(f_1(x), f_2(x))$. Moreover, let C be a unary predicate. If id_C is an invariant of h, then C is an invariant of G.

Proof: We calculate

$$
\begin{array}{ll}
 & (F_C \circ G)(f) \\
= & \{\!\{ \text{ definition } F_C \text{ and } G \text{ } \}\!\} \\
 & g \circ [k, f \circ h] \circ id_C \\
= & \{\!\{ \text{ distributivity } \}\!\} \\
 & g \circ [k \circ id_C, f \circ h \circ id_C] \\
= & \{\!\{ \; id_C \text{ invariant of } h \; \}\!\} \\
 & g \circ [k \circ id_C, f \circ id_C \circ h \circ id_C] \\
= & \{\!\{ \text{ distributivity } \}\!\} \\
 & g \circ [k, f \circ id_C \circ h] \circ id_C \\
= & \{\!\{ \text{ definition } F_C \text{ and } G \text{ } \}\!\} \\
 & (F_C \circ G \circ F_C)(f)(x) \,.
\end{array}
$$

■

Example 4.2 Consider the recursion

$$\mathsf{letrec}\ f(x) = \mathsf{if}\ x \leq 1\ \mathsf{then}\ 0\ \mathsf{else}\ f(x-2) \,.$$

The associated functional is

$$G \stackrel{\text{def}}{=} \lambda f.\lambda x.\,\text{if } x \le 1 \text{ then } 0 \text{ else } f(x-2)\;.$$

Taking $g(x,y) \stackrel{\text{def}}{=} \text{if } x \le 1 \text{ then } 0 \text{ else } y$ and $h(x) \stackrel{\text{def}}{=} x - 2$ we see that by the previous lemma both the predicates *even* and *odd* are invariants of G. ■

This criterion generalizes in a straightforward way to an $n+1$-ary function g and n functions h_i:

Lemma 4.3 Suppose $G(f) = g \circ [k, f \circ h_1, \ldots, f \circ h_n]$ such that id_C is an invariant of h_i for all $i \in \{1, \ldots, n\}$. Then C also is an invariant of G.

Example 4.4 Consider the recursion

$$\text{letrec } oddtree(i) = \langle oddtree(i+2), i, oddtree(i-2)\rangle\;,$$

where $\langle .,.,.\rangle$ is a constructor for node-labelled binary trees. It builds an infinite tree. The predicates *even* and *odd* also are invariants of the associated functional *ODD* with

$$ODD(f)(i) = \langle f(i+2), i, f(i-2)\rangle\;.$$

■

For examples where proving an invariant also involves rule (4) see [12].

Example 4.5 As an example of the use of the invariant rule we choose the improvement of a shortest path algorithm (cf. [13] for the derivation). We start from the version

$$\text{let } shortestpath = \lambda x, y.\, sp(\{x\}, \emptyset, y) \text{ where}$$

$$\begin{array}{l}
\text{letrec } sp = \lambda S, V, y : S \cap V = \emptyset \wedge y \notin V.\\
\quad \text{if } y \in S\\
\quad \text{then } 0\\
\quad \text{else if } succ(S) \cup succ(V) \subseteq S \cup V\\
\qquad\quad \text{then } \infty\\
\qquad\quad \text{else } 1 + sp((succ(S) \cup succ(V)) \backslash (S \cup V)), S \cup V, y)\;.
\end{array}$$

To calculate the length of a shortest path from vertex x to vertex y, where each edge has cost 1, we embed the problem into the function *sp*. It computes the length of a shortest path from vertex set S, the set of vertices to be inspected in the current round, to y along paths not leading through V, the set of vertices already visited. The function *succ* calculates the set of immediate successors of a set of vertices. The algorithm is very inefficient, since it always forms the union $succ(S) \cup succ(V)$. However, a simple analysis shows that the invariant of *sp* can be strengthened by the conjunct $succ(V) \subseteq S \cup V$. Using this, the body can be simplified considerably and we obtain let *shortestpath* $= \lambda x, y.\, sp(\{x\}, \emptyset, y)$ where

letrec $sp = \lambda S, V, y : S \cap V = \emptyset \wedge y \notin V \wedge succ(V) \subseteq S \cup V\,.$
    if $y \in S$
    then 0
    else if $succ(S) \subseteq S \cup V$
        then ∞
        else $1 + ssp(succ(S) \backslash (S \cup V),\, S \cup V,\, y)$.

Afterwards, for passing to a loop we may remove the invariant again. ■

Next we give an example for the case of non-strict semantics.

Example 4.6 Consider a function for constructing the infinite list of squares of integers starting from an integer x:

$$squares = \lambda x.\, [x^2] +\!\!+\; squares(x+1)\ .$$

In this recursion the square of each number is computed separately. We can speed up generation of the numbers in the list using the technique of finite differencing (see e.g. [16]). To this end we define a version of *squares* with an additional parameter and an assertion:

$$esquares = \lambda x, s : s = x^2\, .\; squares(x)\ .$$

A simple unfold/fold transformation shows that we have the recursion

$$esquares = \lambda x, s : s = x^2\, .\; [s] +\!\!+\; esquares(x+1, s+2*x+1)\ ,$$

in which the squares are now computed incrementally. ■

5 Enforcing Relations and Recurrence

Invariants are unary predicates that treat the parameters of recursions in isolation. In some circumstances, however, one wants to relate the parameters of the recursive calls with that of the parent call using a binary relation $R \subseteq T \times T$. For R and $x \in T$ we denote by Rx the residual predicate which holds for $y \in T$ iff $(y, x) \in R$.

Definition 5.1 The *restricting functional* associated with R is

$$F_R \stackrel{\text{def}}{=} \lambda f, x.\, f \circ id_{Rx}\ .$$

We say that a functional $H : [T \to T] \to [T \to T]$ *enforces* R if

$$\forall\, f : \forall\, x : H(f)(x) \;=\; H(F_R(f, x), x)\ .$$

So H enforces R if the arguments of all recursive calls are in relation R with the argument of the parent call.

Example 5.2 Consider the recursion

$$\mathsf{letrec}\ heaptree(n) = \langle heaptree(2*n), n, heaptree(2*n+1)\rangle\ .$$

It builds an infinite tree where the nodes carry the indices of a heap-organized infinite array, starting with index n. The associated functional $HEAP$ with

$$HEAP(f,n) = \langle f(2*n), n, f(2*n+1)\rangle$$

enforces the relation R given by

$$yRx \overset{\text{def}}{\Leftrightarrow} x = y\ div\ 2\ .$$

■

As another application, we consider termination proofs using noetherian relations.

Definition 5.3 A relation $\ll \subseteq T \times T$ is *noetherian* if there is no infinite sequence $(x_i)_{i \in \mathbb{N}}$ with $x_{i+1} \ll x_i$ for all $i \in \mathbb{N}$. We call a functional H *$\ll$-recurrent* if H enforces $\ll$.

For a noetherian relation $\ll$ and a unary predicate P on T one has the following proof principle:

$$\text{(Noetherian Induction)}\quad \frac{\forall\, u \in T : (\forall\, w \in T : w \ll u \Rightarrow P(w)) \Rightarrow P(u)}{\forall\, u \in T : P(u)}$$

The restricting functional associated with a noetherian relation $\ll$ is

$$N_{\ll} \overset{\text{def}}{\Leftrightarrow} \lambda f, x.\ f \circ id_{\ll x}\ ,$$

and so H is $\ll$-recurrent iff

$$\forall\, f : \forall\, x : H(f)(x) = H(N_{\ll}(f,x))(x)\ .$$

A related notion is discussed in [7, 8].

Example 5.4 Consider Ackermann's function

$$\begin{aligned}\mathsf{letrec}\ ack(x,y) = \mathsf{if}\ x = 0\ \mathsf{then}\ y+1 \\ \mathsf{else}\ \mathsf{if}\ y = 0\ \mathsf{then}\ ack(x-1,1) \\ \mathsf{else}\ ack(x-1, ack(x, y-1))\end{aligned}$$

of type $\mathbb{N} \times \mathbb{N} \rightarrow \mathbb{N}$. The corresponding functional is ACK with

$$\begin{aligned} ACK(f)(x,y) = \text{if } x = 0 \text{ then } & y+1 \\ & \text{else if } y = 0 \text{ then } f(x-1,1) \\ & \qquad \text{else } f(x-1, f(x,y-1)) \ . \end{aligned}$$

We want to show that ACK is recurrent. We choose for $\ll$ the lexicographic order on $\mathbb{N} \times \mathbb{N}$, viz.

$$(x_1, y_1) \ll (x_2, y_2) \overset{\text{def}}{\Leftrightarrow} x_1 < x_2 \vee (x_1 = x_2 \wedge y_1 < y_2) \ .$$

We calculate, assuming a strict semantics,

$$\begin{aligned}
& ACK(N_{\ll}(f,(x,y)))(x,y) \\
= \quad & \{\!\![\text{ definition of } ACK \text{ and } N_{\ll} \text{ and (5) }]\!\!\} \\
& \text{if } x = 0 \\
& \quad \text{then } y+1 \\
& \quad \text{else if } y = 0 \\
& \qquad \text{then } (x-1,1) \ll (x,y) \triangleright f(x-1,1) \\
& \qquad \text{else } (x-1, f(x,y-1)) \ll (x,y) \triangleright \\
& \qquad\quad f(x-1, \ (x,y-1) \ll (x,y) \triangleright f(x,y-1)) \\
= \quad & \{\!\![\text{ definition of } \ll \ ]\!\!\} \\
& \text{if } x = 0 \text{ then } y+1 \\
& \quad \text{else if } y = 0 \text{ then } TRUE \triangleright f(x-1,1) \\
& \qquad \text{else } TRUE \triangleright f(x-1, \ TRUE \triangleright f(x,y-1)) \\
= \quad & \{\!\![\text{ by (1) }]\!\!\} \\
& \text{if } x = 0 \text{ then } y+1 \\
& \quad \text{else if } y = 0 \text{ then } f(x-1,1) \\
& \qquad \text{else } f(x-1, f(x,y-1)) \\
= \quad & \{\!\![\text{ definition of } ACK \ ]\!\!\} \\
& ACK(f)(x,y) \ .
\end{aligned}$$

■

6 Properties of Recurrent Recursions

We now want to analyse how certain relations propagate through formation of least fixpoints. To this end, consider a transitive binary relation $\leadsto$ (e.g., refinement of one element by another) on the target domain of the function space under consideration, such that $\perp \leadsto \perp$.

Definition 6.1 We lift $\leadsto$ pointwise to functions by setting

$$f \leadsto g \overset{\text{def}}{\Leftrightarrow} \forall\, x : f(x) \leadsto g(x) \ .$$

A functional F is $\leadsto$-*monotonic* if

$$\forall\, f, g : f \leadsto g \Rightarrow F(f) \leadsto F(g)\ .$$

Now we can show

Theorem 6.2 Let H be a functional that is $\leadsto$-monotonic and $\ll$-recurrent. If, for some function f we have $f \leadsto H(f)$ and g is a fixpoint of H then also $f \leadsto g$.

Proof: We perform a noetherian induction on the predicate

$$P(x) \overset{\text{def}}{\Leftrightarrow} f(x) \leadsto g(x)\ .$$

Suppose $\forall\, y : y \ll x \Rightarrow P(y)$. By $\bot \leadsto \bot$ this implies

$$N_{\ll}(f, x) \leadsto N_{\ll}(g, x)\ . \tag{11}$$

Now,

$$
\begin{array}{ll}
 & f(x) \\
\leadsto & \quad \{\!\!\{ \text{ by the assumption } \}\!\!\} \\
 & H(f)(x) \\
= & \quad \{\!\!\{ H \ll\text{-recurrent } \}\!\!\} \\
 & H(N_{\ll}(f, x))(x) \\
\leadsto & \quad \{\!\!\{ \text{ by (11) and } \leadsto\text{-monotonicity of } H \}\!\!\} \\
 & H(N_{\ll}(g, x))(x) \\
= & \quad \{\!\!\{ H \ll\text{-recurrent } \}\!\!\} \\
 & H(g)(x) \\
= & \quad \{\!\!\{ g \text{ a fixpoint of } H \}\!\!\} \\
 & g(x)\ .
\end{array}
$$

■

This has been used in [4] to prove a declarative analogue of the usual inference rule for while-loops.

Corollary 6.3 (Unique Fixpoint) Let H be a $\ll$-recurrent functional. Then H has at most one fixpoint.

Proof: Take $=$ for $\leadsto$ in the above theorem. ■

Next, we want to talk about preservation of certain properties. For instance, a function is total at an argument x if it preserves definedness of argument x:

$$TOT(f, x) \overset{\text{def}}{\Leftrightarrow} (DEF(x) \Rightarrow DEF(f(x)))\ .$$

In a deterministic setting the definedness predicate *DEF* may be defined as

$$DEF(x) \overset{\text{def}}{\Leftrightarrow} x \neq \text{error} \;;$$

in a non-deterministic setting the precise definition depends on the particular model of choice (erratic, angelic, demonic). Based on *TOT* we define

$$TOTAL(f) \overset{\text{def}}{\Leftrightarrow} \forall\, x : TOT(f, x) \,.$$

Consider now, more generally a unary predicate P. The predicate

$$PRES_P(f, x) \overset{\text{def}}{\Leftrightarrow} (P(x) \Rightarrow P(f(x)))$$

then expresses that f preserves P at argument x. In particular, $TOT = PRES_{DEF}$.

Lemma 6.4 Let H be a functional and f a fixpoint of H. From

$$\forall\, g : \forall\, x : (\forall\, y : y \ll x \Rightarrow PRES_P(g, y)) \Rightarrow PRES_P(H(g), x) \tag{12}$$

we may infer

$$\forall\, x : PRES_P(f, x) \,.$$

Proof: Straightforward noetherian induction on the predicate $Q(x) \overset{\text{def}}{\Leftrightarrow} PRES_P(f, x)$ using the fixpoint property of f. ■

From this we obtain

Theorem 6.5 (Termination Criterion) Let $H : [T \to T] \to [T \to T]$ for flat domain T be $\ll$-recurrent and f be a fixpoint of H. Then from

$$\forall\, g : TOTAL(g) \Rightarrow TOTAL(H(g)) \tag{13}$$

we may infer

$$TOTAL(f) \,.$$

Proof: We show that (12) in the previous lemma is satisfied for $TOT = PRES_{DEF}$. Consider an arbitrary x and assume $\forall\, y : y \ll x \Rightarrow TOT(g, y)$. Choose $u = g(\bot)$ if $g(\bot) \neq \bot$ and arbitrary in $T \backslash \{\bot\}$ otherwise, and define h by

$$h(y) \overset{\text{def}}{=} \begin{cases} g(y) \text{ if } y \ll x \vee y = \bot \,, \\ u \quad\;\; \text{otherwise.} \end{cases}$$

By flatness of T then h is monotonic. Moreover, it satisfies $\forall\, y : TOT(h, y)$ and

$$N_{\ll}(h, x) = N_{\ll}(g, x) \,. \tag{14}$$

Now $\quad TRUE$

$\Leftrightarrow \quad \{\!\!\{ \text{ by (13) } \}\!\!\}$

$TOT(H(h), x)$

$\Leftrightarrow \quad \{\!\!\{ H \ll\text{-recurrent} \}\!\!\}$

$TOT(H(N_{\ll}(h, x)), x)$

$\Leftrightarrow \quad \{\!\!\{ \text{ by (14) } \}\!\!\}$

$TOT(H(N_{\ll}(g, x)), x)$

$\Leftrightarrow \quad \{\!\!\{ H \ll\text{-recurrent} \}\!\!\}$

$TOT(H(g), x)$.

■

As applications of Corollary 6.3 and Theorem 6.5 we obtain that the functional *ACK* associated with Ackermann's function has a unique fixpoint which is a total function.

7 Conclusion

We have given an algebraic characterization of invariants and a general rule for strengthening and weakening them in recursions. The rule covers arbitrary types of recursion, not just linear ones, and is independent of the particular semantic framework used. It applies not only at the level of recursive programs, but also to other recursively defined entities, such as closures in Kleene algebras, which may be seen as data.

The particular approach to assertions leads to a nice calculational framework. Both the proofs of the basic properties and the actual manipulation of assertions largely become elegant (in)equational reasoning. The technique covers also proofs by noetherian induction, in particular, termination proofs, for general recursions.

Acknowledgements. Section 3 was greatly stimulated by [17]. The idea of the functional $N_{\ll}$ and the definition of the notion of being recurrent are due to U. Berger who, in turn, attributes inspiration to H. Schwichtenberg. Helpful comments were provided by R. Backhouse, R. Bird, M. Russling and the anonymous referees.

References

1. R.C. Backhouse, P. de Bruin, G. Malcolm, T.S. Voermans, J. van der Woude: Relational catamorphisms. In: B. Möller (ed.): Constructing programs from specifications. Proc. IFIP TC2/WG 2.1 Working Conference on Constructing Programs from Specifications, Pacific Grove, CA, USA, 13–16 May 1991. Amsterdam: North-Holland 1991, 287–318

2. F.L. Bauer, B. Möller, H. Partsch, P. Pepper: Formal program construction by transformations — Computer-aided, Intuition-guided Programming. IEEE Transactions on Software Engineering **15**, 165–180 (1989)
3. F.L. Bauer, R. Berghammer, M. Broy, W. Dosch, F. Geiselbrechtinger, R. Gnatz, E. Hangel, W. Hesse, B. Krieg-Brückner, A. Laut, T. Matzner, B. Möller, F. Nickl, H. Partsch, P. Pepper, K. Samelson, M. Wirsing, H. Wössner: The Munich project CIP. Volume I: The wide spectrum language CIP-L. Lecture Notes in Computer Science **183**. Berlin: Springer 1985
4. R. Berghammer, H. Ehler, B. Möller: On the refinement of nondeterministic recursive routines by transformation. In: M. Broy, C.B. Jones (eds.): Programming concepts and methods. Amsterdam: North-Holland 1990, 53–71
5. R.S. Bird, O. de Moor: The algebra of programming. Prentice-Hall (to appear)
6. J.H. Conway: Regular algebra and finite machines. London: Chapman and Hall 1971
7. H. Doornbos, R. Backhouse: Induction and recursion on datatypes. In: B. Möller (ed.): Mathematics of Program Construction. Lecture Notes in Computer Science **947**. Berlin: Springer 1995, 242–281
8. H. Doornbos, R. Backhouse: Reductivity. Science of Computer Programming (to appear)
9. B. von Karger, C.A.R. Hoare: Sequential calculus. Information Processing Letters **53**, 123–130 (1995)
10. S.C. Kleene: Introduction to metamathematics. New York: van Nostrand 1952
11. Mathematics of Program Construction Group: Fixed-point calculus. Information Processing Letters **53**, 131-136 (1995)
12. B. Möller: Applicative assertions. In: J.L.A. van de Snepscheut (ed.): Mathematics of Program Construction. Lecture Notes in Computer Science **375**. Berlin: Springer 1989, 348–362
13. B. Möller, M. Russling: Shorter paths to graph algorithms. Science of Computer Programming **22**, 157–180 (1994)
14. B. Möller: Relations as a program development language. In: B. Möller (ed.): Constructing programs from specifications. Proc. IFIP TC2/WG 2.1 Working Conference on Constructing Programs from Specifications, Pacific Grove, CA, USA, 13–16 May 1991. Amsterdam: North-Holland 1991, 373–397
15. C.C. Morgan: Programming from Specifications. Prentice-Hall, 1990
16. H.A. Partsch: Specification and transformation of programs — A formal approach to software development. Berlin: Springer 1990
17. C. Runciman: Subtype constraints as first-class values. Talk given at 48th meeting of IFIP WG 2.1, Günzburg, Germany, October 1995. Paper in preparation
18. M. Russling: A general scheme for breadth-first graph traversal. In: B. Möller (ed.): Mathematics of Program Construction. Lecture Notes in Computer Science **947**. Berlin: Springer 1995, 380–398
19. D. Scott: Data types as lattices. SIAM J. Comp. **5**, 522–587 (1976)

Development Closed Critical Pairs

Vincent van Oostrom*

Information Processing Principles Research Group
3-1 Morinosato Wakamiya, Atsugi-Shi
Kanagawa Prefecture, 243-01, Japan

Abstract. The class of orthogonal rewriting systems (rewriting systems where rewrite steps cannot depend on one another) is the main class of not-necessarily-terminating rewriting systems for which confluence is known to hold. Huet and Toyama have shown that for left-linear first-order term rewriting systems (TRSs) the orthogonality restriction can be relaxed somewhat by allowing *critical pairs* (arising from maximally general ways of dependence between steps), but requiring them to be parallel closed. We extend these results by replacing the parallel closed condition by a *development closed* condition. This also permits to generalise them to higher-order term rewriting, yielding a confluence criterion for Klop's combinatory reduction systems (CRSs), Khasidashvili's expression reduction systems (ERSs), and Nipkow's higher-order pattern rewriting systems (PRSs).

1 Introduction

This paper is concerned with a method to prove confluence of rewriting systems. It's an extension of some confluence results in [CR36, Hue80, Toy88, Klo80, Kha92, Raa93, Tak, MN94, Oos94, ORb] and we refer the reader to these papers and to the handbook chapters [DJ, Klo] for motivation and for standard definitions as well. Here we will mainly be concerned with proving our result:

Left-linear development closed PRSs are confluent.

Let's explain the terminology used. A rewrite system for which the rewrite rules do not depend on one another is called *orthogonal*. Formalising this notion can be quite involved depending on the rewrite formalism it is applied to ([Hue80, Klo80, HL, GLM, MN94, Oos94]), but the intuition to be captured is always the same: an application of a rule replaces some substructure by another one, and in orthogonal systems we moreover have that if two distinct substructures can be replaced then these substructures are independent. Some (non)examples are:

1. The rules $\mathtt{F} \to \mathtt{G}$ and $\mathtt{G} \to \mathtt{H}$ are orthogonal (their left-hand sides $\mathtt{F}$ and $\mathtt{G}$ are independent) hence confluent.

* Partially supported by de Vrije Universiteit, Amsterdam, The Netherlands, and an HCM grant at the Technische Universität München, Germany. Current address: `http://www.cs.vu.nl/~oostrom`.

2. The rules $F \to G$ and $F \to H$ are not orthogonal, since they both depend on the symbol F, and for this reason they're said to be *ambiguous*. The system is not confluent since F can be rewritten to the normal forms G and H.
3. The rules ([Hue80])

$$\begin{aligned} EQ(x,x) &\to T \\ EQ(x,S(x)) &\to F \\ \infty &\to S(\infty) \end{aligned}$$

are not orthogonal, since application of the third rule to $EQ(\infty,\infty)$ destroys the possibility of applying the first rule. The system is not confluent since the term $EQ(\infty,\infty)$ can be rewritten to both T and F. The first and second rule are said to be *non-left-linear*, because of the presence of a repeated variable in its left-hand side.

A fundamental result in rewriting is that forbidding the kind of dependence in the latter two items suffices for orthogonality of rewriting systems:

Left-linear, non-ambiguous term rewriting systems are confluent.

Our aim is to show that for the class of higher-order Pattern Rewriting Systems (PRSs [MN94]), the non-ambiguity condition can be relaxed somewhat without jeopardising confluence. The condition we name ***development closed*** and it expresses that if two steps are dependent, then from the result of performing the inner step, the result of performing the outer step can be reached by one ***development step.***

Remark. The notion of inner and outer are the usual ones obtained from viewing terms as trees. Outer means closer to the root. Observe that steps in disjoint subtrees cannot depend on each other.

1. Adding the rules $G \to H$ and $H \to G$ to the rewriting system in the second item above makes it development closed. The results G and H of two dependent steps can be rewritten to each other in one step, which is a special case of a development step. (Note that in this case the dependent steps are both inner and outer.)
2. The rewrite rules for 'parallel or':

$$\begin{aligned} por(x,T) &\to T \\ por(T,x) &\to T \end{aligned}$$

are development closed since steps only depend on each other in the case of $por(T,T)$. The result then is T for either rewrite rule (so in fact the system is ***weakly orthogonal*** [ORb]) and since the empty step is a special case of a development step, the system is confluent.

Remark. Throughout the text there are many references to [Oos94]. This is only meant for easy reference. Many of the results can be found at many other places.

2 Rewriting

We fix the no(ta)tions for the rewriting systems we're interested in.

Definition 1. An *abstract rewriting system (ARS)* $\rightarrow$ is a binary relation on some set $(a, b \in)A$. The denotation of the 'repeated' notation $\twoheadrightarrow$ is the transitive-reflexive closure of the denotation of $\rightarrow$. Similarly, the denotation of the 'inverse' notation $\leftarrow$ is the inverse of the denotation of $\rightarrow$ and the denotation of the 'union with the inverse' notation $\leftrightarrow$ is the symmetric closure of the denotation of $\rightarrow$. We use infix notation for ARSs. If $a \rightarrow b$, then we say that *a ($\rightarrow$-)rewrites to b*, and the structure $\langle a, \rightarrow, b\rangle$ is called a *$\rightarrow$-step from a to b*. A *rewrite sequence* is a (finite or infinite) sequence of rewrite steps, such that for successive steps the object to which the former rewrites is the same as the object from which the latter rewrites. An ARS $\rightarrow$ has the *diamond property*, if $\leftarrow ; \rightarrow \subseteq \rightarrow ; \leftarrow$. An ARS $\rightarrow$ is *strongly confluent* ([Hue80]), if $\leftarrow ; \rightarrow \subseteq \rightarrow ; \twoheadleftarrow$. An ARS $\rightarrow$ is *confluent*, if $\twoheadrightarrow$ has the diamond property. An object is a *($\rightarrow$)-normal form* if no rewrite steps are possible from it. An ARS is *terminating* if no infinite rewrite sequences are possible.

It is well-known that ARSs having the diamond property (or which are strongly confluent) are confluent ([Hue80]).

Definition 2. Let $\rightarrow$ be an ARS. An *($\rightarrow$-)span* is a pair $(\langle a, \rightarrow, b\rangle, \langle a, \rightarrow, c\rangle)$, which we will usually write as $b \leftarrow a \rightarrow c$.

For cultural reasons we employ the concrete class of higher-order pattern rewriting systems (PRSs [MN94])[2] as a vehicle to present our ideas. The proof-method will be seen to rely on two essential ingredients: trees and the notion of substructure. In PRSs the substructures employed are called *patterns* and the trees arise by viewing the objects of PRSs (λ-terms) as trees in the usual way. Our proofs could be stated in the general framework of HORSs (see [Raa96]) without difficulty, but this would not make them clearer. Stating them for PRSs is a hassle already.

Definition 3. 1. We first define the objects of a higher-order term rewriting system. *Simple types* are defined by the grammar:

$$\sigma ::= o \mid \sigma \rightarrow \sigma$$

Preterms are objects s such that $s\colon\sigma$ for simple type σ can be inferred from
(var) $x^\sigma\colon\sigma$ for termvariables x,
(app) $s\colon\sigma \rightarrow \tau,\ t\colon\sigma \Longrightarrow s(t)\colon\tau$.
(abs) $s\colon\tau \Longrightarrow x^\sigma.s^3\colon\sigma \rightarrow \tau$.

---

[2] Actually we employ a slight variation on PRSs as presented in [ORb] in which all rules are closed, which we find technically and conceptually more convenient.

[3] We omit the usual λ in abstractions.

Higher-order *terms* are obtained from the preterms by quotienting by the theory consisting of α, β and η (that is, $\lambda\eta$ in [Bar84]). To make terms concrete we use their $\beta\overline{\eta}$-normal forms as representatives (unique up to α-conversion) of the equivalence classes. Here, the rewrite relations β and $\overline{\eta}$ are generated by the rules:

$$(x.s)(t) \rightarrow_\beta s[x := t]$$
$$s \rightarrow_{\overline{\eta}} x.s(x)$$

where $\overline{\eta}$ is not allowed to create β-redexes and x does not occur in s.

2. An alphabet of (simply typed) variables $\mathcal{A}$ is distinguished from the set of all variables, and elements of $\mathcal{A}$ will be called *function symbols*. They will be used as constants, i.e. bound externally, in higher-order rewriting. To stress the functional nature of function symbols we use functional notation $\mathsf{F}(s_1, \ldots, s_m)$ instead of applicative notation $\mathsf{F}(s_1)\ldots(s_m)$ in case $m > 1$ (cf. [MN94]).
3. A *pattern rewrite rule* R is a pair $l \rightarrow r$ of closed (note that function symbols are considered bound already, as per the previous item) terms of the same simple type σ, where the *left-hand side* l is a linear pattern. Here a *linear pattern* is a term of the form $\mathbf{x}.s$,[4] such that
   (a) $s{:}\,o$. To understand this it is convenient to think of o as the set of terms.
   (b) s is of the form $\mathsf{F}(s_1, \ldots, s_m)$, F is called the *head-symbol* of the term,
   (c) each x_k among $\mathbf{x}$ occurs exactly once in s and has only ($\overline{\eta}$-normal forms of) pairwise distinct variables not among $\mathbf{x}$ as arguments.

   We'll just say pattern instead of linear pattern and rewrite rule instead of pattern rewrite rule. A higher-order *pattern rewriting systems* (PRS) is a pair $(\mathcal{A},\mathcal{R})$ consisting of an alphabet $\mathcal{A}$ and a set $\mathcal{R}$ of rewrite rules.
4. Let $s =^{\mathrm{def}} C\boxed{l_1,\ldots,l_m}$, $t =^{\mathrm{def}} C\boxed{r_1,\ldots,r_m}$ be preterms, where $R_1 =^{\mathrm{def}} l_1 \rightarrow r_1$, $\ldots$, $R_m =^{\mathrm{def}} l_m \rightarrow r_m$ are rewrite rules, and C is an m-ary *precontext*, i.e. a preterm containing variables $\Box_1, \ldots, \Box_m$. Then we say that t can be obtained from s by *contracting* the *(complete) development redex* $C\boxed{R_1,\ldots,R_m}$.[5] This will be denoted by $s \twoheadrightarrow\!\!\!\!\circ\;_{C\boxed{R_1,\ldots,R_m}} t$. We use u, v, w to denote redexes, and idenify them with their induced steps whenever convenient. This is extended to (the unique representatives of) terms by defining a *development step* $s \twoheadrightarrow\!\!\!\!\circ\; t$ if there exist preterms s' and t' in the same $\beta\overline{\eta}$-equivalence classes as s and t, respectively, such that $s' \twoheadrightarrow\!\!\!\!\circ\; t'$. The ARS associated to the PRS $\mathcal{H}$ is the relation on terms obtained by requiring the precontext employed in a development step to be a unary *context*, i.e. a term containing exactly one occurrence of $\Box$. The so-obtained relation $\rightarrow_{\mathcal{H}}$ (or simply $\rightarrow$) is called the *rewrite step* relation.

Note that we've only defined the linear PRSs, but that suffices for our purposes. For thorough investigations carried out using (also non-linear) PRSs we refer the reader to [MN94, Pre95].

[4] We employ boldface to denote sequences, i.e. $\mathbf{x}$ is a sequence of variables.

[5] This essentially amounts to replacing some derivation subtrees of a simply typed lambda term (in the usual sense) by other ones.

The development rewrite relation defined above is somewhat overly general. Without loss of generality the precontexts can be restricted to contexts, since the precontext can be $\beta\overline{\eta}$-normalised, preserving that there's a development step, i.e. if $C\boxed{\mathbf{l}} \multimap C\boxed{\mathbf{r}}$, then $D\boxed{\mathbf{l}} \multimap D\boxed{\mathbf{r}}$ where D is the $\beta\overline{\eta}$-normal form of C (cf. [Oos94, p. 68]). Furthermore, every step between terms $s \leftrightarrow^*_{\beta\overline{\eta}} s' \multimap t' \leftrightarrow^*_{\beta\overline{\eta}} t$ may be assumed of the form $s \twoheadleftarrow_\beta s' \multimap t' \twoheadrightarrow_\beta t$, by confluence of β and the fact that β-reduction preserves $\overline{\eta}$-normal forms (cf. [Oos94, Prop. 3.2.10]). This justifies restricting attention to β-reduction in the following.

Definition 4. 1. If $C\boxed{l} \twoheadrightarrow_\beta s$ for some context C, pattern l and term s, then we say that l *is a pattern (at C) in s*. A set $l_1,\ldots,l_m$ of patterns (at $C_1,\ldots,C_m$) in s is said to be *independent (at C)*, if C is an m-ary context such that $C\boxed{l_1,\ldots,l_m} \twoheadrightarrow_\beta s$, and such that the head-symbol of l_k descends to the same symbol in s in both $C\boxed{l_1,\ldots,l_m} \twoheadrightarrow_\beta s$ and $C_k\boxed{l_k} \twoheadrightarrow_\beta s$. Here, the descendant relation on function symbols is the usual one for β-reduction ([Bar84, Hue93] or [Oos94, Def. 3.2.12]). It's important to note that due to the linearity of patterns, each function symbol always has exactly one descendant along these β-reductions to s. Independence of sets of redexes is defined via their left-hand sides. We use $\mathcal{U}$, $\mathcal{V}$, $\mathcal{W}$ to range over sets of independent redexes and identify them with their induced development steps whenever convenient.

2. The descendant relation of a development step $s \twoheadleftarrow_\beta C\boxed{l_1,\ldots,l_m} \multimap_{C\boxed{R_1,\ldots,R_m}} C\boxed{r_1,\ldots,r_m} \twoheadrightarrow_\beta t$ of a set of independent redexes is the relation composition of the descendant relations of its three components, the β-expansion, the replacement, and the β-reduction. ([Oos94, Def. 3.1.25]). In the replacement step $C\boxed{l_1,\ldots,l_m} \multimap_{C\boxed{R_1,\ldots,R_m}} C\boxed{r_1,\ldots,r_m}$, function symbols in the context C descend to themselves. Function symbols in the left-(right-)hand sides are said to be *destroyed (created)*.
3. Having defined descendants of function symbols along steps, descendants of patterns can be defined directly via their head-symbol and subsequently descendants of redexes (so called *residuals*) via their left-hand sides ([Oos94, sec. 3.1.1]). A pattern must be independent of the contracted redex to have a descendant. The set of residuals of a set of (implicitly independent) redexes $\mathcal{V}$ *after* a development step contracting $\mathcal{U}$ is written as $\mathcal{V}/\mathcal{U}$.

In the following sections we recapitulate some standard general abstract nonsense about rewriting (in)dependent sets of redexes, paving the technical way for the proof of our main corollaries.

3 Independence

We list the main ingredients of the confluence by developments method and do some shopping in literature to get the results needed for the class of PRSs.

Theorem 5 Prism. *1. Let* $t \;{}_{\mathcal{U}}\!\!\multimap\!\!\!\leftarrow s \multimap_{\mathcal{V}} r$ *be a development span contracting the sets of independent redexes* $\mathcal{U} \subseteq \mathcal{V}$. *Then* $t \multimap_{\mathcal{V}/\mathcal{U}} r$ *and the descendant relations induced by* $\multimap_{\mathcal{U}} ; \multimap_{\mathcal{V}/\mathcal{U}}$ *and* $\multimap_{\mathcal{V}}$ *are the same.*

2. If $\mathcal{U} \cup \mathcal{V}$ is independent, then $\mathcal{V}/\mathcal{U}$ is independent again.
3. Every development step can be serialised, *i.e. if $s \twoheadrightarrow\!\!\!\!\circ\; t$, then there exists a rewrite sequence $s \twoheadrightarrow t$ consisting of rewrite steps, inducing the same descendant relation.*

Proof. 1. This is the Prism Theorem as stated in [Hue93] for the untyped lambda calculus with beta reduction. We will illustrate our proof for PRSs in the case of a set of two rules $l \to r$ and $g \to d$.

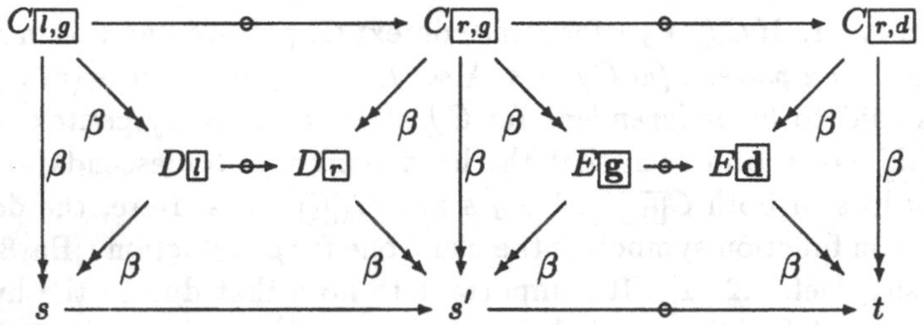

The diagram represents contraction of two independent redexes (l and g) in the term s. The top-line going from $C\boxed{l,g}$ to $C\boxed{r,d}$ corresponds to contracting both in one step, while the middle lines from $D\boxed{l}$ to $D\boxed{r}$ and from $E\boxed{\mathbf{g}}$ to $E\boxed{\mathbf{d}}$ correspond to first contracting l and then the residuals of g. The left-part of the diagram shows how the development step contracting $D\boxed{l}$ can be lifted into a context (C) where also the g-redex is made explicit (we called this the Envelop Lemma [Oos94, Lem. 3.1.42]). Now from this lifted step, it is obvious what should be done next: just contract the residual g in $C\boxed{r,g}$. The problem then is that $C\boxed{r,\ }$ is not a context, only a precontext, but to get the development step we want, we only need to reduce the precontext to its β-normal form E, possibly erasing or duplicating g on the way, and we are done (this we called the Develop Lemma [Oos94, Lem. 3.1.43]). Of course, one needs to check that the descendant relations behave well, but that's a matter of routine and can be found in [Oos94].

2. This is intuitively obvious, since the union being independent, the contraction of a subset may only lead to erasure or duplication of the others. Technically, we've already used this in the proof of the previous item (consider what happened to g in the right-hand side of the diagram above) and is a consequence of the Develop Lemma in [Oos94, Lem. 3.1.43]. Cf. also Section 6.2 in [Hue93], where his compatibility corresponds to our independence.
3. This follows from the Prism Theorem if we can find a particular strategy which decreases the size of the set of independent redexes in every step. This is easy: always contract an innermost redex. This cannot lead to duplication of other redexes. ⊙

From the first two items of the theorem it follows immediately that developments of mutually independent sets of redexes satisfy the diamond property. From serialisation it moreover follows that confluence of $\twoheadrightarrow\!\!\!\!\circ$ is the same as confluence of $\to$ (cf. [Hue80]).

Note that like [Hue93], we didn't employ the ***finiteness of developments theorem*** (FD, i.e. that any serialisation of a development step is finite). How the techniques employed above can be slightly strengthened to yield FD as a bonus can be found in [Oos94], but since we don't need it here we omit it. Moreover, taking development steps for the multi-step derivations in [HL] there's no problem in setting up the theory of ***permutation equivalence***. Again, since we don't need that part of the general abstract theory of independence here, we don't develop it.

In general steps need of course not be independent, but if steps operate on disjoint subterms, they are.

Definition 6. Let l, g be patterns at C, D in s. The patterns are related to each other via the positions (in the usual tree representation, see [Oos94, Def. 3.2.6] for a formal definition of this) of the descendants of their head-symbols in s. If l is on the path to the root of g, then l is *outside* g, or also g is *inside* l. If neither is outside the other, then they're said to be *disjoint*. Redexes are related to each other via their left-hand sides. A redex is said to be *innermost* among a set of (not necessarily independent) redexes if all redexes in the set are either outside or disjoint from it.

In [Hue80] the notation $-\!\!\!\mid\!\mid\!\!\!\rightarrow$ was introduced to denote the development of a set of disjoint ('parallel') redexes.

Lemma 7. *Sets of disjoint redexes are independent.*

Proof. Trivial from the restriction to trees. ⊙

4 Dependence

After having sketched the behaviour of independent steps, we briefly recapitulate some basic theory of dependent steps as known from e.g. [Mil91, Pfe, MN94, Pre95]. The basic results ([Mil91, Pre95]) we need are:

1. *Matching* for patterns is decidable. This yields that the rewrite relation $\to$ of a PRS is decidable.
2. *Unification* of patterns is decidable as well, and *most general unifiers* do exist.

This yields that one can compute so-called critical pairs and that this preserves decidability of matching and unification. Basically, patterns allow you to carry over the techniques from first- to higher-order.

Definition 8. Let $t \leftarrow_{C\boxed{l\to r}} s \to_{D\boxed{g\to d}} r$ be a span of PRS-steps.

1. An *intersector* of the span is a pattern p at C' in l and at D' in g, such that C' at C and D' at D are patterns in the same term (context). This expresses that p is a substructure on which both redexes operate.

2. The intersector is called *critical*, if there does not exist an intersector p' at C'' and D'', such that p is at E in p', and E is a pattern (context) at C'' in C' and at D'' in D'. The term p is said to be the *critical intersection* of the span, denoted sloppywise by $l \sqcap g$. This expresses that p is the maximal substructure on which both redexes operate.
3. The span is *dependent* if an intersector exists for it, that is there is some substructure on which both operate.
4. The dependent span is *critical*, if there does not exist a dependent span $t' \leftarrow_{C'\boxed{l \to r}} s' \rightarrow_{D'\boxed{g \to d}} r'$ such that s' is at E in s, and C' in C and D' in D are both at E. The term s' is said to be the *critical unification* of the span, (imprecisely) denoted by $l \sqcup g$. The critical unification is the largest substructure on which either of the steps operate.

That patterns are closed under all these operations and critical unifications and intersections exist in case of dependent steps follows from results in [Mil91, Pfe, MN94, Pre95].

Lemma 9. *1. A set of redexes is independent if and only if its elements are pairwise independent.*
2. Two redexes are independent if and only if they don't form a dependent span.
3. Every dependent span is obtained by putting a critical span in a context.

Proof. This is tedious. The first item was shown in [Oos94, Prop. 3.1.49]. The second and third item can be dug out from Section 4 in [MN94]. ⊙

Lemma 10. *If two steps u, u' (with left-hand sides l, l') are both not inside v (with left-hand side g) and do both depend on v, then they depend on each other.*

Proof. Since both steps do not depend on v, by Lemma 7 they cannot be disjoint from v and since they're also not inside v, they must be both outside v. Now, because of the tree restriction if some pattern $\tilde{l}$ is outside another one $\tilde{g}$ on which it depends, the head-symbol of $\tilde{g}$ must be in $\tilde{l} \sqcap \tilde{g}$. In particular both $l \sqcap g$ and $l' \sqcap g$ must contain the head-symbol of g, so $l \sqcap l'$ is non-empty. ⊙

For PRSs at least one of the contexts in a critical span must be of the form $\mathbf{x}.\Box(\mathbf{s})$, since if not the contexts would have a common head contradicting minimality ([MN94],[Oos94, p. 102]). A context of the form $\mathbf{x}.\Box(\mathbf{s})$ is called a *substitution context.* Then, the symmetry in the notion of critical span can be avoided by requiring that D is a substitution context, in the definition above. For such critical spans the pair (t,r) is called a *critical pair*.[6] A critical pair is called *trivial*, if $t \equiv r$. The only symmetric case remaining is when both C and D are substitution contexts. In that case the critical pair is called a *root critical pair*, giving rise to a *critical overlay* when both elements of the pair are put into the same context.

Definition 11. A critical pair (t,r) is *development closed*, if $t \multimap\!\!\!\rightarrow r$.

[6] This notion of critical pair is (apart from the initial binder) the usual one! Cf. [Hue80].

5 Development Closed

Huet studied confluence of (first-order) term rewriting systems (TRSs) in [Hue80]. He showed that *parallel closed* left-linear TRSs, i.e. TRSs such that for every critical pair (s,t), we have that $s \xrightarrow{\parallel} t$, are confluent ([Hue80, Lem. 3.3], see the left-hand side of Figure 1).

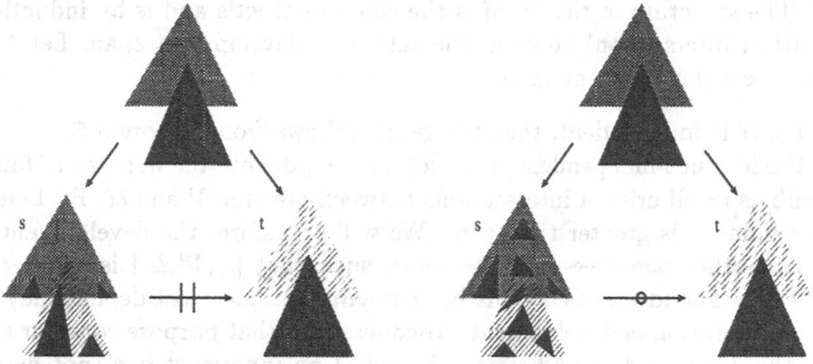

Fig. 1. Parallel closed and Development closed

We show that if a PRS is *development closed*, i.e. for all critical pairs (s,t), $s \xrightarrow{\circ} t$ (see the right-hand side of Figure 1), confluence can be concluded. Since sets of disjoint redexes are independent by Lemma 7, this extends Huet's result. Since left-linear TRSs are PRSs, this moreover generalises it to the higher-order case. Note that our result holds for Klop's combinatory reduction systems ([Klo80]) and Khasidashvili's expression reduction systems ([Kha90]) as well, since these can be embedded into PRSs in a natural way (see [ORa, Raa96]).

Example 1. Consider a PRS having function symbols $\mathbf{app}: o \to (o \to o)$, $\mathbf{abs}: (o \to o) \to o$, $\mathtt{or}: o \to o \to o$, $\mathtt{tt}: o$ and rewrite rules[7]

$$\begin{aligned} y.z.\mathbf{app}(\mathbf{abs}(x.y(x)), z) &\to_{beta} y.z.y(z) \\ y.\mathbf{abs}(x.\mathbf{app}(y, x)) &\to_{eta} y.y \\ x.\mathtt{or}(\mathtt{tt}, x) &\to_{lor} x.\mathtt{tt} \\ x.\mathtt{or}(x, \mathtt{tt}) &\to_{ror} x.\mathtt{tt} \end{aligned}$$

Rewrite steps in this system need not be independent. Critical pairs arise from different ways to unify left-hand sides of rules (cf. [MN94]). The critical pair $(\mathbf{app}(s,t), \mathbf{app}(s,t))$ arises from the critical unification $\mathbf{app}(\mathbf{abs}(x.\mathbf{app}(s,x)), t)$ of *beta* and *eta*, $(\mathbf{abs}(y.s(y)), \mathbf{abs}(x.s(x)))$ arises from the critical unification $\mathbf{abs}(x.\mathbf{app}(\mathbf{abs}(y.s(y)), x))$ of *eta* and *beta*, and $(\mathtt{tt}, \mathtt{tt})$ arises from the critical unification $\mathtt{or}(\mathtt{tt}, \mathtt{tt})$ of *lor* and *ror* (and vice versa). All the critical pairs are trivial so surely development closed, hence the system is confluent.

[7] The beta and eta rules are just the usual beta ($(\lambda x.M)N \to M[x := N]$) and eta ($\lambda x.Mx \to M$, if x not free in M) rules of lambda calculus using higher-order notation (cf. [MN94]).

5.1 Huet

We proceed with our adaptation of the proofmethod in [Hue80].

Lemma 12. *Let $\mathcal{H}$ be a development closed PRS. Then $\multimap\!\!\rightarrow$ satisfies the diamond property.*

Proof. The structure of the proof is the same as Huet's and is by induction on the 'critical intersection' between the sets in a development span. Let $t \leftarrow\!\!\multimap_{\mathcal{V}} s \multimap\!\!\rightarrow_{\mathcal{U}} r$ be a development span.

1. If $\mathcal{V} \cup \mathcal{U}$ is independent, then the result follows from Theorem 5.
2. If $\mathcal{V} \cup \mathcal{U}$ is not independent, then let $|s, \mathcal{V}, \mathcal{U}|$ denote the number of function symbols in all critical intersections between steps in $\mathcal{V}$ and $\mathcal{U}$. By Lemma 9 this number is greater than zero. We will transform the development span into another one $t \leftarrow\!\!\multimap_{\mathcal{V}'} s' \multimap\!\!\rightarrow_{\mathcal{U}'} r$, such that $|s', \mathcal{V}', \mathcal{U}'|$ is smaller than $|s, \mathcal{V}, \mathcal{U}|$. The idea is to do this by removing an innermost dependency using the assumption of development closedness. For that purpose consider among all the dependent steps between $\mathcal{V}$ and $\mathcal{U}$ an innermost one (not necessarily unique), say (wlog) $v =^{\mathrm{def}} D\boxed{l \to r} \in \mathcal{V}$ and $s \to_v t'$. By the innermost assumption, Lemma 10, and the assumption that $\mathcal{U}$ is independent, there exists a unique step $s \to_u r'$, depending on v, with $u =^{\mathrm{def}} E\boxed{g \to d} \in \mathcal{U}$. The resulting dependence span: $t' \leftarrow_v s \to_u r'$ can be written as an F_0-instance of a critical span $t_0 \leftarrow_{D_0\boxed{l\to r}} s_0 \to_{E_0\boxed{g \to d}} r_0$ for some context F_0, by Lemma 9. By assumption the critical pair (t_0, r_0) is development closed, so $t_0 \multimap\!\!\rightarrow_{C_0} r_0$ for some set C_0 of independent redexes. By the Prism Theorem 5, this development can be put inside the context F_0 giving rise to a development $t' \multimap\!\!\rightarrow_C r'$. Define $\mathcal{U}^- =^{\mathrm{def}} \mathcal{U} - \{u\}$ and $\mathcal{V}^- =^{\mathrm{def}} \mathcal{V} - \{v\}$, then we can make the following claim.

   Claim Let $s' =^{\mathrm{def}} t'$, $\mathcal{V}' =^{\mathrm{def}} \mathcal{V}^-/v$, and $\mathcal{U}' =^{\mathrm{def}} C \cup (\mathcal{U}^-/v)$. Then $t \leftarrow\!\!\multimap_{\mathcal{V}'} s' \multimap\!\!\rightarrow_{\mathcal{U}'} r$ is a development span for which $m' =^{\mathrm{def}} |s', \mathcal{V}', \mathcal{U}'|$ is smaller than $m =^{\mathrm{def}} |s, \mathcal{V}, \mathcal{U}|$ (see Figure 2).

   Proof of claim By independence of $\mathcal{V}^-$ and the Prism Theorem, $s' \multimap\!\!\rightarrow_{\mathcal{V}'} t$. By independence of $\mathcal{U}^-$ and the Prism Theorem, $r' \multimap\!\!\rightarrow_{\mathcal{U}^-/u} r$. The critical unification $l \sqcup g$ of u and v at context F in s is independent from $\mathcal{U}^-$ by construction (and since independence is preserved by unification). This means by the Prism Theorem that the descendant relation is obtained from combining the ones on F and $l \sqcup g$. Since performing 'u' from $l \sqcup g$ induces the same (empty) descendant relation as performing '$v\,;C$' this entails that $\mathcal{U}^-/(v\,; C) = \mathcal{U}^-/u$, hence also $(C \cup (\mathcal{U}^-/v))/C = \mathcal{U}^-/u$. Note that the union here is independent since it is obtained by combining independent sets of redexes in independent terms. Using this the result follows from the Prism Theorem since the sequence $C\,; (\mathcal{U}^-/u)$ is a complete development of the set $C \cup (\mathcal{U}^-/v)$.
   It remains to show that $m' < m$. By the innermost assumption redexes inside v are all independent, and by the Prism Theorem their residuals after v are

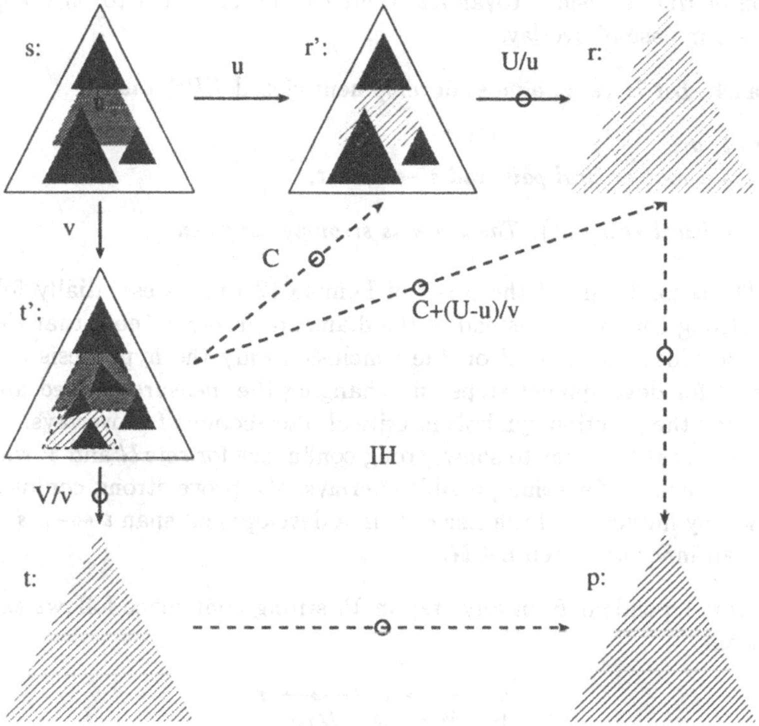

Fig. 2. proof of the claim (only patterns of steps having overlap with u are shown)

independent again. Critical intersections independent from v must therefore be outside or disjoint it, so have unique descendants after v. The only thing which could happen is that they're not longer critical intersections. This can happen when they were critical intersections between redexes in $\mathcal{W} \subseteq \mathcal{V}$ and u. But since every intersection between a redex in C and $\mathcal{W}$ is also an intersection between u and $\mathcal{W}$, per construction, the measure cannot have increased for these intersections either. We conclude by noting that the critical intersection between u and v was non-empty by assumption, contributing to m, but not to m'. $\odot$

As remarked above, the diamond property for $\multimap$ implies its confluence, which in turn implies confluence of $\rightarrow$, so we have our main result:

Corollary 13. *Development closed PRSs are confluent.*

5.2 Toyama

The main result of Toyama's [Toy88] is an improvement of Huet's result by weakening the condition on a critical pair (s,t) in case of overlay (i.e. root overlap) to the existence of a term r such that $s \twoheadrightarrow\!\!\!\!\!\!\!\!\;\;| \; r \twoheadleftarrow t$. Analogous to the above extension of Huet's result, Toyama's result can be extended to only requiring $s \multimap r \twoheadleftarrow t$ in case of overlay.

Lemma 14. *Let $\mathcal{H}$ be an* almost development closed *PRS, that is,*

1. $s \multimap t$, *or*
2. (s,t) *is a root critical pair and* $s \multimap\, ; \twoheadleftarrow t$,

for every critical pair (s,t). Then $\multimap$ is strongly confluent.

Proof. The second part of the proof of Lemma 12 can be essentially followed, proving strong confluence instead of the diamond property (note that this part of the proof does not depend on the conclusion only the hypothesis of strong confluence) for development steps and changing the measure defined above by not counting the function symbols in critical intersections for overlays.

The base case then has to show strong confluence for sets $\mathcal{U}$ and $\mathcal{V}$ which are independent except for some possible overlays. We prove strong confluence for such cases, by induction of the size of $\mathcal{U}$ in a development span $t \,{\circ\!\!-}_{\mathcal{V}}\, s \multimap_{\mathcal{U}} r$. Consider an innermost step $u \in \mathcal{U}$.

1. if u is independent from any step in $\mathcal{V}$, strong confluence follows as in the diagram

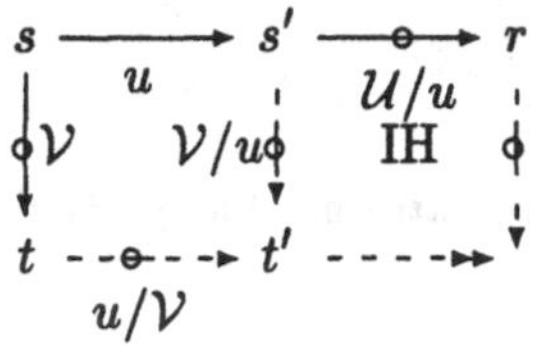

By the innermost assumption u cannot duplicate other redexes, so $\mathcal{U}/u$ has one element less than $\mathcal{U}$.
2. if u has overlay overlap with some step $v \in \mathcal{V}$, the following diagram can be constructed

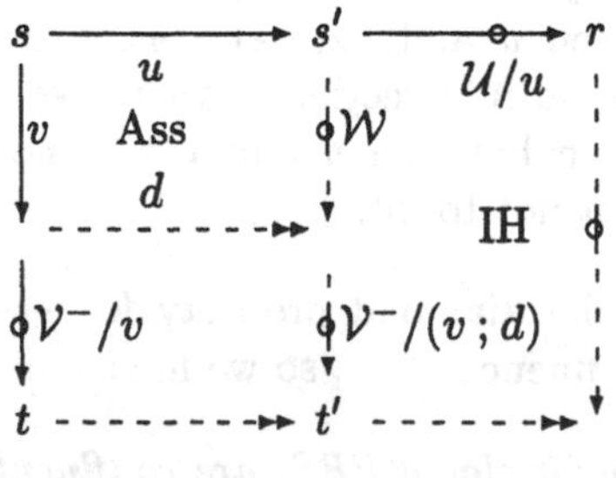

where $\mathcal{V}^- =^{\mathrm{def}} \mathcal{V} - \{v\}$, and *Ass* refers to the overlay case of the definition of almost development closedness. One can construct the critical unification $l \sqcup g$ of u and v at context F in s. Since u and v are overlays, $\mathcal{V}^-$ is independent from both of them, hence also from their critical unification $l \sqcup g$. From this and since the descendant relations induced by the rewrites 'u ; $\mathcal{W}$' and 'v ; d' from the almost development closedness condition are both empty, the rewrites from s obtained by putting them in context F induce the same descendant relation, like in the proof of Lemma 12. Hence $\mathcal{V}^-/(u\,;\mathcal{W}) = \mathcal{V}^-/(v\,;d)$ and $(\mathcal{W} \cup (\mathcal{V}^-/u))/\mathcal{W} = \mathcal{V}^-/(v\,;d)$, and again we see that the sequential composition of the two development steps $\mathcal{W}$ and $\mathcal{V}^-/(v\,;d)$ can be combined into one development step contracting the set $\mathcal{W} \cup (\mathcal{V}^-/u)$ of independent redexes, and the induction hypothesis can be indeed applied for the right part of the diagram, for the same reason as in the previous case. ⊙

As remarked above, strong confluence of $\multimap\!\!\!\rightarrow$ implied its confluence, which in turn implies confluence of $\mathcal{H}$, thereby extending and generalising Toyama's [Toy88, Cor. 3.2].

Corollary 15. *Almost development closed PRSs are confluent.*

Although our results don't seem to have a great number of killer-applications, they do support our feeling that many first-order techniques based on orthogonality carry over to the higher-order case.

Example 2. 1. Consider the non-orthogonal, non-terminating, non-right-linear, highly-artificial TRS having rules:

$$\mathtt{A} \to \mathtt{G(B)}$$
$$\mathtt{A} \to \mathtt{H(K)}$$
$$x.\mathtt{G}(x) \to x.\mathtt{F}(x,x)$$
$$x.\mathtt{H}(x) \to x.\mathtt{F}(x,x)$$
$$\mathtt{B} \to \mathtt{A}$$
$$\mathtt{K} \to \mathtt{A}$$

This TRS is confluent by Lemma 14.

2. A more useful higher-order example to which the lemma also applies, but not immediately since the system is not a PRS (because the pattern condition is violated) is the following definitional expansion rule:

$$y.z.\mathbf{app}(\mathbf{abs}(x.y(x,x)),z) \to_{betadelta} y.z.\mathbf{app}(\mathbf{abs}(x.y(x,z)),z)$$

This rule allows to 'expand' some 'abbreviations' x to their 'definition' z in the 'text' y.

6 Conclusion

In this paper we have proven all development closed PRSs to be confluent, thereby obtaining a critical pair based confluence result, for non-terminating higher-order term rewriting systems.

The idea of our confluence proof for development closed rewriting system is the same as in [Hue80, Toy88], but for the refined measure function. The reason for this adaptation is that the confluence results obtained in those papers were based on the diamond property for 'parallel rewriting'. Since parallel rewriting doesn't satisfy the diamond property for higher-order rewriting systems, not even for such simple ones as the lambda calculus, it has to be replaced by the notion of 'development rewriting'. Trying this back in 1992, I only observed that the measure function used by Huet didn't work. It took two years before looking at it again and realising that a straightforward adaptation of the measure function did the trick.

The proof is modular in the following sense. The basis of the result is formed by a more or less abstract theory of independence of redexes as found at many places ([CR36, Klo80, HL, Kha92, Hue93, Oos94, Mel95]) and briefly recapitulated here. For orthogonal systems this immediately yields confluence. The development closed condition requires on top of that also a term structure of the objects of the rewriting system.

The known relaxation of orthogonality to weak orthogonality (having only trivial critical pairs, [ORb]) puts weaker demands on the structure of objects than the development closed condition does and in particular it doesn't require them to be trees. So, for term rewriting, development closedness is better, but for graph rewriting our proof fails and weak orthogonality is the strongest result available at this moment. Of course, this should be formalised in some formal system, and we believe that the framework as introduced in [Oos94, ORb, Raa96] is appropriate for that purpose.

One can also choose not to work with a notion of descendants at all, and to prove the diamond property directly for an inductively defined notion of complete development, by induction over that definition. This method (and variations) one can find described in [Acz78, Raa93, Tak, Nip, MN94, Raa96] and in many ad hoc confluence proofs in literature as well.

The advantages of keeping the descendants around is that it yields such nice byproducts as the theory of permutation equivalence. The advantage of not using descendants is that it yields short confluence proofs. A paper displaying both for weakly orthogonal rewriting system in a general setting is [ORb].

Some random questions remaining are:

1. Can the Tait & Martin-Löf method be employed to yield our main results? This is a methodological question (there's no sense an sich in reproving known results by known methods).
2. Do the results carry over to the term graph rewriting world? From the abstract nature of the proof I expect the answer to be affirmative. The problem is more that existing notions of term graph rewriting and concepts such as 'critical pairs' for these are rather ad hoc.
3. This brings us immediately to an important issue: unlike the theory of independence there's not much been done for dependence. We expect that our setting is convenient for formulating such a theory.
4. Can the requirement for development closedness that for each critical pair (s,t), t can be reached from s by a *complete* development of some set of re-

dexes can be relaxed by omitting the completeness requirement?[8] I would say yes, but could imagine an out-of-sync phenomenon as in the counterexample against confluence of non-ambiguous TRSs as well.

5. There's no reason why one should restrict attention to critical *pairs*. It seems worthwhile to study such intuitive notions as critical *clusters*.

Acknowledgements I thank Femke van Raamsdonk for collaboration on the general setting, Mizuhito Ogawa, Aart Middeldorp, Jan Willem Klop and Fer-Jan de Vries for feedback, Paul Taylor for his diagrams and Hideki Isozaki for colour-support. Finally, thanks to the referees for useful comments.

References

[Acz78] Peter Aczel. A general Church-Rosser theorem. Technical report, University of Manchester, July 1978.

[AGM92] S. Abramsky, Dov M. Gabbay, and T. S. E. Maibaum, editors. *Handbook of Logic in Computer Science*, volume 2, Background: Computational Structures. Oxford University Press, New York, 1992.

[Bar84] H. P. Barendregt. *The Lambda Calculus, Its Syntax and Semantics*, volume 103 of *Studies in Logic and the Foundations of Mathematics*. Elsevier Science Publishers B.V., Amsterdam, revised edition, 1984.

[BG93] M. Bezem and J. F. Groote, editors. *Proceedings of the International Conference on Typed Lambda Calculi and Applications, TLCA'93, March 1993, Utrecht, The Netherlands*, volume 664 of *Lecture Notes in Computer Science*, Berlin Heidelberg, 1993. Springer-Verlag.

[CR36] Alonzo Church and J. B. Rosser. Some properties of conversion. *Transactions of the American Mathematical Society*, 39:472–482, January to June 1936.

[DJ] Nachum Dershowitz and Jean-Pierre Jouannaud. Rewrite systems. In [Lee90, Ch. 6 pp. 243–320].

[GLM] Georges Gonthier, Jean-Jacques Lévy, and Paul-André Melliès. An abstract standardisation theorem. pp. 72–81 in LICS'92.

[HL] Gérard Huet and Jean-Jacques Lévy. Computations in orthogonal rewriting systems. Chch. 11,12 in [LP91].

[Hue80] Gérard Huet. Confluent reductions: Abstract properties and applications to term rewriting systems. *Journal of the Association for Computing Machinery*, 27(4):797–821, October 1980.

[Hue93] Gérard Huet. Residual theory in λ-calculus : A formal development. Rapport de Recherche 2009, INRIA, Août 1993.

[Kha90] Z. O. Khasidashvili. Expression reduction systems. In *Proceedings of I. Vekua Institute of Applied Mathematics*, volume 36, pages 200–220, Tbilisi, 1990.

[Kha92] Zurab Khasidashvili. The Church-Rosser theorem in orthogonal combinatory reduction systems. Rapports de Recherche 1825, INRIA-Rocquencourt, December 1992.

[8] Question asked by Aart Middeldorp.

[Kir93] Claude Kirchner, editor. *Rewriting Techniques and Applications, 5th International Conference, RTA-93, Montreal, Canada, June 16–18, 1993*, volume 690 of *Lecture Notes in Computer Science*. Springer-Verlag, Berlin Heidelberg, 1993.

[Klo] J. W. Klop. Term rewriting systems. In [AGM92, pp. 1–116].

[Klo80] J. W. Klop. *Combinatory Reduction Systems*. PhD thesis, Rijksuniversiteit Utrecht, June 1980. Mathematical Centre Tracts 127.

[Lee90] Jan van Leeuwen, editor. *Handbook of Theoretical Computer Science*, volume B: Formal Models and Semantics. Elsevier Science Publishers B.V., Amsterdam, 1990.

[LP91] Jean-Louis Lassez and Gordon Plotkin, editors. *Computational Logic: Essays in Honor of Alan Robinson*. The MIT Press, Cambridge, Massachusetts, 1991.

[Mel95] Paul-André Melliès. *Description Abstraite des Systèmes de Réécriture*. Thèse de doctorat, Université Paris VII, Preliminary version of September 28, 1995.

[Mil91] Dale Miller. A logic programming language with lambda-abstraction, function variables, and simple unification. in [SH91, pp. 253–281], 1991.

[MN94] Richard Mayr and Tobias Nipkow. Higher-order rewrite systems and their confluence. On `ftp.informatik.tu-muenchen.de` as `local/lehrstuhl/nipkow/hrs.dvi.gz`, November 1994. To appear in TCS.

[Nip] Tobias Nipkow. Orthogonal higher-order rewrite systems are confluent. In [BG93, pp. 306–317].

[Oos94] Vincent van Oostrom. *Confluence for Abstract and Higher-Order Rewriting*. PhD thesis, Vrije Universiteit, Amsterdam, March 1994. Available at `http://www.cs.vu.nl/~oostrom`.

[ORa] Vincent van Oostrom and Femke van Raamsdonk. Comparing combinatory reduction systems and higher-order rewrite systems. pp. 276–304 in HOA'93, LNCS 816.

[ORb] Vincent van Oostrom and Femke van Raamsdonk. Weak orthogonality implies confluence: the higher-order case. pp. 379–392 in LFCS'94, LNCS 813.

[Pfe] Frank Pfenning. Unification and anti-unification in the calculus of constructions. pp. 74–85 in LICS'91.

[Pre95] Christian Prehofer. *Solving Higher-Order Equations: From Logic to Programming*. PhD thesis, Technische Universität München, February 1995. Report TUM-19508.

[Raa93] Femke van Raamsdonk. Confluence and superdevelopments. in [Kir93, pp. 168–182], 1993.

[Raa96] Femke van Raamsdonk. *Confluence and Normalisation for Higher-Order Rewriting*. PhD thesis, Vrije Universiteit, Amsterdam, 1996. Preliminary version of 19 January 1996.

[SH91] P. Schroeder-Heister, editor. *Extensions of Logic Programming, International Workshop, Tübingen, FRG, December 8–10, 1989*, volume 475 of *Lecture Notes in Artificial Intelligence*. Springer-Verlag, 1991.

[Tak] Masako Takahashi. λ-calculi with conditional rules. In [BG93, pp. 406–417].

[Toy88] Yoshihito Toyama. Commutativity of term rewriting systems. In F. Fuchi and L. Kott, editors, *Programming of Future Generation Computer*, volume II, pages 393–407. North-Holland, 1988.

Two *Different* Strong Normalization Proofs?

— computability versus functionals of finite type —

Jaco van de Pol

Department of Philosophy, Utrecht University
Heidelberglaan 8, 3584 CS Utrecht, The Netherlands
E-mail: jaco@phil.ruu.nl

Abstract. A proof of $\forall t \exists n \mathrm{SN}(t, n)$ (term t performs at most n reduction steps) is given, based on strong computability predicates. Using modified realizability, a bound on reduction lengths is extracted from it. This upper bound is compared with the one Gandy defines, using strictly monotonic functionals. This reveals a remarkable connection between his proof and Tait's. We show the details for simply typed λ-calculus and Gödel's T. For the latter system, program extraction yields considerably sharper upper bounds.

1 Introduction

The purpose of this paper is to compare two different methods to prove strong normalization (SN). The first method uses the notion of ***strong computability predicates***. This method is attributed to Tait [9], who used convertibility predicates to prove a normal form theorem for various systems. The other method to prove strong normalization uses functionals of finite type. To each typed term a functional of the same type is associated. This functional is measured by a natural number. In order to achieve that a rewrite step gives rise to a decrease of the associated number, the notion ***strictly monotonic functional*** is developed. The number is an upper bound for the length of reduction sequences starting from a certain term. This method was invented by Gandy [2].

In the literature, these two methods are often put in contrast (e.g. [2, § 6.3] and [3, § 4.4]). Using functionals seems to be more transparent and economizes on proof theoretical complexity; strong computability should generalize to more complex systems. On the other hand, seeing the two proofs one gets the feeling that "somehow, the same thing is going on". Indeed De Vrijer [11, § 0.1] remarks that a proof using strong computability can be seen as abstracting from concrete information in the functionals that is not strictly needed in a termination proof, but which provides for an estimate of reduction lengths.

In this paper we will substantiate this feeling. First we will *decorate* the proof à la Tait with concrete numbers. This is done by introducing binary predicates $\mathrm{SN}(t, n)$, which mean that the term t may perform at most n reduction steps. A formal, constructive proof of $\exists n \mathrm{SN}(t, n)$ is given for any t. From this proof, we extract a program, via the ***modified realizability interpretation***. Remarkably, this program equals (more or less) the functional assigned to the term t in the proof à la Gandy.

The paper is organized as follows. In Section 3, we decorate Tait's SN-proof for simply typed λ-calculus. Modified realizability is introduced in Section 4. In Section 5 the proofs of Section 3 are formalized; also the program extraction is carried out there. In Section 5.3, the extracted functionals are compared with those used by Gandy. The same project is carried out for Gödel's T in Section 6. Other possible extensions are considered in Section 7.

The idea of using a realizability interpretation to extract functionals from Tait's SN-proof already occurs in [1]. In that paper, the *normal form* of a term is extracted. The contribution of this paper is, that by extracting numerical upper bounds for the length of reduction sequences, a comparison with Gandy's proof can be made. Furthermore, we also deal with Gödel's T, which yields a sharper upper bound than provided by Gandy's proof. The author is grateful to Ulrich Berger for discussions on the subject, and to Marc Bezem and Jan Springintveld for reading and improving preliminary versions of the paper.

2 Simply Typed λ-Calculus

The set of *simple types* contains a certain set of base types and is closed under the binary operator →. By convention, metavariables $\iota, \iota_1, \cdots$ range over base types; $\rho, \sigma, \tau, \cdots$ over arbitrary types. The adjective "simple" will often be dropped.

The set of *simply typed λ-terms* contains a certain set of typed variables and is closed under typed application and λ-binding. We reserve $r, s, t, \cdots$ for arbitrary simply typed terms, and $x^\sigma, y^\tau, \cdots$ for typed variables. Constants can be seen as variables that will not occur after a λ. With *terms* we will mean simply typed λ-terms. To indicate that r has type σ, we write r^σ. The typing rules are as follows:

1. A variable (or constant) x^σ is of type σ.
2. If s and t are of type $\rho \to \sigma$ and ρ, respectively, then (st) is of type σ.
3. If x^σ is a variable and s is of type τ, then $(\lambda x^\sigma s)$ is of type $\sigma \to \tau$.

Type decoration and outer brackets are often omitted.

Standard notions of bound and free variables (FV(s)) will be used. We will identify α-convertible terms (i.e. terms that are equal up to the names of the bound variables). Substituting a term t for the free occurrences of x in a term s is denoted by $s[x := t]$. Renamings to avoid unintended capture of free variables are performed automatically.

The binary rewrite relation $s \to_\beta t$ is defined as the compatible closure of the β-rule: $(\lambda x s)t \mapsto s[x := t]$. We write $s \to_\beta^n t$, if there is a reduction sequence from s to t of n steps.

Definition 1. (SN for simply typed λ-calculus)

1. A term t is *strongly normalizing*, denoted by SN(t), if every reduction sequence $t \equiv s_0 \to_\beta s_1 \to_\beta \cdots$ is finite.

2. A term is *strongly normalizing in at most n steps* (SN(t,n)) if every reduction sequence out of t is finite, and has length at most n.

In the next section, we present the proof à la Tait, that every simply typed λ-term has an upper bound n such that it is strongly normalizing in at most n steps. By König's Lemma, this is equivalent to strong normalization, because the β-reduction relation is finitely branching.

We will often abbreviate a sequence of terms $t_1, \ldots, t_n$ by $\bar{t}$. In the same way, sequences of variables $\bar{x}$ or types $\bar{\sigma}$ will occur frequently. The length of such sequences is implicitly known, or unimportant. The empty sequence is denoted by ϵ. The *simultaneous* substitution of the variables $\bar{x}$ by the terms $\bar{t}$ in a term s is denoted by $s[\bar{x} := \bar{t}]$.

By convention, $\bar{\sigma} \rightarrow \tau$ means $\sigma_1 \rightarrow (\sigma_2 \rightarrow \cdots \rightarrow (\sigma_n \rightarrow \tau))$ and $s\bar{t}$ means $(((st_1)t_2) \cdots t_n)$ and $\lambda\bar{x}.t$ means $(\lambda x_1(\lambda x_2 \cdots (\lambda x_n t)))$. We also use the following (less standard) notation for sequences of variables, terms and types:

- $\bar{x}^{\bar{\sigma}} \equiv x_1^{\sigma_1}, \cdots, x_n^{\sigma_n}$
- $\bar{\sigma} \rightarrow \bar{\tau} \equiv \bar{\sigma} \rightarrow \tau_1, \ldots, \bar{\sigma} \rightarrow \tau_n$
- $\bar{s}\bar{t} \equiv s_1\bar{t}, \ldots, s_n\bar{t}$
- $\lambda\bar{x}.\bar{t} \equiv \lambda\bar{x}.t_1, \ldots, \lambda\bar{x}.t_n$

Note that by this convention $\bar{\sigma} \rightarrow \epsilon \equiv \epsilon$ (in particular, $\epsilon \rightarrow \epsilon \equiv \epsilon$) and $\epsilon \rightarrow \sigma \equiv \sigma$.

3 Informal SN-Proof à la Tait

Tait's method to prove strong normalization starts with defining a "strong computability" predicate which is stronger than "strong normalizability". The proof consists of two parts: One part stating that strongly computable terms are strongly normalizing, and one part stating that any term is strongly computable. The first is proved with induction on the types (simultaneously with the statement that every variable is strongly computable). The second part is proved with induction on the term structure (in fact a slightly stronger statement is proved). We will present a version of this proof that contains information about reduction lengths.

Definition 2. The set of *strongly computable* terms is defined inductively as follows:

(i) $\mathrm{SC}_\iota(t)$ iff there exists an n such that SN(t,n).
(ii) $\mathrm{SC}_{\sigma\rightarrow\tau}(t)$ iff for all s with $\mathrm{SC}_\sigma(s)$, $\mathrm{SC}_\tau(ts)$.

Lemma 3 (SC Lemma). *(a) For all terms t, if* SC(t) *then there exists an n with* SN(t,n).
(b) For all terms t of the form $x\bar{t}$, if there exists an n with SN(t,n)*, then* SC(t).

In (b), $\bar{t}$ may be the empty sequence.

Proof. (Simultaneous induction on the type of t)

(a) Assume SC(t).
If t is of base type, then SC(t) just means that there exists an n with SN(t,n). If t is of type $\sigma \to \tau$, we take a variable x^σ, which is of the form $x\bar{t}$. Note that x is in normal form, hence SN(x, 0) holds. By IH(b), SC(x); and by the definition of SC(t), SC(tx). By IH(a) we have that there exists an n such that SN(tx, n). We can take this n, because any reduction sequence from t gives rise to a sequence from tx of the same length. Hence SN(t,n) holds.

(b) Assume that $t = x\bar{t}$ and SN(t,n) for some n.
If t is of base type, then the previous assumption forms exactly the definition of SC(t).
If t has type $\sigma \to \tau$, assume SC(s) for arbitrary s^σ. By IH(a), SN(s,m) for some m. Because reductions in $x\bar{t}s$ can only take place inside $\bar{t}$ or s, we have SN($x\bar{t}s, m+n$). IH(b) yields that SC($x\bar{t}s$). This proves SC(t).

□

Lemma 4 (Abstraction Lemma). *For all terms s, t and $\bar{r}$ and variables x, it holds that if* SC($s[x := t]\bar{r}$) *and* SC(t), *then* SC($(\lambda x.s)t\bar{r}$).

Proof. (Induction on the type of $s\bar{r}$.) Let s, x, t and $\bar{r}$ be given, with SC($s[x := t]\bar{r}$) and SC(t). Let σ be the type of $s\bar{r}$.

If $\sigma = \iota$, then by definition of SC, we have an n such that SN($s[x := t]\bar{r}, n$). By Lemma 3(a) we obtain the existence of m, such that SN(t,m). We have to show, that there exists a p with SN($(\lambda x.s)t\bar{r}, p$). We will show that we can put $p := m + n + 1$. Consider an arbitrary reduction sequence of $(\lambda x.s)t\bar{r}$. Without loss of generality, we assume that it consists of first a steps in s (yielding s'), b steps in $\bar{r}$ (yielding $\bar{r}'$) and c steps in t (yielding t'). After this the outermost redex is contracted, yielding $s'[x := t']\bar{r}'$, and finally d steps occur. Clearly, $c \leq m$. Notice that we also have a reduction sequence $s[x := t]\bar{r} \to^* s'[x := t']\bar{r}'$ of at least $a + b$ steps (we cannot count reductions in t, because we do not know whether x occurs free in s). So surely, $a + b + d \leq n$. Summing this up, we have that any reduction sequence from $(\lambda x.s)t\bar{r}$ has length at most $m + n + 1$.

Let $\sigma = \rho \to \tau$. Assume SC(r), for arbitrary r^ρ. Then by definition of SC($s[x := t]\bar{r}$), we have $\text{SC}_\tau(s[x := t]\bar{r}r)$, and by IH SC($(\lambda x.s)t\bar{r}r$). This proves SC($(\lambda x.s)t\bar{r}$). □

In the following lemma, θ is a substitution, i.e. a finite mapping from variables into terms.

Lemma 5 (Main Lemma). *For all terms t and substitutions θ, if* SC(x^θ) *for all free variables x of t, then* SC(t^θ).

Proof. (Induction on the structure of t.) Let t and θ be given, such that SC(x^θ) for all $x \in \text{FV}(t)$.

If $t = x$, then the last assumption yields SC(t^θ).

If $t = rs$, we have SC(r^θ) and SC(s^θ) by IH for r and s. Then by definition of SC(r^θ), we have SC($r^\theta s^\theta$), hence by equality of $r^\theta s^\theta$ and $(rs)^\theta$, SC(t^θ) follows.

If $t = \lambda x.s$, assume that SC(r) for an arbitrary r. By IH for s, applied on the substitution $\theta[x := r]$, we see that SC($s^{\theta[x:=r]}$), hence by equality SC($(s^\theta)[x := r]$). Now we can apply Lemma 4, which yields that SC($(\lambda x.s^\theta)r$). Again by using equality, we see that SC($(\lambda x.s)^\theta r$) holds. This proves SC($(\lambda x.s)^\theta$). (Note that implicitly renaming of bound variables is required.) □

Theorem 6. *For any term t there exists an n, such that* SN(t, n).

Proof. Let θ be the identity substitution, with as domain the free variables of t. By Lemma 3(b), SC(x) is guaranteed. Now we can apply Lemma 5, yielding SC(t^θ). Because $t^\theta = t$, we obtain SC(t). Lemma 3(a) yields the existence of an n with SN(t, n). □

4 A Variant of Modified Realizability

As mentioned before, we want to extract the computational content from the SN-proof of Section 3. To this end we use *modified realizability*, introduced by Kreisel [4]. In [10, § 3.4] modified realizability is presented as a translation of HA^ω into itself. This interpretation eliminates existential quantifiers, at the cost of introducing functions of finite type (functionals), represented by λ-terms.

Following Berger [1], we present modified realizability as an interpretation of a first order fragment (MF) into a higher-order, negative (i.e. $\exists$-free) fragment (NH). We will also take over a refinement by Berger, which treats specific parts of a proof as *computationally irrelevant*.

4.1 The Modified Realizability Interpretation

A formula can be seen as the specification of a program. E.g. $\forall x \exists y.P(x, y)$ specifies a program f of type $o \to o$, such that $\forall x.P(x, f(x))$ holds. In general a sequence of programs is specified.

A refinement by Berger enables to express that existentially quantified variables are independent of certain universal variables, by underlining the universal ones. In $\underline{\forall x} \exists y.P(x, y)$ the underlining means that y is not allowed to depend on x. So a number m is specified, with $\forall x.P(x, m)$. This could of course also be specified by the formula $\exists y \forall x.P(x, y)$, but in specifications of the form $\underline{\forall x}.P(x) \to \exists y.Q(x, y)$ the underlining cannot be eliminated that easily. This formula specifies a number m, such that $\forall x.P(x) \to Q(x, m)$ holds. The $\underline{\forall x}$ cannot be pushed to the right, nor can the $\exists y$ be pulled to the left, without changing the intuitionistic meaning.

Specifications are expressed in minimal many-sorted first-order logic (MF). This logic is based upon a many-sorted first-order signature. Terms over such a signature are defined as usual ($a, b, c, \ldots$ denote arbitrary terms). The formulae of MF are either atomic ($P\vec{a}$), or of the form $\varphi \to \psi$, $\forall x^\iota \varphi$, $\underline{\forall x^\iota} \varphi$ or $\exists x^\iota \varphi$. Here $\varphi, \psi, \ldots$ denote arbitrary MF formulae. This logic is Minimal, because negation is not included, and it deals with First-order objects only.

As programming language, we use the simply typed λ-calculus. Because programs are higher-order objects, MF cannot talk about them. To express correctness of programs, we introduce **N**egative **H**igher-order logic (NH). The terms of NH are simply typed λ-terms *considered modulo β*, with the MF sorts as base types, MF function symbols as constants and with the MF predicate symbols. The formulae are atomic ($P\overline{s}$), or composed with $\varphi \to \psi$ or $\forall x^\rho \varphi$. Here $\varphi, \psi, \ldots$ denote arbitrary NH formulae. Negative means that there are no existential quantifiers, and Higher-order refers to the objects.

Below we define $\tau(\varphi)$, the sequence of types of the programs specified by the MF formula φ. This operation is known as "forgetting dependencies" (of types on terms). Furthermore, if $\overline{s}$ is a sequence of programs of type $\tau(\varphi)$, we define an NH formula $\overline{s}$ **mr** φ (**m**odified **r**ealizes). This NH formula expresses correctness of $\overline{s}$ with respect to the specification φ.

Definition 7. (modified realizability interpretation)

$$
\begin{aligned}
\tau(P\overline{a}) &:= \epsilon & \epsilon \;\mathbf{mr}\; P\overline{a} &:= P\overline{a} \\
\tau(\varphi \to \psi) &:= \tau(\varphi) \to \tau(\psi) & \overline{s} \;\mathbf{mr}\; \varphi \to \psi &:= \forall \overline{x}^{\tau(\varphi)}(\overline{x} \;\mathbf{mr}\; \varphi) \to (\overline{s}\overline{x} \;\mathbf{mr}\; \psi) \\
\tau(\forall x^\iota \varphi) &:= \iota \to \tau(\varphi) & \overline{s} \;\mathbf{mr}\; \forall x^\iota \varphi &:= \forall x^\iota (\overline{s}x \;\mathbf{mr}\; \varphi) \\
\tau(\underline{\forall x^\iota} \varphi) &:= \tau(\varphi) & \overline{s} \;\mathbf{mr}\; \underline{\forall x^\iota} \varphi &:= \forall x^\iota (\overline{s} \;\mathbf{mr}\; \varphi) \\
\tau(\exists x^\iota \varphi) &:= \iota, \tau(\varphi) & r, \overline{s} \;\mathbf{mr}\; \exists x^\iota \varphi(x) &:= \overline{s} \;\mathbf{mr}\; \varphi(r)
\end{aligned}
$$

In the **mr**-clauses, x^ι should not occur in $\overline{s}$ and $\overline{x}$ should be fresh. Note that only existential quantifiers give rise to a longer sequence of types. In particular, if φ has no existential quantifiers, then $\tau(\varphi) \equiv \epsilon$. (We use that $\overline{\sigma} \to \epsilon \equiv \epsilon$). Nested implications give rise to arbitrarily high types. In $\underline{\forall x^\iota} \varphi$, the program specified by φ may not depend on x, so the "$\iota \to$" is discarded in the τ-clause. In the **mr**-clause, the programs $\overline{s}$ do not get x as input, as intended. But to avoid that x becomes free in φ, we changed Berger's definition by adding $\forall x^\iota$.

By induction on the MF formula φ one sees that if $\overline{s}$ is of type $\tau(\varphi)$, then $\overline{s}$ **mr** φ is a correct formula of NH, so in particular, it will not contain $\exists$- and $\underline{\forall}$-quantifiers (nor of course the symbol **mr**).

4.2 Derivations and Program Extraction

In the previous section we introduced the formulae of MF, the formulae of NH and a translation of the former into the latter. In this section we will introduce proofs for MF and for NH. The whole point will be, that from an MF proof of φ a program can be extracted, together with an NH proof that this program meets its specification φ.

Proofs are formalized by derivation terms, a linear notation for natural deduction. Derivation terms are defined as the least set containing assumption variables ($u^\varphi, v^\psi, \ldots$) and closed under certain syntactic operations. To express some side conditions, the sets of assumption variables (FA(d)) and of computational relevant variables (CV(d)) are defined simultaneously. By convention, x and y range over object variables. We let d, e range over derivations.

The introduction rule for the $\underline{\forall}$-quantifier has an extra proviso: we may only extend a derivation d of φ to one of $\underline{\forall x}\varphi$, if x is not *computationally relevant* in d. Roughly speaking, all free object variables of d occurring as argument of a $\forall$-elimination or as witness in an $\exists$-introduction are computationally relevant.

Definition 8. (derivations, free assumptions, computational relevant variables)

$$
\begin{array}{lll}
\mathit{ass}: u^{\varphi} & \mathrm{FA}(u) = \{u\} & \mathrm{CV}(u) = \emptyset \\
\to^{+}: (\lambda u^{\varphi} d^{\psi})^{\varphi\to\psi} & \mathrm{FA}(\lambda u d) = \mathrm{FA}(d) \setminus \{u\} & \mathrm{CV}(\lambda u d) = \mathrm{CV}(d) \\
\to^{-}: (d^{\varphi\to\psi} e^{\varphi})^{\psi} & \mathrm{FA}(de) = \mathrm{FA}(d) \cup \mathrm{FA}(e) & \mathrm{CV}(de) = \mathrm{CV}(d) \cup \mathrm{CV}(e) \\
\forall^{+}: (\lambda x^{\sigma} d^{\varphi})^{\forall x^{\sigma}\varphi} & \mathrm{FA}(\lambda x d) = \mathrm{FA}(d) & \mathrm{CV}(\lambda x d) = \mathrm{CV}(d) \setminus \{x\} \\
\qquad \text{provided (1)} & & \\
\forall^{-}: (d^{\forall x^{\sigma}\varphi(x)} a^{\sigma})^{\varphi(a)} & \mathrm{FA}(da) = \mathrm{FA}(d) & \mathrm{CV}(da) = \mathrm{CV}(d) \cup \mathrm{FV}(a) \\
\underline{\forall}^{+}: (\underline{\lambda x^{\sigma}} d^{\varphi})^{\underline{\forall x^{\sigma}}\varphi} & \mathrm{FA}(\underline{\lambda x} d) = \mathrm{FA}(d) & \mathrm{CV}(\underline{\lambda x} d) = \mathrm{CV}(d) \\
\qquad \text{provided (2)} & & \\
\underline{\forall}^{-}: (d^{\underline{\forall x^{\sigma}}\varphi(x)} \underline{a^{\sigma}})^{\varphi(a)} & \mathrm{FA}(d\underline{a}) = \mathrm{FA}(d) & \mathrm{CV}(d\underline{a}) = \mathrm{CV}(d) \\
\exists^{+}: (\exists^{+}[a^{\sigma}; d^{\varphi(a)}])^{\exists x^{\sigma}\varphi(x)} & \mathrm{FA}(\exists^{+}[a;d]) = \mathrm{FA}(d) & \mathrm{CV}(\exists^{+}[a;d]) \\
 & & \quad = \mathrm{CV}(d) \cup \mathrm{FV}(a) \\
\exists^{-}: (\exists^{-}[d^{\exists x^{\sigma}\varphi(x)}; y; u^{\varphi(y)}; e^{\psi}])^{\psi} & \mathrm{FA}(\exists^{-}[d;y;u;e]) & \mathrm{CV}(\exists^{-}[d;y;u;e]) \\
\qquad \text{provided (3)} & \quad = \mathrm{FA}(d) \cup (\mathrm{FA}(e) \setminus \{u\}) & \quad = \mathrm{CV}(d) \cup (\mathrm{CV}(e) \setminus \{y\})
\end{array}
$$

where the provisos are:

(1) $x \notin \mathrm{FV}(\psi)$ for any $u^{\psi} \in \mathrm{FA}(d)$.
(2) $x \notin \mathrm{FV}(\psi)$ for any $u^{\psi} \in \mathrm{FA}(d)$ and moreover, $x \notin \mathrm{CV}(d)$.
(3) $y \notin \mathrm{FV}(\psi)$ and $y \notin \mathrm{FV}(\chi)$ for all $v^{\chi} \in \mathrm{FA}(e) \setminus \{u\}$.

An MF-derivation is a derivation with all quantifier rules restricted to base types. An NH-derivation is a derivation without the $\underline{\forall x}$ and the $\exists$-rules. We will write $\Phi \vdash_{\mathrm{MF}} \psi$ if there exists a derivation d^{ψ}, with all free assumptions among Φ. Likewise for $\vdash_{\mathrm{NH}}$.

From MF-derivations, we can read off a program and a correctness proof for this program. This is best illustrated by the $\exists^{+}$ rule: If we use this rule to prove $\exists x\varphi(x)$, then we immediately see the witness a and a proof d of $\varphi(a)$. In general, we can define $\mathbf{ep}(d)$, the sequence of extracted programs from a derivation d. To deal with assumption variables in d, we fix for every assumption variable u^{φ} a sequence of object variables $\overline{x}_u^{\tau(\varphi)}$. The extracted program is defined with respect to this choice.

Definition 9. (extracted program from MF-derivations)

$$
\begin{array}{rr}
\mathbf{ep}(u^{\varphi}) := \overline{x}_u^{\tau(\varphi)} & \mathbf{ep}(d^{\forall x^{\iota}\varphi(x)} a^{\iota}) := \mathbf{ep}(d)a \\
\mathbf{ep}(\lambda u^{\varphi} d^{\psi}) := \lambda \overline{x}_u^{\tau(\varphi)} \mathbf{ep}(d) & \mathbf{ep}(\underline{\lambda x^{\iota}} d^{\varphi}) := \mathbf{ep}(d) \\
\mathbf{ep}(d^{\varphi\to\psi} e^{\varphi}) := \mathbf{ep}(d)\mathbf{ep}(e) & \mathbf{ep}(d^{\underline{\forall x^{\iota}}\varphi(x)} \underline{a^{\iota}}) := \mathbf{ep}(d) \\
\mathbf{ep}(\lambda x^{\iota} d^{\varphi}) := \lambda x^{\iota} \mathbf{ep}(d) & \mathbf{ep}(\exists^{+}[a^{\iota}; d^{\varphi(a)}]) := a, \mathbf{ep}(d)
\end{array}
$$

$$
\mathbf{ep}(\exists^{-}[d; y; u^{\varphi(y)}; e^{\psi}]) := \mathbf{ep}(e)[y := s][\overline{x}_u := \overline{t}\,], \text{where } s, \overline{t} = \mathbf{ep}(d^{\exists x^{\iota}\varphi(x)})
$$

The whole enterprise is justified by the following

Theorem 10 (Correctness [1]). *If d is an* MF *derivation of φ, then there exists an* NH *derivation $\mu(d)$ of* $\mathbf{ep}(d)$ $\mathbf{mr}$ φ. *Moreover, the only free assumptions in $\mu(d)$ are of the form $\overline{x}_u$* $\mathbf{mr}$ ψ, *for some assumption u^ψ occurring in d already.*

Proof. First the following facts are verified by induction on d:

1. $\mathrm{FV}(\mathbf{ep}(d)) \subseteq \bigcup\{\overline{x}_u | u \in \mathrm{FA}(d)\} \cup \mathrm{CV}(d)$.
2. $\mathbf{ep}(d^\varphi)$ is a sequence of terms of type $\tau(\varphi)$.

Then the existence of $\mu(d)$ can be proved by induction on d. We only deal with one typical case. See e.g. [1] for the other cases.

$$\mu(\underline{\lambda x^\iota} d) := \lambda x^\iota(\mu(d))$$

By induction hypothesis, we have $\mu(d)$ proves $\mathbf{ep}(d)$ $\mathbf{mr}$ φ. By the proviso of $\underline{\forall}^+$, $x \notin \mathrm{CV}(d)$, hence (by fact 1) $x \notin \mathrm{FV}(\mathbf{ep}(d))$. Furthermore, x does not occur in free assumptions of d, hence not in assumptions of $\mu(d)$, so $\lambda x \mu(d)$ is a correct derivation of $\forall x(\mathbf{ep}(d)\ \mathbf{mr}\ \varphi)$, which is equivalent (because $x \notin \mathbf{ep}(d)$, and using β-equality) to $\mathbf{ep}(\underline{\lambda x} d)\ \mathbf{mr}\ \underline{\forall x}\varphi$. □

4.3 On Using Axioms

If we use an axiom $\mathbf{ax}^\varphi$ (as open assumption) in a proof d of MF, then the extracted program $\mathbf{ep}(d)$ will probably contain the free variables $\overline{x}_{\mathbf{ax}}^{\tau(\varphi)}$ (as holes) and the correctness proof $\mu(d)$ may contain a free assumption $\overline{x}_{\mathbf{ax}}$ $\mathbf{mr}$ φ (according to Theorem 10).

The goal is to complete the program in a correct way. More specifically, we look for realizers $\overline{t}_{\mathbf{ax}}$, such that the NH formula $\overline{t}_{\mathbf{ax}}$ $\mathbf{mr}$ φ holds in our intended interpretation. If this succeeds, then the extracted program can be completed by taking $\mathbf{ep}(d)[\overline{x}_{\mathbf{ax}} := \overline{t}_{\mathbf{ax}}]$. This is correct, because the assumptions in the correctness proof $\mu(d)$ hold. We conclude that the justification of postulated principles should be given in terms of NH, because in this logic the correctness proofs live. We will now justify some typical principals. (See also [10] and [1].)[1]

∃-free axioms. Let φ be such a formula. Then $\tau(\varphi) \equiv \epsilon$. The only candidate realizer is the empty sequence of programs. Note that the formula ϵ $\mathbf{mr}$ φ is obtained from φ by removing all underlinings. So a formula without existential quantifiers may be used as axiom, whenever it is true after removing all underlinings.

Equality axioms. Assume that $=$ is a binary predicate symbol. The usual axioms for symmetry, transitivity and reflexivity of $=$ are existential-free and are justified as above. The replacement scheme requires another justification. We introduce the axiom scheme

$$\mathbf{repl} : s = t \rightarrow \varphi(s) \rightarrow \varphi(t) \ .$$

[1] In the full version we regard some principles that depend on the underlining.

Note that $\tau(s = t \to \varphi(s) \to \varphi(t)) \equiv \tau(\varphi) \to \tau(\varphi)$. The identity can be taken as realizer, as the following calculation shows:

$$\begin{aligned} &\lambda \overline{x}^{\tau(\varphi)}.\overline{x}\ \mathbf{mr}\ (s = t \to \varphi(s) \to \varphi(t)) \\ &\equiv s = t \to \forall \overline{x}.\overline{x}\ \mathbf{mr}\ \varphi(s) \to \overline{x}\ \mathbf{mr}\ \varphi(t), \end{aligned}$$

The latter NH formula is true, if we interpret = by the identity relation. This means that we can use the replacement scheme in MF proofs. Its realizer is the identity on sequences.

5 Formalized Proofs and Extracted Programs

In this section the proof of Section 3 will be formalized in first-order predicate logic, as introduced in Section 4. This is not unproblematic as the informal proof contains induction on types and terms, which is not a part of the framework. This is solved by defining a series of proofs, by recursion over types or terms. In this way the induction is shifted to the metalevel. There is a price to be paid: instead of a uniform function U, such that $U(t)$ computes the desired upper bound for a term t, we only extract for any t an expression Upper[t], which computes an upper bound for term t only. So here we lose a kind of uniformity. It is well known that the absence of a uniform first-order proof is essential, because the computability predicate is not arithmetizable [10, § 2.3.11].

Another incompleteness arises, because some combinatorial results will be plugged in as axioms. This second incompleteness is harmless for our purpose, because all these axioms are formulated without using existential quantifiers. Hence they are realized by the empty sequence (and finding formal proofs for these facts would be waste of time).

5.1 Fixing Signature and Axioms

As to the language, we surely have to represent λ-terms. To this end, we adopt for each type ρ new sorts $\mathcal{V}_\rho$ and $\mathcal{T}_\rho$, that interpret variables and terms *modulo α-conversion* of type ρ, respectively. Constants of sort $\mathcal{V}_\rho$ are added to represent variables (written $\ulcorner x \urcorner$). Function symbols for typed application and abstraction are included as well. With $\ulcorner s \urcorner$, we denote the representation of a λ-term s in this first-order language, using the following function symbols:

$\mathsf{V}_\rho : \mathcal{V}_\rho \to \mathcal{T}_\rho$, to inject variables into terms;
$\_ \bullet_{\rho,\sigma} \_ : \mathcal{T}_{\rho\to\sigma} \times \mathcal{T}_\rho \to \mathcal{T}_\sigma$, denoting typed application;
$\boldsymbol{\lambda}_{\rho,\sigma} : \mathcal{V}_\rho \times \mathcal{T}_\sigma \to \mathcal{T}_{\rho\to\sigma}$, denoting typed abstraction.

Note that e.g. $\ulcorner \lambda x.x \urcorner \equiv \boldsymbol{\lambda}(\ulcorner y \urcorner, \mathsf{V}(\ulcorner y \urcorner))$, for some arbitrary but fixed choice of y. Although the terms in the intended model are taken modulo α-conversion, the first-order terms cannot have this feature. We will also need function symbols to represent simultaneous substitution: for any sequence of types $\sigma, \tau_1, \ldots, \tau_n$, a symbol $\_(\_,\_,\ldots := \_,\_,\ldots)$ of arity $\mathcal{T}_\sigma \times \mathcal{V}_{\tau_1} \times \cdots \times \mathcal{V}_{\tau_n} \times \mathcal{T}_{\tau_1} \times \cdots \times \mathcal{T}_{\tau_n} \to \mathcal{T}_\sigma$. The intended meaning of $s(\overline{x} := \overline{t})$ is the simultaneous substitution in s of x_i by

t_i. If for some i and j, x_i and x_j happen to be the same, the first occurrence from left to right takes precedence (so the other substitution is simply discarded).

In order to represent upper bounds for reduction sequences, we introduce a sort **nat**, denoting the natural numbers, with constants 0^{nat}, 1^{nat} and $\_+\_$ of arity $\mathsf{nat} \times \mathsf{nat} \to \mathsf{nat}$, with their usual meaning.

We let r, s and t range over terms of sorts $\mathcal{T}_\rho$; x and y are variables of sorts $\mathcal{V}_\rho$; m and n range over sort **nat**. We abbreviate $((s \bullet t_1) \bullet \cdots \bullet t_n)$ by $s \bullet \bar{t}$. Type decoration is often suppressed.

Finally, we add binary predicate symbols $\_ =_\rho \_$ for equality on sort $\mathcal{T}$ and SN_σ of arity $\mathcal{T}_\sigma \times \mathsf{nat}$, representing the relation of Definition 1(2).

We can now express the axioms that will be used in the formal proof. We will use the axiom schema **repl** : $s = t \to \varphi(s) \to \varphi(t)$ to replace equals by equals. Furthermore, we use all well typed instances of the following axiom schemata.

1. $\underline{\forall x}.\ \mathrm{SN}_\rho(\mathsf{V}(x), 0)$
2. $\underline{\forall x, \bar{t}, s, m, n}.\ \mathrm{SN}_{\rho\to\sigma}(\mathsf{V}(x) \bullet \bar{t}, m) \to \mathrm{SN}_\rho(s, n) \to \mathrm{SN}_\sigma(\mathsf{V}(x) \bullet \langle \bar{t}, s\rangle, m+n)$
3. $\underline{\forall s, x, m}.\ \mathrm{SN}_\sigma(s \bullet \mathsf{V}(x), m) \to \mathrm{SN}_{\rho\to\sigma}(s, m)$
4. $\underline{\forall r, y, \bar{x}, s, \bar{t}, \bar{r}, m, n}.\ \mathrm{SN}_\iota(r(y, \bar{x} := s, \bar{t}\,) \bullet \bar{r}, m) \to \mathrm{SN}_\rho(s, n) \to$
   $\mathrm{SN}_\iota(\boldsymbol{\lambda}(y, r)(\bar{x} := \bar{t}\,) \bullet \langle s, \bar{r}\,\rangle, m+n+1)$
5. $\underline{\forall \bar{t}}.\ t_i = \mathsf{V}(\ulcorner x_i \urcorner)(\ulcorner \bar{x} \urcorner := \bar{t}\,)$, provided i is the first occurrence of $\ulcorner x_i \urcorner$ in $\ulcorner \bar{x} \urcorner$.
6. $\underline{\forall r, s, \bar{x}, \bar{t}}.\ r(\bar{x} := \bar{t}\,) \bullet s(\bar{x} := \bar{t}\,) = (r \bullet s)(\bar{x} := \bar{t}\,)$
7. $\underline{\forall s, \bar{x}}.\ s(\bar{x} := \mathsf{V}(\bar{x})) = s$, where $\mathsf{V}(\bar{x})$ stands for $\mathsf{V}(x_1), \ldots, \mathsf{V}(x_m)$

In the formal proofs, we will refer to these axioms by number (e.g. ax_5). Axioms 1–3 express simple combinatorial facts about SN. The equations 5–7 axiomatize substitution. Axiom 4 is a mix, integrating a basic fact about reduction and an equation for substitution. The reason for this mixture is that we thus avoid variable name clashes. This is the only axiom that needs some elaboration.

In the intended model, $(\lambda x r)[\bar{x} := \bar{t}\,]$ equals $\lambda x(r[\bar{x} := \bar{t}\,])$, because we can perform an α-conversion, renaming x. However, we cannot postulate the similar equation

$$\forall x, \bar{x}, \bar{t}, r.\ \boldsymbol{\lambda}(x, r)(\bar{x} := \bar{t}\,) = \boldsymbol{\lambda}(x, r(\bar{x} := \bar{t}\,))$$

as an axiom, because we cannot avoid that e.g. t_1 gets instantiated by a term containing the free variable x, such that the same x would occur both bound and free[2]. Now in the proof of Lemma 4 it is shown how the reduction length of $(\lambda y.t)s\bar{r}$ can be estimated from the reduction lengths of s and $t[y := s]\bar{r}$. After substituting $r[\bar{x} := \bar{t}\,]$ for t, and using the abovementioned equation (thus avoiding that variables in $\bar{t}$ become bound), we get Axiom 4.

5.2 Proof Terms and Extracted Programs

As in the informal proof, we define formulae $\mathrm{SC}_\rho(t)$ by induction on the type ρ. These will occur as abbreviations in the formal derivations.

$$\begin{cases} \mathrm{SC}_\iota(t) := \exists n^{\mathsf{nat}} \mathrm{SN}_\iota(t, n) \\ \mathrm{SC}_{\rho\to\sigma}(t) := \underline{\forall s^{\mathcal{T}_\rho}} \mathrm{SC}_\rho(s) \to \mathrm{SC}_\sigma(t \bullet s) \end{cases}$$

[2] Strictly speaking, [1] erroneously ignores this subtlety.

Due to the underlined quantifier, $\tau(\mathrm{SC}_\sigma(s)) \equiv \sigma'$, where σ' is obtained from σ by renaming base types ι to **nat**. The underlined quantifier takes care that numerical upper bounds only use numerical information about subterms: the existential quantifier hidden in $\mathrm{SC}(t \bullet s)$ can only use the existential quantifier in $\mathrm{SC}(s)$; not s itself. In fact, this is the reason for introducing the underlined quantifier.

Formalizing the SC Lemma. We proceed by formalizing Lemma 3. We will define proofs

$$\begin{array}{l} \Phi_\rho : \underline{\forall t}.\, \mathrm{SC}_\rho(t) \to \exists n \mathrm{SN}_\rho(t,n) \quad \text{and} \\ \Psi_\rho : \underline{\forall x \bar{t}}.\, (\exists m \mathrm{SN}_\rho(\mathsf{V}(x) \bullet \bar{t}, m)) \to \mathrm{SC}_\rho(\mathsf{V}(x) \bullet \bar{t}\,) \end{array}$$

with simultaneous induction on ρ:

$$\begin{array}{l} \Phi_\iota := \underline{\lambda t} \lambda u^{\mathrm{SC}(t)} u \\ \Phi_{\rho\to\sigma} := \underline{\lambda t} \lambda u^{\mathrm{SC}(t)} \\ \qquad\qquad \exists^-[\underline{\Phi_\sigma(t \bullet \mathsf{V}(x))}\Big(u \underline{\mathsf{V}(x)} (\Psi_\rho \underline{x} \exists^+[0; (ax_1\underline{x})])\Big); \\ \qquad\qquad\quad m; v^{\mathrm{SN}(t\bullet\mathsf{V}(x),m)}; \\ \qquad\qquad\quad \exists^+[m; (ax_3\underline{txm}v)]] \end{array}$$

$$\begin{array}{l} \Psi_\iota := \underline{\lambda x \bar{t}} \lambda u^{\exists m \mathrm{SN}(\mathsf{V}(x)\bullet\bar{t},m)} u \\ \Psi_{\rho\to\sigma} := \underline{\lambda x \bar{t}}\ \lambda u^{\exists m \mathrm{SN}(\mathsf{V}(x)\bullet\bar{t},m)} \underline{\lambda s} \lambda v^{\mathrm{SC}(s)} \\ \qquad\qquad \exists^-[u; m; u_0^{\mathrm{SN}(\mathsf{V}(x)\bullet\bar{t},m)}; \\ \qquad\qquad\quad \exists^-[(\Phi_\rho \underline{s} v); n; v_0^{\mathrm{SN}(s,n)}; \\ \qquad\qquad\qquad \Psi_\sigma \underline{x \bar{t} s}\ \exists^+[(m+n); (ax_2\underline{x\bar{t}smn}u_0 v_0)]]] \end{array}$$

Having the concrete derivations, we can extract the computational content, using the definition of **ep**. Note that the underlined parts are discarded, and that an $\exists$-elimination gives rise to a substitution. The resulting functionals are $\mathbf{ep}(\Phi_\rho) : \rho \to \mathbf{nat}$ and $\mathbf{ep}(\Psi_\rho) : \mathbf{nat} \to \rho$,

$$\begin{aligned} \mathbf{ep}(\Phi_\iota) &= \lambda x_u x_u \\ \mathbf{ep}(\Phi_{\rho\to\sigma}) &= \lambda x_u m[m := \mathbf{ep}(\Phi_\sigma)(x_u(\mathbf{ep}(\Psi_\rho)0))] \\ &= \lambda x_u \mathbf{ep}(\Phi_\sigma)(x_u(\mathbf{ep}(\Psi_\rho)0)) \\ \mathbf{ep}(\Psi_\iota) &= \lambda x_u x_u \\ \mathbf{ep}(\Psi_{\rho\to\sigma}) &= \lambda x_u \lambda x_v \mathbf{ep}(\Psi_\sigma)(m+n)[n := \mathbf{ep}(\Phi_\rho)x_v][m := x_u] \\ &= \lambda x_u \lambda x_v \mathbf{ep}(\Psi_\sigma)(x_u + (\mathbf{ep}(\Phi_\rho)x_v)) \end{aligned}$$

Formalizing the Abstraction Lemma. We proceed by formalizing Lemma 4, which deals with abstractions. Let r have sort $\mathcal{T}_{\bar{\rho}\to\rho}$, and each r_i sort $\mathcal{T}_{\rho_i}$ (so $r \bullet \bar{r}$ has sort $\mathcal{T}_\rho$). Let s have sort $\mathcal{T}_\sigma$, y sort $\mathcal{V}_\sigma$, each t_i sort $\mathcal{T}_{\tau_i}$ and each x_i sort $\mathcal{V}_{\tau_i}$. We construct proofs

$$\Lambda_{\rho,\sigma,\bar{\rho},\bar{\tau}} : \underline{\forall r, y, \bar{x}, s, \bar{t}, \bar{r}}.\, \mathrm{SC}_\rho(r(y, \bar{x} := s, \bar{t}\,) \bullet \bar{r}\,) \to \mathrm{SC}_\sigma(s) \to \mathrm{SC}_\rho(\boldsymbol{\lambda}(y, r)(\bar{x} := \bar{t}\,) \bullet \langle s, \bar{r}\,\rangle)$$

by induction on ρ. This corresponds to the induction on ρ in the informal proof. The base case uses Axiom 4. Only the first two subscripts will be written in the sequel.

$$
\begin{aligned}
\Lambda_{\iota,\sigma} = \underline{\lambda r, y, \overline{x}, s, \overline{t}, \overline{r}}\, \lambda u^{\mathrm{SC}_\iota(r(y,\overline{x}:=s,\overline{t})\bullet\overline{r})} \lambda v^{\mathrm{SC}(s)} \\
\exists^-[u; m; u_0^{\mathrm{SN}(r(y,\overline{x}:=s,\overline{t})\bullet\overline{r},m)}; \\
\exists^-[(\Phi_\sigma \underline{s} v); n; v_0^{\mathrm{SN}(s,n)}; \\
\exists^+[m+n+1; (ax_4 \underline{ry\overline{x}s\overline{t}\overline{r}mnu_0 v_0})]]]
\end{aligned}
$$

$$
\begin{aligned}
\Lambda_{\rho\to\tau,\sigma} = \underline{\lambda r, y, \overline{x}, s, \overline{t}, \overline{r}}\, \lambda u^{\mathrm{SC}(r(y,\overline{x}:=s,\overline{t})\bullet\overline{r})} \lambda v^{\mathrm{SC}(s)} \\
\underline{\lambda r'} \lambda w^{\mathrm{SC}_\rho(r')} (\Lambda_{\tau,\sigma} \underline{ry\overline{x}s\overline{t}\overline{r}r'} (u \underline{r'} w) v)
\end{aligned}
$$

Having these proofs, we can extract their programs, using the definition of **ep**. In this way we get $\mathbf{ep}(\Lambda_{\rho,\sigma}) : \rho \to \sigma \to \rho$,

$$
\begin{aligned}
\mathbf{ep}(\Lambda_{\iota,\sigma}) &= \lambda x_u \lambda x_v (m+n+1)[n := \mathbf{ep}(\Phi_\sigma)x_v][m := x_u] \\
&= \lambda x_u \lambda x_v (x_u + (\mathbf{ep}(\Phi_\sigma)x_v) + 1) \\
\mathbf{ep}(\Lambda_{\rho\to\tau,\sigma}) &= \lambda x_u \lambda x_v \lambda x_w (\mathbf{ep}(\Lambda_{\tau,\sigma})(x_u x_w) x_v)
\end{aligned}
$$

Formalizing the Main Lemma. The main lemma (5) states that every term s is strongly computable, even after substituting strongly computable terms for variables. The informal proof of Lemma 5 is with induction on s. Therefore, we can only give a formal proof for each s separately. Given a term s with all free variables among $\overline{x}$, we construct by induction on the term structure a proof

$$\Pi_{s,\overline{x}} : \forall \underline{t_1, \ldots, t_n}. \mathrm{SC}(t_1) \to \cdots \to \mathrm{SC}(t_n) \to \mathrm{SC}(\ulcorner s \urcorner (\ulcorner \overline{x} \urcorner := \overline{t})).$$

$$
\begin{aligned}
&\Pi_{x_i,\overline{x}} := \underline{\lambda \overline{t}} \lambda \overline{u} (\mathbf{repl}\ (ax_5\, \underline{\overline{t}})\ u_i) \\
&\Pi_{rs,\overline{x}} := \underline{\lambda \overline{t}} \lambda \overline{u} (\mathbf{repl}\ (ax_6\, \underline{\ulcorner r \urcorner \ulcorner s \urcorner \ulcorner \overline{x} \urcorner \overline{t}})\ (\Pi_{r,\overline{x}} \underline{\overline{t}} \overline{u}\, \underline{\ulcorner s \urcorner (\ulcorner \overline{x} \urcorner := \overline{t})}\ (\Pi_{s,\overline{x}} \underline{\overline{t}} \overline{u}))) \\
&\Pi_{\lambda x r,\overline{x}} := \underline{\lambda \overline{t}} \lambda \overline{u} \underline{\lambda s} \lambda v^{\mathrm{SC}(s)} (\Lambda_{\rho,\sigma} \underline{r' \ulcorner y \urcorner \ulcorner \overline{x} \urcorner s \overline{t}}\ (\Pi_{r',y,\overline{x}} \underline{s \overline{t}} v \overline{u}) v),
\end{aligned}
$$

where in the last equation, we assume that $\ulcorner \lambda x r \urcorner = \boldsymbol{\lambda}(\ulcorner y \urcorner, r')$, with $x : \sigma$ and $r : \rho$.

Again we extract the programs from these formal proofs. Because the realizer of **repl** is the identity, we can safely drop it from the extracted program. For terms s^σ with free variables among $\overline{x}$, each $x_i : \tau_i$, we get $\mathbf{ep}(\Pi_{s,\overline{x}}) : \overline{\tau} \to \sigma$,

$$
\begin{aligned}
\mathbf{ep}(\Pi_{x_i,\overline{x}}) &= \lambda \overline{x}_u x_{u,i} \\
\mathbf{ep}(\Pi_{rs,\overline{x}}) &= \lambda \overline{x}_u (\mathbf{ep}(\Pi_{r,\overline{x}}) \overline{x}_u (\mathbf{ep}(\Pi_{s,\overline{x}}) \overline{x}_u)) \\
\mathbf{ep}(\Pi_{\lambda x r,\overline{x}}) &= \lambda \overline{x}_u \lambda x_v (\mathbf{ep}(\Lambda_{\rho,\sigma})(\mathbf{ep}(\Pi_{r',y,\overline{x}}) x_v \overline{x}_u) x_v),
\end{aligned}
$$

where again it is assumed that $\ulcorner \lambda x r \urcorner = \boldsymbol{\lambda}(\ulcorner y \urcorner, r')$, $x : \sigma$ and $r : \rho$.

Formalization of the Theorem. Now we are able to give a formal proof of $\exists n\mathrm{SN}(\ulcorner s\urcorner, n)$, for any term s. Extracting the computational content of this proof, we get an upper bound for the length of reduction sequences starting from s. We will define formal proofs $\Omega_s : \exists n\mathrm{SN}(\ulcorner s\urcorner, n)$ for each term s ($\ulcorner s\urcorner$ denotes the representation of s). Let $\overline{x}$ be the sequence of free variables in $s : \sigma$, each $x_i : \tau_i$.
$\Omega_s := (\Phi_\sigma \underline{\ulcorner s\urcorner}(\mathbf{repl}\ (ax_7 \underline{\ulcorner s\urcorner \ulcorner \overline{x}\urcorner})\ (\Pi_{s,\overline{x}} \underline{\mathsf{V}(\ulcorner \overline{x}\urcorner)\Psi_1 \cdots \Psi_n}))),$

where $\Psi_i := (\Psi_{\tau_i} \underline{\mathsf{V}(\ulcorner x_i\urcorner)} \exists^+[0; (ax_1 \ulcorner x_i\urcorner)])$ is a proof of $\mathrm{SC}(\mathsf{V}(\ulcorner x_i\urcorner))$ (Ψ is defined in Section 5.2) and $\mathsf{V}(\ulcorner\overline{x}\urcorner)$ stands for $\mathsf{V}(\ulcorner x_1\urcorner), \cdots, \mathsf{V}(\ulcorner x_n\urcorner)$. As extracted program, we get $\mathbf{ep}(\Omega_s) : \mathsf{nat}$,

$$\mathbf{ep}(\Omega_s) = \mathbf{ep}(\Phi_\sigma)(\mathbf{ep}(\Pi_{s,\overline{x}})(\mathbf{ep}(\Psi_{\tau_1})0) \cdots (\mathbf{ep}(\Psi_{\tau_n})0))$$

5.3 Comparison with Gandy's Proof

In order to compare the extracted programs from the formalized proofs with the strictly monotonic functionals used by Gandy [2], we recapitulate these programs and introduce a readable notation for them.

$$\begin{array}{rll} M_\sigma : \sigma \to \mathsf{nat} & := \mathbf{ep}(\Phi_\sigma) \\ S_\sigma : \sigma & := \mathbf{ep}(\Psi_\sigma)0 \\ L_{\rho,\sigma} : \rho \to \sigma \to \rho & := \mathbf{ep}(\Lambda_{\rho,\sigma}) \\ [\![s^\sigma]\!]_{\overline{x}\mapsto\overline{t}} : \sigma & := \mathbf{ep}(\Pi_{s,\overline{x}})\overline{t} \\ \mathsf{Upper}[\mathsf{t}] : \mathsf{nat} & := \mathbf{ep}(\Omega_t). \end{array}$$

Function application is written more conventionally as $f(x)$ and some recursive definitions are unfolded. Assuming that $\sigma = \sigma_1 \to \cdots \to \sigma_n \to \mathsf{nat}$, these functionals obey the following equations:

$$\begin{aligned} M_\sigma(f) &= f(S_{\sigma_1}, \ldots, S_{\sigma_n}) \\ S_\sigma(\overline{x}) &= M_{\sigma_1}(x_1) + \cdots + M_{\sigma_n}(x_n) \\ L_{\sigma,\tau}(f, y, \overline{x}) &= f(\overline{x}) + M_\tau(y) + 1 \\ [\![x_i]\!]_{\overline{x}\mapsto\overline{t}} &= t_i \\ [\![rs]\!]_{\overline{x}\mapsto\overline{t}} &= [\![r]\!]_{\overline{x}\mapsto\overline{t}}([\![s]\!]_{\overline{x}\mapsto\overline{t}}) \\ [\![\lambda x^\sigma r^\rho]\!]_{\overline{x}\mapsto\overline{t}}(y) &= L_{\rho,\sigma}([\![r]\!]_{x,\overline{x}\mapsto y,\overline{t}}, y) \\ \mathsf{Upper}[t^\tau] &= M_\tau([\![t]\!]_{\overline{x}\mapsto\overline{S}}). \end{aligned}$$

The Correctness Theorem 10 guarantees that $\mathrm{SN}(\ulcorner t\urcorner, \mathsf{Upper}[t])$ is provable in NH, so $\mathsf{Upper}[t]$ puts an upper bound on the length of reduction sequences from t. This expression can be compared with the functionals in the proof of Gandy.

First of all, the ingredients are the same. In [2] a functional (say G) is defined playing the rôle of both S and M (and indeed, $S_{\sigma\to\mathsf{nat}} = M_\sigma$). S is a special strictly monotonic functional and M serves as a measure on functionals. Then Gandy gives a non-standard interpretation t^* of a term t, by assigning the special strict functional to the free variables, and interpreting λ-abstraction by a λI term, so that reductions in the argument will not be forgotten. This corresponds to our $[\![t]\!]_{\overline{x}\mapsto\overline{S}}$, where in the λ-case the argument is remembered by $L_{\rho,\sigma}$ and eventually added to the result. Finally, Gandy shows that in each reduction step the measure of the assigned functionals decreases. So the measure of the non-standard interpretation serves as an upper bound.

Looking into the details, there is one slight difference. The bound Upper[t] is sharper than the upper bound given by Gandy. The reason is that Gandy's special functional (resembling S and M by us) is inefficient. It obeys the equation (with $\sigma \equiv \sigma_1 \to \cdots \to \sigma_n \to \mathsf{nat}$)

$$G_\sigma(x_1, \ldots, x_n) := G_{\sigma_1 \to \mathsf{nat}}(x_1) + 2^0 G_{\sigma_2 \to \mathsf{nat}}(x_2) + \cdots + 2^{n-2} G_{\sigma_n \to \mathsf{nat}}(x_n).$$

Gandy defines G_σ with a + functional on all types and a peculiar induction. By program extraction, we found functionals defined by simultaneous induction, using an extra argument as accumulator (see the definition of $\mathbf{ep}(\Phi)$ and $\mathbf{ep}(\Psi)$), thus avoiding the + functional and the implicit powers of 2.

We conclude this section by stating that program extraction provides a tool to compare the two SN-proofs in the case of simply typed λ-calculus.

6 Application to Gödel's T

Gödel's T extends simply typed λ-calculus with higher-order primitive recursion. The set of base types is extended with a type o of natural numbers. We let p and q range over terms of type o. Constants 0° and $S^{o \to o}$ are added. For each type σ, we add a constant $R_\sigma : \sigma \to (o \to \sigma \to \sigma) \to o \to \sigma$. The following rules express higher-order primitive recursion:

$$R_\sigma st0 \mapsto s \qquad \text{and} \qquad R_\sigma st(Sp) \mapsto tp(R_\sigma stp) \ .$$

With $\to_{\beta R}$ we denote the compatible closure of the β rule and the two recursion rules. It is a well known fact that $\to_{\beta R}$ is a terminating rewrite relation.

The proof à la Tait of this fact (see e.g. [10, 2.2.31]) extends the case of the β-rule, by proving that the new constants are strongly computable. We will present a version with concrete upper bounds. It turns out to be rather cumbersome to give a concrete number. Some effort has been put in identifying and proving the right "axioms" (Lemma 12–15) from which the decorated proof can be constructed (Lemma 16, 17). The extracted upper bounds are compared with the functionals used by Gandy (Section 6.3).

6.1 Changing the Interpretation of SN(t, n)

Consider the following consequence of $\mathrm{SC}_{o \to o}(r)$ for fixed r. This formula is equivalent to $\underline{\forall p}\forall m.\mathrm{SN}(p, m) \to \exists n \mathrm{SN}(rp, n)$. So we can bound the reduction length of rp uniformly in the upper bound for p. More precisely, if $\mathrm{SN}(p, m)$ then $\mathrm{SN}(rp, [\![r]\!](m))$. A stronger uniformity principle appears in [11, § 2.3.4]).

The uniformity principle does not hold if we substitute $R_o st$ for r: Although $\mathrm{SN}(S^k 0, 0)$ holds for each k, $Rst(S^k 0)$ can perform k reduction steps. So $\mathrm{SC}(Rst)$ cannot hold. This shows that it is impossible to prove $\mathrm{SC}(R)$ with SC as in

Definition 2. Somehow, the numerical value (k) has to be taken into account too.

To proceed, we have to change the interpretation of the predicate SN(t,n). We have to be a bit careful here, because speaking about *the* numerical value of a term s would mean that we assume the existence of a unique normal form. The following definition avoids this assumption:

Definition 11. 1. Second interpretation of SN: SN(t,n) holds if and only if for all reduction sequences of the form $t \equiv s_0 \to_{\beta R} s_1 \to_{\beta R} \cdots \to_{\beta R} s_m \equiv S^k(r)$, we have $m + k \leq n$. Note that k can only be non-zero for terms of type o.
2. A finite reduction sequence $s_0 \to_{\beta R} \cdots \to_{\beta R} s_n$ is *maximal* if s_n is normal (i.e. there is no term t with $s_n \to_{\beta R} t$). An infinite reduction sequence is always maximal.

So SN(t,n) means that for any reduction sequence from t to some s, n is at least the length of this sequence plus the number of leading S-symbols in s. Note that SN(t,n) already holds, if n bounds the length plus value of all *maximal* reduction sequences from t.

We settle the important question to what extent the proofs of Section 5 remain valid. Because these are formal proofs, with SN just as a predicate symbol, the derivation terms remain correct. These derivation terms contain axioms, the validity of which was shown in the intended model. But we have changed the interpretation of the predicate symbol SN. So what we have to do, is to verify that the axioms of Section 5.1 remain correct in the new interpretation.

The axiom schema **repl**, ax_5, ax_6 and ax_7 are independent of the interpretation of SN. Axioms 1, 2 and 3 remain true, because the terms in their conclusion have no leading S-symbols (note that 1 and 2 have a leading variable; 3 is of arrow type). Axiom 4 is proved by a slight modification of the proof of Lemma 4. The following observation is used: If $(\lambda x.s)t\overline{r} \to^*_{\beta R} S^\ell(q)$, then at some point we contract the outermost β-redex, say $(\lambda x.s')t'\overline{r'} \to_\beta s'[x := t']\overline{r'}$. The latter is also a reduct of $s[x := t]\overline{r}$, so ℓ is bounded by the upper bound for the numerical value of this term.

6.2 Informal Decorated Proof

To prove SC(0), SC(S) and SC(R), we need some axioms, expressing basic truths about SN. In this section, $\to$ is written for $\to_{\beta R}$. For 0 and S we have:

Lemma 12. *1.* SN($0^o, 0^{\mathsf{nat}}$)
2. For all terms p and numbers m, $\text{SN}_o(p,m)$ *implies* $\text{SN}_o((Sp), m+1)$.

Proof. 0 is normal and has no leading S-symbols. If $(Sp) \to^n S^k(r)$ for some n, k and r, then $p \to^n S^{k-1}(r)$. From SN(p,m) we obtain $k + n \leq m + 1$. This holds for every reduction sequence, so SN($(Sp), m+1$) holds. □

It is less clear which facts we need for the recursion operator. To prove SC(R_σ) (see Lemma 17), we need to prove $\text{SC}_\sigma(Rstp)$ for strongly computable

s, t and p. If p is strongly computable, then $\mathrm{SN}(p,m)$ holds for some m. With induction on m, we will prove $\forall p(\mathrm{SN}(p,m) \to \mathrm{SC}_\sigma(Rstp))$. We need two axioms to establish the base case and the step case of this induction. For the base case, we need (schematic in the type σ):

Lemma 13.

$$\forall s,t,\overline{r},p,\ell,n.\ \mathrm{SN}_\iota(s\overline{r},\ell) \to \mathrm{SN}_{o\to\sigma\to\sigma}(t,n) \to \mathrm{SN}_o(p,0) \to \mathrm{SN}_\iota(R_\sigma stp\overline{r},\ell{+}n{+}1)$$

Proof. Assume $\mathrm{SN}(s\overline{r},\ell)$, $\mathrm{SN}(t,n)$ and $\mathrm{SN}(p,0)$. The latter assumption tells that p is normal and cannot be a successor. If $p \not\equiv 0$, then reductions in $Rstp\overline{r}$ can only occur inside s, t and $\overline{r}$, and these are bounded by $\ell+n$. If $p \equiv 0$, then a maximal reduction of $Rstp\overline{r}$ will consist of first some steps within s, t and $\overline{r}$ (of respectively a, b and c steps, say) followed by an application of the first recursion rule, and finally d more steps. This gives a reduction of the form:

$$Rst0\overline{r} \to^{a+b+c} Rs't'0\overline{r'} \to s'\overline{r'} \to^d S^i(r)$$

We can construct a reduction sequence from $s\overline{r}$ via $s'\overline{r'}$ to $S^i(r)$ of length $a{+}c{+}d$. By the first assumption, $a+c+d+i \leq \ell$, by the second assumption $b \leq n$, so $a+b+c+1+d+i \leq \ell+n+1$. As this upper bound holds for an arbitrary maximal reduction sequence, it holds for all reduction sequences, so we get $\mathrm{SN}(Rstp\overline{r},\ell+n+1)$. □

The next lemma is needed for the step case. Note that if $\mathrm{SN}(p,m+1)$ holds, then p may reduce to either 0 (in at most $m+1$ steps) or to (Sp') (in at most m steps). This explains the first two hypotheses of the following lemma.

Lemma 14.

$$\forall s,t,\overline{r},\ell,m,n.\ \mathrm{SN}_\iota(s\overline{r},\ell) \to \Big(\forall q.\mathrm{SN}_o(q,m) \to \mathrm{SN}_\iota(tq(Rstq)\overline{r},n)\Big) \to \Big(\forall p.\mathrm{SN}_o(p,m+1) \to \mathrm{SN}_\iota(Rstp\overline{r},\ell+m+n+2)\Big)$$

Proof. Assume $\mathrm{SN}_\iota(s\overline{r},\ell)$, $\forall q.\mathrm{SN}(q,m) \to \mathrm{SN}_\iota(tq(Rstq)\overline{r},n)$ and $\mathrm{SN}(p,m+1)$, for arbitrary $s,t,\overline{r},\ell,m,n$ and p. Consider an arbitrary maximal reduction sequence from $Rstp\overline{r}$. It consists of reduction steps inside s, t, p and $\overline{r}$ (of a, b, c and d steps to the terms s', t', p' and $\overline{r'}$, respectively), possibly followed by an application of a recursion rule, concluded by some more steps. We make a case distinction to the shape of the reduct p' after these steps:

Case A: $p' \equiv 0$ Then the maximal reduction has the following shape:

$$Rstp\overline{r} \to^{a+b+c+d} Rs't'0\overline{r'} \to s'\overline{r'} \to^e S^i(r)$$

We can construct a reduction from $s\overline{r}$ to $S^i(r)$ of $a+d+e$ steps, hence, by the first assumption, $a{+}d{+}e{+}i \leq \ell$. From the third assumption, we get $c \leq m{+}1$. To bound b, we can only use the second hypothesis. Note that $\mathrm{SN}(0,0)$ and hence $\mathrm{SN}(0,m)$ holds. The second assumption applied to 0 yields $\mathrm{SN}(t0(Rst0)\overline{r},n)$, so necessarily $b \le n$. Now the reduction sequence can be bounded, viz. $a+b+c+d+1+e+i \le \ell+m+n+2$.

Case B: $p' \equiv (Sq)$ Then the maximal reduction has the following shape:

$$Rstp\overline{r} \rightarrow^{a+b+c+d} Rs't'(Sq)\overline{r'} \rightarrow t'q(Rs't'q)\overline{r'} \rightarrow^{e} S^i(r)$$

First, $\mathrm{SN}(q,m)$ holds, because if $q \rightarrow^j S^k(q')$, then $p \rightarrow^{c+j} S^{k+1}(q')$, so $c+j+k+1 \le m+1$, hence $j+k \le m$. Next note, that there is a reduction from $tq(Rstq)\overline{r}$ to $S^i(r)$ of $a+2b+d+e$ steps. Now the second assumption can be applied, which yields that $a+2b+d+e+i \le n$. Finally, $c \le m$. Adding up all information, we get $a+b+c+d+1+e+i \le m+n+1$.

Case C: If cases A and B do not apply, then p' is normal (because a maximal reduction sequence is considered), and no recursion rule applies. The reduction sequence has length $a+b+c+d$ and the result has no leading S-symbols. Now $c \le m+1$, $a+d \le \ell$ and $b \le n$ can be obtained as in Case A. Clearly $a+b+c+d \le \ell+m+n+1$.

In all cases, the length of the maximal reduction plus the number of leading S-symbols is bounded by $\ell+m+n+2$, so indeed $\mathrm{SN}(Rstp\overline{r}, \ell+m+n+2)$ holds. □

The nice point is that this lemma is $\exists$-free, so it hides no computational content. Unfortunately, it is not strong enough to enable the induction step. We have $\forall q.\mathrm{SN}(q,m) \rightarrow \mathrm{SC}(Rstq)$ as induction hypothesis, and we may assume $\mathrm{SN}(p, m+1)$. In order to apply Lemma 14, we are obliged to give an n, such that $\forall q.\mathrm{SN}(q,m) \rightarrow \mathrm{SN}(tq(Rstq)\overline{r}, n)$ holds, but using the induction hypothesis we can only find an n for each q separately.

The solution of this problem relies on the fact that the upper bound n above does not really depend on q. In the formalism of Section 4, this is expressed by the $\underline{\forall q}$-quantifier. We change the lemma accordingly:

Lemma 15.

$$\begin{aligned}\underline{\forall s,t,\overline{r},\ell,m.}\ & \mathrm{SN}_\iota(s\overline{r},\ell) \rightarrow \\ & \Big(\underline{\forall q}.\mathrm{SN}_o(q,m) \rightarrow \exists n \mathrm{SN}_\iota(tq(Rstq)\overline{r},n)\Big) \rightarrow \\ & \Big(\underline{\forall p}.\mathrm{SN}_o(p,m+1) \rightarrow \exists n \mathrm{SN}_\iota(Rstp\overline{r},\ell+m+n+2)\Big)\end{aligned}$$

The justification of this lemma has to be given in terms of NH, as pointed out in Section 4.3. Lemma 15 contains existential quantifiers, so we have to insert a realizer. Of course we take as realizer $\lambda n.n$. Now it can be verified that

$$\lambda n.n \ \mathbf{mr}\ (\text{Lemma 15}) \equiv (\text{Lemma 14})\ .$$

Eventually, we can prove that the new constants are strongly computable. The Numeral Lemma is a direct consequence of Lemma 12. The Recursor Lemma uses Lemmas 13, 15 and 3. The SC-formula is an abbreviation introduced in Section 5.2. The proofs below are in MF, so the underlining is important.

Lemma 16 (Numeral Lemma). SC(0) *and* SC(S).

Lemma 17 (Recursor Lemma). *For all* σ, SC(R_σ) *is strongly computable.*

Proof. Note that R_σ has type $\sigma \to (o \to \sigma \to \sigma) \to o \to \sigma$. We assume SC($s$), SC($t$) and SC($p$) for arbitrary terms s, t and p. We have to show $\mathrm{SC}_\sigma(R_\sigma stp)$. From the definition of $\mathrm{SC}_o(p)$ we obtain $\exists m \mathrm{SN}(p,m)$. Now $\forall m \underline{\forall p}.\mathrm{SN}(p,m) \to \mathrm{SC}(Rstp)$ is proved by induction on m, which finishes the proof.

Case 0: Let SN(p, 0). Let arbitrary, strongly computable $\overline{r}$ be given. We have to prove $\exists k \mathrm{SN}(Rstp\overline{r}, k)$. From SC($s$) and SC($\overline{r}$) we get SC($s\overline{r}$), hence SN($s\overline{r}, \ell$) for some ℓ (using the definition of SC repeatedly). Lemma 3 and the assumption SC(t) imply SN(t, n) for some n. Now Lemma 13 applies, yielding $\mathrm{SN}(Rstp\overline{r}, \ell + n + 1)$. So we put $k := \ell + n + 1$.

Case $m+1$: Assume $\underline{\forall q}.\mathrm{SN}(q,m) \to \mathrm{SC}(Rstq)$ (IH) and $\mathrm{SN}(p, m+1)$. Let arbitrary, strongly computable $\overline{r}$ be given. We have to prove $\exists k \mathrm{SN}(Rstp\overline{r}, k)$. As in Case 0, we obtain SN($s\overline{r}, \ell$) for some ℓ. In order to apply Lemma 15, we additionally have to prove $\underline{\forall q}.\mathrm{SN}(q,m) \to \exists n \mathrm{SN}(tq(Rstq)\overline{r}, n)$.

So assume SN(q, m) for arbitrary q. This implies SC(q) and, by IH, SC($Rstq$). Now by definition of SC(t), we have SC($tq(Rstq)\overline{r}$), i.e. SN($tq(Rstq)\overline{r}, n$) for some n. Now Lemma 15 applies, yielding $\mathrm{SN}(Rstp\overline{r}, \ell + m + n' + 2)$ for some n'. We put $k := \ell + m + n' + 2$. □

6.3 Extracted Programs compared with Gandy's Functionals

The informal proof of the previous section can be formalized in the extension of MF, obtained by adding induction axioms to it. System NH has to be extended with induction axioms accordingly. We also have to add the (simultaneous) primitive recursors, in order to get realizers of the induction axioms (see [10, § 1.6.16, § 3.4.5]). Objects in the extended NH are regarded modulo βR-equality.

We will omit the formal proofs here, due to lack of space[3]. Instead, we directly give the program that can be extracted from the formalized proof. This program will contain the primitive recursor R_σ, because Lemma 17 contains induction to a formula φ with $\tau(\varphi) = \sigma$. Using the notation of Section 5.3, the extracted functionals read:

$$\begin{aligned} [\![0]\!] &= 0 \\ [\![S]\!](m) &= m+1 \\ [\![R_\sigma]\!](x, f, 0, \overline{z}) &= x(\overline{z}) + M_{o\to\sigma\to\sigma}(f) + 1 \\ [\![R_\sigma]\!](x, f, m+1, \overline{z}) &= x(\overline{z}) + m + f(m, [\![R_\sigma]\!](x, f, m), \overline{z}) + 2 \end{aligned}$$

These clauses can be added to the definition of $[\![\_]\!]$ (Section 5.3), which now assigns a functional to each term of Gödel's T. This also extends Upper[_], which now computes the upper bound for reduction lengths of terms in Gödel's T. But, due to the changed interpretation of the SN-predicate, we know even more. In

[3] The formal proofs are in the full version.

fact, Upper[t] puts an upper bound on the length ***plus the numerical value*** of each reduction sequence. More precisely, if $t \to^i S^j(t')$ then $i + j \leq$ Upper[t].

Gandy's SN-proof can be extended by giving a strictly monotonic interpretation R^* of R, such that the recursion rules are decreasing. The functional used by Gandy resembles the one above, but gives larger upper bounds. It obeys the following equations:

$$R^*(x, f, 0, \overline{z}) = x(\overline{z}) + G(f) + 1$$
$$R^*(x, f, m+1, \overline{z}) = f(m, R^*(x, f, m), \overline{z}) + R^*(x, f, m, \overline{z}) + 1.$$

Here G is Gandy's version of the functional M (see Section 5.3). Clearly, the successor step of R^* uses the previous result twice, whereas $[\![R]\!]$ uses it only once. Both are variants of the usual recursor. In the base case, the step function f is remembered by both. This is necessary, because the first recursor rule drops its second argument, while reductions in this argument may not be discarded. In step $m + 1$ the two versions are really different; R^* adds the results of the steps $0, \cdots, m$, while $[\![R]\!]$ only adds the result of step 0 and the numerical argument m. The addition of the result of step 0 is necessary to achieve strict monotonicity of $[\![R]\!]$ in its third argument.

7 Conclusion

With two case studies we showed, that modified realizability is a useful tool to reveal the similarity between SN-proofs using strong computability and SN-proofs using strictly monotonic functionals. The extra effort for Gödel's T has paid off, because we found sharper upper bounds than in [2, 7]. Moreover, the new upper bound puts a bound on the sum of the length and numerical value of reduction sequences. This information helps to improve the proof that uses strictly monotonic functionals.

We think that our method can be applied more often. In a typical computability proof SC-predicates are defined with induction on types. It is then proved by induction on terms, that any term satisfies SC. By induction on the types, SN follows. After decorating such a proof with an administration for reduction lengths, the appropriate modified realizability interpretation maps SC-predicates to functionals of the original type and SN-predicates to numbers. The extracted program follows the induction on terms to obtain a non-standard interpretation of the term. This object is mapped to an upper bound by the proof that SC implies SN.

The realizability interpretation follows the type system closely. To deal with Gödel's T, induction was added. In the same way, conjunction and disjunction can be added to deal with products and coproducts (see also [2]). Recently, Loader [5] extended Gandy's proof to System F. As he points out, Girard's SN proof for System F (using reducibility candidates, see e.g. [3]) can be decorated, after which modified realizability yields the same upper bound expressions. Another extension could deal with recursion over infinitely branching trees.

A problem arises with the permutative conversions for existential quantifiers in first order logic. Prawitz [8] gives an SN-proof using strong validity (SV). In [7] an SN-proof is given based on strict functionals. The SV-predicate is defined using a general inductive definition, hence the computational contents of Prawitz' proof is not clear. Consequently, the two SN-proofs cannot be related with our method.

The latter system, and also Gödel's T, can be seen as instances of higher-order term rewrite systems. In [6, 7] a method is given to use strict functionals in termination proofs for such rewrite systems. The connection with computability should help in finding strict functionals for such proofs. One could for example extract functionals from a computability proof for a core system, and then change them by hand to obtain termination of a richer system.

The connection between computability and functionals gives rise to the following questions: Are the functionals extracted from SN-proofs always strictly monotonic? What are right notions of strict monotonicity for higher type systems? Can some "easy" classes of strictly monotonic functionals be identified?

References

1. U. Berger. Program extraction from normalization proofs. In M. Bezem and J.F. Groote ed., *Proc. of TLCA '93*, Utrecht, volume 664 of *LNCS*, pages 91–106. Springer Verlag, 1993.
2. R.O. Gandy. Proofs of strong normalization. In J.R. Hindley and J.P. Seldin ed., *To H.B. Curry: Essays on Combinatory Logic, Lambda Calculus and Formalism*, pages 457–477. Academic Press, London, 1980.
3. J.-Y. Girard, Y. Lafont, and P. Taylor. *Proofs and Types*, volume 7 of *Cambridge tracts in theoretical computer science*. Cambridge University Press, 1989.
4. G. Kreisel. Interpretation of analysis by means of constructive functionals of finite types. In A. Heyting ed., *Constructivity in Mathematics*, pages 101–128. North-Holland, 1959.
5. R. Loader. Normalisation by translation. http://sable.ox.ac.uk/ loader/, 1995.
6. J.C. van de Pol. Termination proofs for higher-order rewrite systems. In J. Heering et al ed., *Proc. of HOA '93*, volume 816 of *LNCS*, pages 305–325. Springer Verlag, 1994.
7. J.C. van de Pol and H. Schwichtenberg. Strict functionals for termination proofs. In M. Dezani-Ciancaglini and G. Plotkin ed., *Proc. of TLCA'95*, volume 902 of *LNCS*, pages 350–364. Springer Verlag, 1995.
8. D. Prawitz. Ideas and results in proof theory. In J.E. Fenstad ed., *Proc. of the 2nd Scandinavian Logic Symposium*, pages 235–307, Amsterdam, 1971. North-Holland.
9. W.W. Tait. Intensional interpretation of functionals of finite types I. *JSL*, 32:198–212, 1967.
10. A.S. Troelstra. *Metamathematical Investigation of Intuitionistic Arithmetic and Analysis*. Number 344 in LNM. Springer Verlag, Berlin, 1973. A 2nd corrected edition appeared as ILLC X-93-05, University of Amsterdam.
11. R. de Vrijer. Exactly estimating functionals and strong normalization. *Proc. of the Koninklijke Nederlandse Akademie van Wetenschappen*, 90(4):479–493, Dec 1987.

Third-Order Matching in the Polymorphic Lambda Calculus

Jan Springintveld*

Computing Science Institute
Department of Mathematics and Computer Science
University of Nijmegen
P.O. Box 9010, 6500 GL Nijmegen, The Netherlands
email: jans@cs.kun.nl

Abstract. We show that it is decidable whether a third-order matching problem in the polymorphic lambda calculus has a solution. We give an algorithm that, given such a problem, returns a substitution if it has a solution and *fail* otherwise.

1 Introduction

This paper concerns the study of higher-order matching. It is well-known that higher-order matching is a central technique for higher-order rewriting and proof development in type theory. In this paper, we study higher-order matching in the polymorphic lambda calculus ($\lambda 2$) [Gir72, Rey74]. From [Dow93b], it follows that higher-order matching is undecidable in this calculus. It is important to understand why higher-order matching is undecidable and to what extent subcases of the full problem still *are* decidable. In particular, one would like to know if the full power of polymorphism is required for the undecidability. In this paper, we study matching in a fragment of the polymorphic lambda calculus and prove that matching is decidable in case the types of the variables have order at most 3.

The proof is a reduction to third-order matching in the simply typed lambda calculus ($\lambda\tau$), which is decidable, as proven in [Dow94] (recently this result was extended to the fourth-order case [Pad94b, Pad94a]). It consists of two parts. One part is a translation that maps third-order matching problems P in $\lambda 2$ of a certain restricted format (and their solutions) to third-order matching problems $[\![P]\!]$ in $\lambda\tau$ (and their solutions). We derive from $[\![P]\!]$ a natural number n, depending only on P, with the following property: if P has a solution then it has a solution with number of symbols less than n. The existence of such an upper bound suffices to establish decidability of third-order matching in the

* The research for this paper has been carried out at the Department of Philosophy of Utrecht University. The preparation of this paper took for a large part place at CWI, Amsterdam. This work is supported by the Netherlands Computer Science Research Foundation (SION) with financial support of the Netherlands Organisation for Scientific Research (NWO).

restricted format. The other part of our proof is a reduction of the general format to the restricted format.

The present proof only works for matching problems of order 3; it remains to generalise the algorithm to higher finite orders. Another feature of the proof is that it relies on the use of fresh variables (declared in auxiliary contexts in substitutions). In other words, it relies on the assumption that all types are inhabited. If this is not assumed, third-order matching is not decidable, because one can encode the undecidable problem of inhabitation in $\lambda 2$ (see [Löb76]) as a third-order matching problem (for details, see Chapter 4 of [Spr95b]).

We recall some results concerning higher-order matching in the subsystems of the Calculus of Constructions [CH88], as given by the lambda cube [Bar92]. Second-order matching is decidable in all systems of the lambda cube [Dow91d]. Third- and higher-order matching is undecidable in systems with dependent types [Dow91b]. According to the erratum to [Dow91b], fourth-order matching is undecidable in $\lambda\omega$, the extension of $\lambda 2$ with type constructors. An open problem is whether third-other matching is decidable in $\lambda\omega$; it is decidable in $\lambda\underline{\omega}$, a system with only type constructors [Spr95a]. Furthermore, the decidability of matching of order $n \geq 4$ in $\lambda\underline{\omega}$ and of matching of order $n \geq 5$ in the simply typed lambda calculus are still open. More information on higher-order matching in the simply typed lambda calculus can be found in [Wol93].

This paper is organised as follows. In Section 2 we recapitulate the systems $\lambda\tau$ and $\lambda 2$. In Section 3 we present the notions of order, matching problems and solutions. We will also give an explicit description of third-order types. In Section 4, we describe and motivate the restricted format. In Section 5, we reduce third-order matching to restricted third-order matching. Finally, we present in Section 6, our translation and prove decidability of restricted third-order matching. The full version of this paper is Chapter 4 of [Spr95b].

2 The systems $\lambda\tau$ and $\lambda 2$

In this section we introduce two typed lambda calculi: $\lambda\tau$, the simply typed lambda calculus with one base type O, and $\lambda 2$, the polymorphic lambda calculus (the Π-fragment of System F [Gir72, Rey74]). These systems are presented in the PTS format [Bar92] with quantified contexts [Mil92]. In Example 2.4 we give some simple typings. For more information on System F, we refer to [Gir89].

Definition 2.1. *Terms* are given by the following abstract syntax: $\mathcal{T} ::= \mathcal{C} \mid \mathcal{V} \mid \mathcal{T}\mathcal{T} \mid \lambda\mathcal{V}{:}\mathcal{T}.\mathcal{T} \mid \Pi\mathcal{V}{:}\mathcal{T}.\mathcal{T}$. Here $\mathcal{C}$ is an infinite set of constants and $\mathcal{V}$ is an infinite set of variables; x, y, y', y_1, ... range over $\mathcal{V}$. Among the constants, three elements are singled out: O, $*$ and $\square$. Their role will become clear in Definition 2.3. Roman letters range over $\mathcal{T}$. In examples we will sometimes use greek letters for type variables. When $x \notin FV(A_2)$, we write $\Pi x{:}A_1.A_2$ as $A_1 \rightarrow A_2$. We apply the usual conventions concerning brackets; so ABC means $(AB)C$ and $A \rightarrow B \rightarrow C$ means $A \rightarrow (B \rightarrow C)$. Using the brackets convention we can write every application term $A_1 A_2$ as $B_1 \ldots B_n A_2$ (for some $n \geq 1$) with B_1 not an application term. When A has a subterm of the form $\lambda x_1{:}A_1.A_2$ then A_1 is called a *domain in* A. The set of free variables of A is defined as usual and denoted by $FV(A)$. Also the substitution of A for x in B (denoted by $B[x := A]$) and the relations $\rightarrow_\beta$, $\twoheadrightarrow_\beta$, $\rightarrow_\eta$, $\twoheadrightarrow_\eta$, $\twoheadrightarrow_{\beta\eta}$, $=_\beta$ and $=_{\beta\eta}$ are defined on terms as usual. Syntactic equality (modulo α-conversion) is denoted by $\equiv$.

Next we introduce, in the style of [Mil92], quantified declarations and contexts. Usually, a declaration of a variable in a context is of the form $x : B$, meaning that the variable x has type B. In quantified contexts, the declaration is modified by a universal or existential quantifier, e.g. $\exists x : B$. This is needed for the proper definition of (solutions for) matching problems. The intuition for the quantification of variables is that a universally quantified variable is considered to be constant, in the sense that solutions for matching problems are not allowed to substitute terms for it; substitutions are only allowed to substitute terms for variables that are quantified existentially. We stress that the quantification does not affect the identity of variables or the typing rules, but only serves to mark variables in matching problems as constant or substitutable.

Definition 2.2. In this paper Q, Q_1, ... range over $\{\exists, \forall\}$. $Qx : B$ is called a *(quantified) declaration.* A *(quantified) context* is a finite ordered sequence of quantified declarations $Q_i x_i : C_i$, where the x_i are pairwise distinct. Contexts are denoted by capital Greek letters Δ, Γ, Γ_0, ... The empty context is denoted by $\langle\rangle$. If $\exists x : C$ occurs in Γ, then x is said to be *existential in* Γ. If $\forall x : C$ occurs in Γ, then x is said to be *universal in* Γ. If every declaration in Γ is of the form $\exists x : C$, then Γ is an *existential context.*

If $\Gamma \equiv \langle Q_1 x_1 : A_1, \ldots, Q_n x_n : A_n\rangle$, then $dom(\Gamma) = \{x_1, \ldots, x_n\}$, $FV(\Gamma) = \{x_1, \ldots, x_n\} \cup \bigcup_{1 \le i \le n} FV(A_i)$, and $\Gamma, Qx : B$ denotes $\langle Q_1 x_1 : A_1, \ldots, Q_n x_n : A_n, Qx : B\rangle$. (In general, we denote the concatenation of Γ and Δ by Γ, Δ.) Furthermore, for $1 \le i \le n$, Γ_{x_i} denotes $\langle Q_1 x_1 : A_1, \ldots, Q_{i-1} x_{i-1} : A_{i-1}\rangle$ and $Ex(\Gamma)$ denotes $\langle \exists x_1 : A_1, \ldots, \exists x_n : A_n\rangle$. If the declaration $Qx : C$ occurs in Γ, then $\Gamma(x)$ denotes C. Finally, $\Gamma \sqcup \Delta$ denotes Γ, Δ, but with declarations $Qx : C$ in Δ such that $x \in dom(\Gamma)$ removed from Δ.

A *judgement* in a system $\lambda\_$ is of the form $\Gamma \vdash_{\lambda\_} A : B$, where Γ is a quantified context and A and B are terms. We use the abbreviation $\Gamma \vdash A : B : C$ for $\Gamma \vdash A : B$ and $\Gamma \vdash B : C$. A term A is called *legal* when there exist a context Γ and a term B such that $\Gamma \vdash A : B$ or $\Gamma \vdash B : A$. A context Γ is called legal when there exist terms A and B such that $\Gamma \vdash A : B$. If $\Gamma \vdash_{\lambda\_} A : B$, then we call A *closed in* Γ if all free variables x in A are universal in Γ and moreover $\Gamma(x)$ is closed in Γ_x.

Definition 2.3. The systems $\lambda\tau$ and $\lambda 2$ are defined using the rules in Table 1 and the specification $(\mathcal{S}, \mathcal{A})$. Here $\mathcal{S}$ is a collection of *sorts*, and $\mathcal{A}$ is a set of *axioms* of the form $c : s$ with $c \in \mathcal{C}$ and $s \in \mathcal{S}$. In Table 1, s ranges over $\mathcal{S}$. The system $\lambda\tau$ can be obtained by taking $\mathcal{S} = \{*\}$ and $\mathcal{A} = \{O : *\}$. The system $\lambda 2$ can be obtained by taking $\mathcal{S} = \{*, \Box\}$ and $\mathcal{A} = \{* : \Box\}$. In this paper we let $\lambda\_$ range over $\lambda\tau$ and $\lambda 2$ and (except in Table 1) s over $\{*, \Box\}$.

Axiom	$\langle\rangle \vdash c : s$	if $c : s \in \mathcal{A}$
Start	$\dfrac{\Gamma \vdash B : s}{\Gamma, Qx : B \vdash x : B}$	if $x \notin FV(\Gamma)$
Weakening	$\dfrac{\Gamma \vdash A : B \quad \Gamma \vdash B' : s}{\Gamma, Qx : B' \vdash A : B}$	if $x \notin FV(\Gamma)$
Product	$\dfrac{\Gamma \vdash A_1 : s \quad \Gamma, Qx : A_1 \vdash A_2 : *}{\Gamma \vdash \Pi x{:}A_1.A_2 : *}$	
Application	$\dfrac{\Gamma \vdash A_1 : \Pi x{:}B_1.B_2 \quad \Gamma \vdash A_2 : B_1}{\Gamma \vdash A_1 A_2 : B_2[x := A_2]}$	
Abstraction	$\dfrac{\Gamma, Qx : A_1 \vdash A_2 : B_2 \quad \Gamma \vdash \Pi x{:}A_1.B_2 : *}{\Gamma \vdash \lambda x{:}A_1.A_2 : \Pi x{:}A_1.B_2}$	

Table 1. The rules

Example 2.4. In $\lambda\tau$ one can type, e.g., the (closed) identity function for base type O: $\langle\rangle \vdash_{\lambda\tau} \lambda x{:}O.x : O{\rightarrow}O$. In $\lambda 2$ the polymorphic identity function can be typed: $\langle\rangle \vdash_{\lambda 2} \lambda\alpha{:}*.\lambda x{:}\alpha.x : \Pi\alpha{:}*.\alpha{\rightarrow}\alpha$. In the usual System F notation, this would look like $\langle\rangle \vdash_{\lambda 2} \Lambda\alpha.\lambda x^{\alpha}.x : \forall\alpha.\alpha{\rightarrow}\alpha$. Judgements with quantified variables will appear in the next two sections.

The main difference between the PTS versions and the original versions of the simply typed and polymorphic lambda calculus is that in the PTS version the well-formedness of contexts and types is checked inside the system. The judgement $\Gamma \vdash_{\lambda\_} A : *$ means that A is a *type*. If $\Gamma \vdash_{\lambda\_} A : B : *$, A is called an *object*. Types of $\lambda\tau$ are given by the syntax $O \mid O{\rightarrow}O$. If $A \equiv \Pi x{:}B.C$ and $x \in FV(C)$ it holds that $\lambda\_ = \lambda 2$ and $B \equiv *$. Types which contain such a type as subexpression are called *polymorphic*. When x is declared in Γ and its type in Γ is polymorphic, we say that x is a *polymorphic variable* (PV) in Γ. *Object variables in* Γ are variables x such that $\Gamma \vdash_{\lambda\_} \Gamma(x) : *$; *type variables* are variables x such that $\Gamma(x) \equiv *$. Note that polymorphic variables are object variables and that type variables do not occur in legal contexts of $\lambda\tau$. When A is of the form $xA_1 \dots A_n$ with x a PV in Γ, then we say that *A starts with a PV*. In $\lambda\tau$ and $\lambda 2$, object variables do not occur in types. In $\lambda\tau$, types occur only as domains in objects; in $\lambda 2$, a type can also occur as a subterm A_i $(2 \leq i \leq n)$ in application terms $A_1 \dots A_n$. In this case A_i is called a *polymorphic argument of* A_1.

The systems $\lambda\tau$ and $\lambda 2$ enjoy many well-known properties such as Strong Normalisation, Confluence, Subject Reduction (all w.r.t. $\beta\eta$-reduction) and Unicity of Types. See [Bar92] or [Geu93b]. By Confluence and Strong Normalisation, each legal term A has a unique $\beta\eta$-normal form. We denote it by $nf(A)$. By Unicity of Types we know that if a term A has a type in a context Γ then this type is unique. We denote it by $\tau(\Gamma; A)$. It will often be convenient to assume that terms are in η-long-β-normal form (LNF). This notion is defined e.g. in [Spr95a]. Types are in LNF and objects in LNF can be described as follows.

Lemma 2.5. *Suppose $\Gamma \vdash_{\lambda\_} A : \Pi x_1{:}A_1 \ldots \Pi x_n{:}A_n.C : *$, where all terms are in* LNF. *Then $A \equiv \lambda x_1{:}A_1 \ldots \lambda x_n{:}A_n.xB_1 \ldots B_m$, for some variable x (possibly among $\{x_1, \ldots, x_n\}$) and terms $B_1, \ldots, B_m$ (in* LNF*).*

Let $\Gamma \vdash_{\lambda\_} \lambda x_1{:}S_1 \ldots \lambda x_n{:}S_n.xt_1 \ldots t_m : T$. When we analyse the subterm $xt_1 \ldots t_m$ or a subterm t_i (for instance in an induction on the LNF-structure of well-typed terms) we do this in the context $\Gamma, \forall x : S_1, \ldots, \forall x_n : S_n$. This context is called the *current context*. So bound variables are treated as ("locally") universal variables. This involves a choice of the variables $x_1, \ldots, x_n$. When we do this for two terms $\lambda x_1{:}S_1 \ldots \lambda x_n{:}S_n.xt_1 \ldots t_m$ and $\lambda x_1{:}S_1 \ldots \lambda x_n{:}S_n.x't'_1 \ldots t'_m$ simultaneously and we say, e.g., that $xt_1 \ldots t_m$ is syntactically equal to $x't'_1 \ldots t'_m$ then we assume that the choice of variables is done consistently. A *free type in $xt_1 \ldots t_m$ w.r.t. Γ* is a subterm T of $xt_1 \ldots t_m$ such that $\Gamma \vdash_{\lambda 2} T : *$ (so T does not contain $x_1, \ldots, x_n$).

3 Matching problems

We present the notions of a *matching problem*, the *order of such a problem* and a *solution* for such a problem. An example of a matching problem will be given in Section 4.

Definition 3.1. Suppose $\Gamma \vdash_{\lambda\_} A : s$. We define $ord_\Gamma(A)$, the *order of A in Γ*, as follows: $ord_\Gamma(A)$ equals 2 if $A \equiv *$; 1 if $A \equiv O$ or $A \equiv x$ and x is universal in Γ; ∞ if $A \equiv x$ and x is existential in Γ; $\max(\{1 + ord_\Gamma(A_1), ord_{\Gamma,\exists x:A_1}(A_2)\})$ if $A \equiv \Pi x{:}A_1.A_2$. By convention, $\max(\{n, \infty\}) = \infty$ and $n + \infty = \infty$. When Γ is clear from the context, we simply speak about the order of A. For a motivation of these definitions, see Chapter 2 of [Spr95b].

The following lemma gives the form of types of the first three orders in $\lambda 2$. Note that third-order types are built from arrows, variables and the constant $*$. The presence of $*$ in types allows for polymorphic arguments in terms.

Lemma 3.2. *Let $\Gamma \vdash_{\lambda 2} A : *$.*

1. *If $ord_\Gamma(A) = 1$ then A is a variable, universal in Γ.*
2. *If $ord_\Gamma(A) = 2$ then for some $n \geq 1$, $A \equiv A_1 \to \cdots \to A_n \to A'$ where A' and, for all $1 \leq i \leq n$, A_i are universal variables.*

3. *If $ord_\Gamma(A) = 3$ then for some $n \geq 1$, $A \equiv A_1 \to \cdots \to A_n \to A'$ where A' is universal variable and, for all $1 \leq i \leq n$, A_i is either equal to $*$ or of the form described in 2.*

Example 3.3. Let Γ be $\langle \forall C : *, \forall D : * \rangle$. Then, e.g., $\Gamma \vdash_{\lambda 2} * \to D : *$ and $\Gamma \vdash_{\lambda 2} (C \to C) \to * \to D : *$. We have $ord_\Gamma(* \to D) = 3$ and $ord_\Gamma((C \to C) \to * \to D) = 3$.

Definition 3.4.

1. A ***substitution*** is a finite set of triples $\langle x_i\,;\, \gamma_i\,;\, M_i \rangle$, such that the x_i are pairwise distinct, γ_i is an existential context and $dom(\gamma_i)$ consists of fresh variables, possibly occurring in the M_j's. We let σ, σ', τ, ... range over substitutions.
2. If $\langle x\,;\, \gamma\,;\, M \rangle \in \sigma$, then we say that σ ***binds*** x. M is called a ***substitution term***, γ a ***substitution context***. To indicate that the substitution context is an "auxiliary" context, we denote it by a small Greek letter. Put $dom(\sigma) = \{x \mid x \text{ bound by } \sigma\}$.
3. A substitution σ operates on a term A by replacing x in A by M, for $\langle x\,;\, \gamma\,;\, M \rangle \in \sigma$. σ is extended to a function on contexts as follows: $\sigma(\langle\rangle) = \langle\rangle$; $\sigma(\Gamma, Qx : C) = \sigma(\Gamma), \gamma$ if $Q = \exists$ and $\langle x\,;\, \gamma\,;\, M \rangle \in \sigma$; otherwise $\sigma(\Gamma, Qx : C) = \sigma(\Gamma), Qx : \sigma(C)$.
4. Let Γ be a legal context in λ_-. Then we call σ ***well-typed in*** Γ if (i) σ binds no variables that are universal in Γ; (ii) $\sigma(\Gamma)$ is legal in λ_-; (iii) for all existential variables x in Γ that are bound by σ we have that $\sigma(\Gamma_x), \gamma \vdash_{\lambda_-} M : \sigma(\Gamma(x))$, where $\langle x\,;\, \gamma\,;\, M \rangle$ is the unique triple in σ that binds x. We have that if $\Gamma \vdash_{\lambda_-} A : B$ and σ is well-typed in Γ, then $\sigma(\Gamma) \vdash_{\lambda_-} \sigma(A) : \sigma(B)$.
5. Let σ be well-typed in some context Γ which is legal in λ_- and suppose $dom(\sigma) \subseteq dom(\Gamma)$. For fixed sort s, we call σ an ***s-substitution*** if for every $x \in dom(\sigma)$ we have: $\Gamma \vdash_{\lambda_-} \Gamma(x) : s$.

Definition 3.5.

1. A ***matching problem in*** λ_- is a triple $\langle \Gamma\,;\, A\,;\, B \rangle$, where Γ is a quantified context such that (i) there exists a term C such that $\Gamma \vdash_{\lambda_-} A : C$ and $\Gamma \vdash_{\lambda_-} B : C$; (ii) B is closed in Γ. If $\Gamma \vdash_{\lambda_-} C : *$, then we say that $\langle \Gamma\,;\, A\,;\, B \rangle$ is a matching problem ***for objects***; if $\Gamma \vdash_{\lambda_-} C : \Box$, then we say that $\langle \Gamma\,;\, A\,;\, B \rangle$ is a matching problem ***for types***. A matching problem $\langle \Gamma\,;\, A\,;\, B \rangle$ is ***of order*** n if the types of the existential variables in Γ have order at most n in Γ. *In this paper, we assume that the type of every variable in Γ is closed. In the full version [Spr95a] we show that this can be assumed without loss of generality.*
2. A ***solution*** for a matching problem $\langle \Gamma\,;\, A\,;\, B \rangle$ in λ_- is a substitution σ, well-typed in Γ, such that $\sigma(A) =_{\beta\eta} B$. For fixed sort s, we call σ an ***s-solution*** when σ is an s-substitution. σ is called a solution for a collection of matching problems $\{P_i \mid i \in I\}$ if σ is a solution for every P_i. By a standard argument, a set of matching problems $\{\langle \Gamma\,;\, A_i\,;\, B_i \rangle \mid 1 \leq i \leq n\}$ (for some $n \in \mathbb{N}$) can be encoded as a single matching problem. We call the matching problems in such a set ***Γ-compatible***.

4 Restricted matching problems

In this section we will define several notions concerning $\lambda 2$ terms, culminating in the notion "restricted". For motivation, we sketch our translation and explain problems that occur when defining it, which lead us to the restricted format. To be useful, the translation should at least preserve typing judgements, substitution and reduction.

First we translate the types of $\lambda 2$ to simple types in a straightforward way: replace type variables and $*$ by the base type O, preserving the arrow structure of types and interpreting $\Pi\alpha{:}*.B$ as $*\to B$. For example, if A and B are type variables, then $\Pi\alpha{:}*.(\alpha\to A)\to B$ is translated to $O\to(O\to O)\to O$.

Next we turn to the translation of terms. In the polymorphic lambda calculus, types also occur as arguments of terms. Such occurrences can be encoded as terms of type O: type variables are interpreted as variables of type O, the type forming operator $\to$ is represented by a constant $c^{\to}$ of type $O\to O\to O$, and the operator Π is represented by a constant c^{Π} of type $(O\to O)\to O$. The binding of type variables by Π is simulated by the λ-binder. For example, $\Pi\alpha{:}*.(\alpha\to A)\to B$ is translated to $c^{\Pi}(\lambda\alpha{:}O.c^{\to}(c^{\to}\alpha A)B)$.

The first problem can now be described as follows. Consider the judgement (omitting quantifiers)

$$\langle A:*, B:*, x:\Pi\alpha{:}*.\alpha\rangle \vdash_{\lambda 2} x(A\to B)(xA) : B.$$

We see that the types $A\to B$ and A occur as argument of the polymorphic variable x. But, given the encoding described above, it is hopeless to find a $\lambda\tau$-type that corresponds to $\Pi\alpha{:}*.\alpha$ such that when we assign this type to x, the translated judgement is valid in $\lambda\tau$. For each occurrence of x we would have to assign a different $\lambda\tau$-type to x. Instead, we add, for each occurrence of x, a fresh variable which fixes the resulting type inequalities. The translated judgement becomes:

$$\Gamma \vdash_{\lambda\tau} p_1 x(c^{\to}AB)(p_2 xA) : O,$$

where $\Gamma \equiv \langle c^{\to} : O\to O\to O, p_1 : (O\to O)\to O\to O\to O, p_2 : (O\to O)\to O\to O, A : O, B : O, x : O\to O\rangle$.

This solves the problem that concerns the "static" aspect of polymorphism, i.e. the aspect related to the fact that types of terms may depend on polymorphic arguments in that term. We now have achieved that the translation preserves typing judgements. Next, we turn to the second problem, which concerns the "dynamic" aspect of polymorphism, i.e. the fact that, by β-reduction, types may be moved around. This concerns the preservation of substitution and reduction. Consider the third-order matching problem P given by the triple $\langle\Gamma, A, B\rangle$

$$\begin{aligned}
\Gamma &\equiv \langle \forall C:*, \forall D:*, \forall x:\Pi\alpha{:}*.(\alpha\to\alpha)\to D, \exists f:*\to D\rangle,\\
A &\equiv f(C\to D),\\
B &\equiv x(C\to D)(\lambda y{:}C\to D.\lambda z{:}C.yz)
\end{aligned}$$

One possible solution for P substitutes $\lambda\alpha{:}*.x\alpha(\lambda y{:}\alpha.y)$ for f. To see that this is a solution for P, check that $(\lambda\alpha{:}*.x\alpha(\lambda y{:}\alpha.y))(C\to D)$ β-reduces in one step to $x(C \rightarrow D)(\lambda y{:}C \rightarrow D.y)$ and that the latter term is the $\beta\eta$-normal form of B. We see that in the β-reduction step, $C \rightarrow D$ is substituted for α in $\lambda y{:}\alpha.y$, changing the arrow structure of the domain in that term. In the spirit of the translation sketched above, we would translate $\lambda y{:}\alpha.y$ to $\lambda y{:}O.y$ and $\lambda y{:}C \rightarrow D.y$ to $\lambda y{:}O \rightarrow O.y$. Now the problem is: how do we represent in $\lambda\tau$ the substitution of types for variables in types?

Fortunately, we do not have to solve the problem in its full form. Note that if the argument of $\lambda\alpha{:}*.x\alpha(\lambda y{:}\alpha.y)$ in $\sigma(A)$ is a type variable E rather than $C \rightarrow D$ the problem would not occur. Then the β-reduction step would change $\lambda y{:}\alpha.y$ into $\lambda y{:}E.y$ and both terms are translated to $\lambda y{:}O.y$. This turns out to be the key idea. For we can define a translation with the desired properties for third-order matching problems in the so-called *restricted* format. A matching problem $P = \langle \Gamma, A, B \rangle$ is called restricted when (among other things) for every existential object variable f in A all polymorphic arguments of all occurrences of f are in fact universal type variables. The effect of this is that when we substitute a term for f, the polymorphic arguments of an occurrence of the substituted term are all universal type variables. For the terms occurring in restricted third-order matching problems and their solutions (the so-called *simple terms*), one can prove that this property is preserved under substitution of terms and reduction. Thus we have obtained for the restricted formatted, a translation that preserves typing judgements, substitution and reduction. The reduction of the general case to the restricted case is the subject of the next section.

Definition 4.1. Suppose $\Gamma \vdash_{\lambda 2} A : B$. Let $n \in \mathrm{N} \cup \{\infty\}$. An *n-redex in A* is a β-redex $(\lambda x{:}A_1.A_2)A_3$ in A such that the order of the type of $\lambda x{:}A_1.A_2$ (in the current context) is at most n. A term is called *finite-redexed* if there exists an $n \in \mathrm{N}$ such that all β-redexes in A are n-redexes. A *polymorphic β-redex* in A is a β-redex in A of the form $(\lambda x{:}*.A_2)A_3$.

Definition 4.2. Suppose $\Gamma \vdash_{\lambda 2} A : B$. We say that A is *polymorphic-atomic-redexed* (PAR, for short) if for all subterms in A of the form $(\lambda x{:}C_1.C_2)D_1 \ldots D_n$ with $n \geq 1$ the following holds for every $1 \leq i \leq n$: if D_i is a type then D_i is a universal variable (in the current context).

One easily checks that both the property of being PAR and the property that every β-redex is an n-redex (for finite n) hold for β-normal terms but are not preserved under β-reduction. Fortunately, the *conjunction* of the property of being PAR and the property that every β-redex is a 3-redex is preserved under β-reduction.

Definition 4.3. Suppose $\Gamma \vdash_{\lambda 2} A : B$. We say that A is *simple* if A is PAR and every β-redex in A is a 3-redex.

Simple terms have the important property that the translation of Section 6 preserves β-reduction on simple terms. To prove this, it is essential to prove that simplicity is closed under β-reduction.

Lemma 4.4. *Let $\Gamma \vdash_{\lambda 2} A : B$, where A is simple. Suppose $A \twoheadrightarrow_\beta A'$. Then A' is simple.*

Note that the conjunction of the property of being PAR and the property that every β-redex is a 4-redex is *not* preserved under β-reduction. Consider for example, $A \equiv (\lambda x{:}\, * \rightarrow B.x(\Pi\alpha{:}*.\alpha))\lambda\beta{:}*.y\beta$ where B is a universal type variable and y is a variable of type $*\rightarrow B$. Then A is PAR and its only β-redex is a 4-redex. But A β-reduces in one step to $(\lambda\beta{:}*.y\beta)(\Pi\alpha{:}*.\alpha)$ which is not PAR. Here the restriction to *third*-order matching is essential.

Now we describe the situation in which simple terms appear in our analysis of matching problems and solutions.

Definition 4.5. Suppose $\Gamma \vdash_{\lambda 2} A : B$, with A in LNF. Let $X = \{x_1, \ldots, x_n\} \subseteq FV(A)$ be a set of object variables in Γ. Then A is called ***simplified** w.r.t.* X if for all $x \in X$, for all occurrences $xt_1 \ldots t_m$ of x in A and for all $1 \leq i \leq m$ we have that if t_i is a type then it is a universal variable (in the current context). We call a matching problem $\langle \Gamma\, ;\, A\, ;\, B\rangle$ simplified when A is simplified w.r.t. the set of free object variables in A that are existential in Γ.

Lemma 4.6. *Let $\langle \Gamma\, ;\, A\, ;\, B\rangle$ be a simplified third-order matching problem. Let σ be a $*$-substitution, well-typed in Γ. Then $\sigma(A)$ is simple and the β-normal form of $\sigma(A)$ is in LNF. So, if σ is a solution for P, then $\sigma(A) \twoheadrightarrow_\beta B$.*

A consequence of this result is that when encoding solutions to simplified matching problems, we only have to consider β-reduction, instead of full $\beta\eta$-reduction. The following specialisation of simplicity will also be useful.

Definition 4.7. Suppose $\Gamma \vdash_{\lambda 2} A : B$. We call A ***redex-monomorphic*** if for all subterms in A of the form $(\lambda x{:}C_1.C_2)D_1 \ldots D_n$ with $n \geq 1$ we have that no D_i $(1 \leq i \leq n)$ is a type. Note that such A are also PAR. We call A ***ultrasimple*** when A is redex-monomorphic and every β-redex in A is a 3-redex. Note that, indeed, every ultrasimple term is simple. Terms that are β-normal are ultrasimple and ultrasimplicity is preserved under $\beta\eta$-reduction.

The following two notions finish the definition of restrictedness. The first one concerns the arguments of existential object variables in matching problems; the second one essentially amounts to the clause that all types occurring in matching problems for objects are closed. Their role will become clear in the next section.

Definition 4.8. Suppose $\Gamma \vdash_{\lambda 2} A_1 \rightarrow \cdots \rightarrow A_n \rightarrow A : *$, with $n \geq 0$ and A a variable. Suppose that $ord_\Gamma(A_1 \rightarrow \cdots \rightarrow A_n \rightarrow A)$ is finite. We call $A_1 \rightarrow \cdots \rightarrow A_n \rightarrow A$ ***structured*** when there exists $1 \leq i \leq n$ such that $A_1, \ldots, A_i$ are all equal to $*$ and none of $A_{i+1}, \ldots, A_n$ are equal to $*$. (Note that $A \not\equiv *$.) We call a matching problem $\langle \Gamma\, ;\, A\, ;\, B\rangle$ of finite order in $\lambda 2$ structured when all the types of the existential object variables that occur in A are structured.

Definition 4.9. Let $P = \langle \Gamma\, ;\, A\, ;\, B\rangle$ be a matching problem in $\lambda 2$. We call P ***type-closed*** when either P is a matching problem for types, or the type of every variable in Γ is closed in Γ and every type T in A is closed in its current context.

Finally, we can define the notion "restricted".

Definition 4.10. We call P *restricted* when P is structured, type-closed and simplified.

5 Reduction to restricted third-order matching

In Section 6, we will establish that it is decidable whether a restricted third-order matching problem $P = \langle \Gamma\,;\, A\,;\, B\rangle$ in $\lambda 2$ has a solution or not. Since every third-order matching problem for types is restricted, this will settle the case of matching problems for types. In this section we will reduce third-order matching for objects to restricted third-order matching for objects. First, we prove that we may w.l.o.g. assume that P is structured and type-closed.

Lemma 5.1 *(Reduction to the structured type-closed case). Let $P = \langle \Gamma\,;\, A\,;\, B\rangle$ be a third-order matching problem for objects in $\lambda 2$. From P we can (effectively) construct a finite set $\Pi(P)$ of structured, type-closed matching problems (of finite order in $\lambda 2$) such that P has a solution iff some problem in $\Pi(P)$ has a solution.*

Proof *(Sketch).* First, we show that we may w.l.o.g. assume that P is structured. Since the types of the existential variables in A are of finite order they are not polymorphic. Hence in a subterm of the form $xA_1 \ldots A_n$ with x an existential variable, the type of $xA_1 \ldots A_i$ $(1 \le i \le n)$ depends solely on the type of x, not on a polymorphic argument of x among $A_1, \ldots, A_{i-1}$. Thus we may as well permute the arguments of the existential variables to obtain a structured matching problem (changing the types of the existential variables accordingly.)

As to type-closed. We define a finite set of substitutions by taking every possible substitution obtained by the following scenario: for every subset S of the existential type variables in A, for every $x \in S$, substitute some free type (see below Lemma 2.5) in B (w.r.t. Γ) for x, and substitute $\Pi\alpha{:}{*}.\alpha$ for the existential type variables not in S. Now a finite set $\Pi(P)$ of matching problems is obtained from $\langle \Gamma\,;\, A\,;\, B\rangle$ by replacing A by $\sigma(A)$ with σ taken from the set of substitutions defined above. Since free types in B are closed, it is easily seen that each such problem is type-closed. If P has solution σ then the type part of σ (possibly augmented with some substitutions of $\Pi\alpha{:}{*}.\alpha$ for type variables) will be among the substitutions considered and from the object part of σ a solution can be constructed for the resulting matching problem in $\Pi(P)$. Conversely, if one of the matching problems in $\Pi(P)$ has a solution, then a solution for P can be pieced together from the substitution considered and the solution for the resulting matching problem. □

The motivation for reducing matching to type-closed matching is as follows. A solution for a type-closed matching problem P may be assumed to be $*$-substitution that will leave the polymorphic arguments in A unaffected. This allows us to replace w.l.o.g. polymorphic arguments of existential variables in A

by fresh universal type variables, obtaining a simplified matching problem. The motivation for structuring third-order matching problems for objects in $\lambda 2$ is that if σ is a solution for a structured problem $P = \langle \Gamma ; A ; B \rangle$ then there exists a very surveyable $\beta\eta$-reduction path from $\sigma(A)$ to B, enabling us to analyse what happens to the subterms of $\sigma(A)$. This analysis is given in the following lemma.

Lemma 5.2. *Let $\langle \Gamma ; A ; B \rangle$ be a structured, type-closed third-order matching problem for objects in $\lambda 2$. Let σ be a $*$-substitution, well-typed in Γ. Suppose that $\sigma(A) \twoheadrightarrow_{\beta\eta} A'$ via a reduction path that begins with contracting, exhaustively, all polymorphic β-redexes according to the Leftmost Innermost Reduction Order [FH89, p. 119]. Consider an occurrence of an existential variable in A and a polymorphic argument C_i of that occurrence of x. We can underline $\underline{C_i} \equiv \sigma(C_i)$ in $\sigma(A)$ and follow the underlined term in the reduction sequence described above.*

Write $E \xrightarrow{\theta} F$ if there is some (β- or η-) step from E to F in the reduction sequence described above. We have that the underlined terms in F (if any) are syntactically equal to the underlined terms in E. For $T \in \mathcal{T}$, write $T \xrightarrow{-} T$ if T' is the result of replacing all underlined terms in T (if any) by a fresh type variable z. The following diagram commutes.

$$\begin{array}{ccc} E & \xrightarrow{\theta} & F \\ {\scriptstyle -}\downarrow & & \downarrow{\scriptstyle -} \\ E' & \xrightarrow{\theta} & F' \end{array}$$

Proof *(Sketch)*. When we, initially and exhaustively, contract all polymorphic β-redexes in $\sigma(A)$, then in this part of the reduction sequence, C_i may be removed and may be moved around, but is not changed. This results in a term A'' that is ultrasimple. Contraction of a $\beta\eta$-redex in a ultrasimple term does not involve types except as domains, which, again, may be removed or moved around but are not changed. □

Using this result, we will show that we can without loss of generality assume that third-order matching problems for objects are restricted.

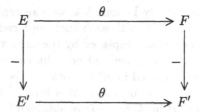

Definition 5.3. Let $\langle \Gamma ; A ; B \rangle$ be a structured, type-closed third-order matching problem for objects in $\lambda 2$. Suppose $\Gamma \vdash_{\lambda 2} A : C$ (and so $\Gamma \vdash_{\lambda 2} B : C$). Write A' for the result of the following operation: for every occurrence of an existential variable x in A, replace every polymorphic argument T of that occurrence of x by a fresh variable M_T. We demand that all occurrences in A of a polymorphic argument that are replaced, are replaced by the same variable and that no (occurrences of) two syntactically unequal polymorphic arguments are replaced by the same variable. Let $\mathsf{M} = \{M_1, \ldots, M_n\}$ be the set of fresh variables thus obtained. Write Δ for the context consisting of declarations in the

set $\{\forall M : * \mid M \in \mathsf{M}\}$ (in any order). Let Γ' be Γ, Δ. Then $\Gamma' \vdash_{\lambda 2} A' : C$. Note that, for a suitable choice of variables M, a substitution σ is well-typed in Γ iff σ is well-typed in Γ'.

Let B' be $nf(B)$. Let $\Psi(B)$ be the set of all terms D such that $\Gamma' \vdash_{\lambda 2} D : C$ and D is the result of the following operation: for every type in B' that also occurs in A and is replaced by a variable in the transformation of A into A', replace some (possibly none) occurrences of this type in B' by the same variable. Note that $\Psi(B)$ is a finite set. Define $\Theta(P) = \{\langle \Gamma' ; A' ; \mathit{lnf}_{\Gamma'}(D)\rangle \mid D \in \Psi(P)\}$. Note that $\Theta(P)$ is a finite set of Γ'-compatible, restricted third-order matching problems for objects in $\lambda 2$.

Lemma 5.4 *(Reduction to the restricted case). Let $P = \langle \Gamma ; A ; B\rangle$ be a structured, type-closed third-order matching problem for objects in $\lambda 2$. Then P has a solution $\Leftrightarrow$ some $Q \in \Theta(P)$ has a solution.*

Proof *(Sketch).* Let B' be $nf(B)$. We treat the "$\Rightarrow$" case; the remaining case is similar. Suppose P has solution σ; then $\sigma(A) \twoheadrightarrow_{\beta\eta} B'$. As remarked above, σ is also well-typed in Γ' and so the $\beta\eta$-normal form of $\sigma(A')$ exists; call it D. We are done when we can show that $Q = \langle \Gamma' ; A' ; \mathit{lnf}_{\Gamma'}(D)\rangle \in \Theta(P)$. Since we may assume that σ is a $*$-substitution (so does not affect types) we can write $\sigma(A')$ as the result of replacing occurrences of polymorphic arguments in $\sigma(A)$ by variables from M. By Lemma 5.2, we can replace some types in B' by variables from M to obtain D, with each such replaced type syntactically equal to a (replaced) type in $\sigma(A)$ and replaced by the same variable as its counterpart in $\sigma(A)$. Since σ is a $*$-substitution and no subterms of substitution terms of σ are replaced, each type replaced in B' to obtain D is in fact syntactically equal to a type replaced in the transformation of A into A' and replaced by the same variable as its counterpart in A. Since D is obtained from B', we can prove that every variable in $FV(D)$ is universal and of closed type in Γ' and $\Gamma' \vdash_{\lambda 2} D : C$. We may conclude that $Q = \langle \Gamma' ; A' ; \mathit{lnf}_{\Gamma'}(D)\rangle \in \Theta(P)$. □

Corollary 5.5. *Third-order matching for objects in $\lambda 2$ reduces to restricted third-order matching for objects in $\lambda 2$.*

6 Encoding restricted third-order matching problems

In this section, we will encode restricted third-order matching problems in $\lambda 2$ (and their solutions) as third-order matching problems in $\lambda\tau$ (and their solutions). To this end, we define a translation $[\![-]\!]_{-}$ from terms in $\lambda 2$ to terms in $\lambda\tau$. Its most important properties are that it preserves judgements on finite-redexed terms and β-reduction on simple terms (Lemma 4.6 states that we do not have to consider full $\beta\eta$-reduction). We assume the existence of a function P that assigns to each $\lambda\tau$-type A a variable p_A of type A. We will assume that all these variables are mutually distinct. Moreover we will assume that when we translate terms in $\lambda 2$ to terms in $\lambda\tau$ the new variables p_D are distinct from all variables that occur in the $\lambda 2$-terms.

The translations are defined on well-typed terms (and contexts). This means that the definitions depend on the current context in which a term is typed. For readability, we leave this dependency implicit.

We begin with a translation of types in $\lambda 2$ to types in $\lambda\tau$.

Definition 6.1. Suppose $\Gamma \vdash_{\lambda 2} A : s$. We define $\ll A \gg$ as follows.

$$\ll A \gg = \begin{cases} O & \text{if } A \equiv * \\ \ll A_1 \gg \to \ll A_2 \gg & \text{if } A \equiv \Pi x{:}A_1.A_2 \\ O & \text{otherwise} \end{cases}$$

Next, we define $[\![\cdot]\!]$ on legal contexts.

Definition 6.2. Let Γ be legal in $\lambda 2$. We define $[\![\Gamma]\!]$ by induction on the length of Γ, using fresh variables c^{Π}, $c^{\to}$.

$$[\![\langle\rangle]\!] = \langle c^{\Pi} : (O{\to}O){\to}O, c^{\to} : O{\to}O{\to}O\rangle$$
$$[\![\Gamma, Qx : A]\!] = [\![\Gamma]\!], Qx : \ll A \gg$$

Lemma 6.3. *If $\Gamma \vdash_{\lambda 2} A : s$, then $[\![\Gamma]\!] \vdash_{\lambda\tau} \ll A \gg : *$ and $ord_{[\![\Gamma]\!]}(\ll A \gg) \leq ord_{\Gamma}(A)$.*

We proceed to define $[\![-]\!]_{-}$ on finite-redexed objects. This function returns pairs of the form $([\![A]\!]^1 \mid [\![A]\!]^2)$, where $[\![A]\!]^1$ is a term and $[\![A]\!]^2$ is a context containing declarations of variables that are added in the construction of $[\![A]\!]^1$ (we use a slightly non-standard comma). For an example of the translation we refer to Section 4.

Definition 6.4. Suppose $\Gamma \vdash_{\lambda 2} A : B : s$, where A is finite-redexed. We define $[\![A]\!]$ by induction on the structure of A.

$$[\![x]\!] = (x \mid \langle\rangle)$$

$$[\![A_1 A_2]\!] = \begin{cases} (p_D y'[\![C_1]\!]^1 \ldots [\![C_n]\!]^1 [\![A_2]\!]^1 \mid \\ \langle \forall p_D : D\rangle \sqcup [\![C_1]\!]^2 \sqcup \ldots \sqcup [\![C_n]\!]^2 \sqcup [\![A_2]\!]^2) \\ \text{if } A_1 \equiv y C_1 \ldots C_n, \text{ with } y \text{ a PV} \qquad (\dagger) \\ \\ ([\![A_1]\!]^1 [\![A_2]\!]^1 \mid [\![A_1]\!]^2 \sqcup [\![A_2]\!]^2) \\ \text{if } A_1 \text{ does not start with a PV} \end{cases}$$

$$[\![\lambda x{:}A_1.A_2]\!] = (\lambda x{:} \ll A_1 \gg .[\![A_2]\!]^1 \mid [\![A_2]\!]^2)$$

$$[\![\Pi x{:}A_1.A_2]\!] = \begin{cases} (c^{\Pi}(\lambda x{:}O.[\![A_2]\!]^1) \mid \langle\rangle) & \text{if } x \in FV(A_2) \\ (c^{\to}[\![A_1]\!]^1[\![A_2]\!]^1 \mid \langle\rangle) & \text{otherwise} \end{cases}$$

($\dagger$): $y' \equiv \mathit{lnf}_{[\![\Gamma]\!]}(y)$ and $D \equiv \ll \Gamma(y) \gg \to$
$\ll \tau(\Gamma; C_1) \gg \to \cdots \to \ll \tau(\Gamma; C_n) \gg \to \ll \tau(\Gamma; A_2) \gg \to \ll \tau(\Gamma; A_1 A_2) \gg$.

We collect some useful facts. Suppose $\Gamma \vdash_{\lambda 2} B : C$. It is easy to see that there exists a linear function (which we denote by F) such that if m is the length of $[\![B]\!]^1$ then the length of B is bounded by $\mathsf{F}(m)$. This will be used in the proof of Theorem 6.11.

Lemma 6.5. *Suppose $\Gamma \vdash_{\lambda 2} A : B : s$, where A is finite-redexed. If A is β-normal (resp. in* LNF*) then $[\![A]\!]^1$ is β-normal (resp. in* LNF*). We have*

$$[\![A]\!]^2, [\![\Gamma]\!] \vdash_{\lambda\tau} [\![A]\!]^1 : \ll B \gg .$$

Lemma 6.6. *Suppose $\Gamma \vdash_{\lambda 2} A : B : s$, where A is simple. If $A \twoheadrightarrow_\beta A'$, then $[\![A]\!]^1 \twoheadrightarrow_\beta [\![A']\!]^1$.*

Proof. By induction on the generation of $\twoheadrightarrow_\beta$. By Lemma 4.4, simplicity is preserved by β-reduction. For the β-step, we use that for terms that are PAR and finite-redexed, the appropriate substitution lemmas hold. □

Definition 6.7. Let $P = \langle \Gamma ; A ; B \rangle$ be a matching problem of finite order n in $\lambda 2$. Define $[\![P]\!] = \langle [\![A]\!]^2 \sqcup [\![B]\!]^2, [\![\Gamma]\!] ; [\![A]\!]^1 ; [\![B]\!]^1 \rangle$. By Lemma 6.5 and Lemma 6.3, $[\![P]\!]$ is a matching problem of order n in $\lambda\tau$.

The $[\![-]\!]_-$ translation is extended to substitutions by applying the translation to substitution contexts and substitution terms. Some care is needed. Consider a solution σ for a matching problem $\langle \Gamma ; A ; B \rangle$. While translating a substitution term of a triple in σ we may have to add auxiliary variables p_D to the substitution context of that triple. Because our function P produces precisely one variable for each $\lambda\tau$-type, such a variable should be added to a substitution context at most once in the process of translating σ as a whole. Otherwise the resulting substitution would not be well-typed in $[\![\Gamma]\!]$, because p_D would be declared twice. Also, we must take care that the variables added to the substitution contexts are not already added to $[\![\Gamma]\!]$ in the construction of $[\![A]\!]^1$ and $[\![B]\!]^1$. There are various ways to ensure that these constraints are met. We will assume that for a matching problem $P = \langle \Gamma ; A ; B \rangle$, $[\![\sigma]\!]_P$ is defined and well-typed in $[\![\Gamma]\!]$. As before, the subscript P will be omitted.

Theorem 6.8. *Let $\langle \Gamma ; A ; B \rangle$ be a restricted third-order matching problem for objects in $\lambda 2$ and let σ be a $*$-solution for P. (It is no restriction to assume that $dom(\sigma) \subseteq dom(\Gamma)$.) Then $[\![\sigma]\!]$ is a solution for $[\![P]\!]$.*

Proof *(Sketch).* We have taken care that $[\![\sigma]\!]$ is well-typed in $[\![A]\!]^2 \sqcup [\![B]\!]^2, [\![\Gamma]\!]$. We need to show that $[\![\sigma]\!]([\![A]\!]^1) \twoheadrightarrow_\beta [\![B]\!]^1$. This follows from the fact that $\sigma(A)$ is simple (Lemma 4.6) and the fact that $[\![\ ]\!]$ preserves substitutions and β-reduction on simple terms (Lemma 6.6). □

The following lemma is a kind of converse to Theorem 6.8. It states that for σ to be a solution for restricted $P = \langle \Gamma ; A ; B \rangle$ in $\lambda 2$, it suffices that σ is well-typed in Γ and $[\![\sigma]\!]([\![A]\!]^1)$ reduces to $[\![B]\!]^1$.

Lemma 6.9. *Let $P = \langle \Gamma ; A ; B \rangle$ be a restricted third-order matching problem for objects in $\lambda 2$. Let σ be a $*$-substitution, well-typed in Γ, $dom(\sigma) \subseteq dom(\Gamma)$. Suppose that $[\![\sigma]\!]([\![A]\!]^1) \twoheadrightarrow_\beta [\![B]\!]^1$. Then σ is a solution for P.*

Results similar to the last two results hold for matching problems for types and their solutions. The proof of our main theorem uses the following result concerning third-order matching in $\lambda\tau$.

Theorem 6.10 *(Dowek [Dow94]). Let P be a third-order matching problem in $\lambda\tau$. There exists an $n \in \mathbb{N}$, that can be effectively computed from P only, such that the following holds. From a solution τ for P one can construct a solution $\tau^\star$ for P such that the length of each substitution term in $\tau^\star$ is less than n. So if P has a solution then it has a solution such that the length of its substitution terms is less than n. We denote this n by bound(P).*

The transition from an arbitrary solution τ for a matching problem in $\lambda\tau$ to $\tau^\star$ uses a variable e of type O which is assumed to be in the context. It proceeds by replacing certain "superfluous" subterms of substitution terms in τ by terms either of the form $\lambda x_1{:}S_1 \ldots \lambda x_n{:}S_n.e$, or of the form $\lambda x_1{:}S_1 \ldots \lambda x_n{:}S_n.x_j$ (for some $1 \leq j \leq n$ depending on the subterm in question). Instead of using one fixed variable e, one can equivalently take a fresh variable each time a replacement occurs. It is easy to see that for each substitution term we need at most $bound([\![P]\!])$ new variables. So for the substitution as a whole the number of new variables need not exceed the number of existential variables in A times $bound([\![P]\!])$. In our-set up, the new variables are added to the corresponding substitution contexts when necessary.

The point of this adaptation is that we can mimick the adapted transformation on $\lambda 2$-substitutions σ in such a way that it yields well-typed substitutions and that it commutes with our $[\![\,]\!]$-translation, i.e. $([\![\sigma]\!])^\star = [\![\sigma^\star]\!]$.

Decidability of third-order matching in $\lambda 2$ follows from Corollary 5.5 and the following theorem.

Theorem 6.11. *It is decidable whether a restricted third-order matching problem in $\lambda 2$ has a solution or not.*

Proof. Let $P = \langle \Gamma\,;\, A\,;\, B\rangle$ be given. We present an algorithm which returns a substitution if P has a solution and returns *fail* otherwise.

1. Translate P to $[\![P]\!]$ and fix a set $\mathcal{E}$ of $n \cdot bound([\![P]\!])$ fresh variables, where n is the number of existential variables in A.
2. Enumerate all substitutions consisting of substitution terms whose length is smaller than $\mathsf{F}(bound([\![P]\!]))$ and whose free variables are in $dom(\Gamma) \cup \mathcal{E}$. (For F, see below Definition 6.4).
3. For each such substitution σ, check whether it is a solution for P. If so, return σ and stop.
4. If none of the substitutions is a solution for P, return *fail* and stop.

We have to show that this algorithm is sound and complete and that it always terminates. Soundness is trivial and termination is easy, given the considerations preceding this theorem. As to completeness. We treat the case where P is a matching problem for objects. Suppose P has solution σ. Since P is restricted, it is no restriction to assume that σ is a $*$-substitution. By Theorem 6.8, $[\![\sigma]\!]$ is a solution for $[\![P]\!]$. By Theorem 6.10, $([\![\sigma]\!])^\star$ is also a solution for $[\![P]\!]$. We have shown that the transition from $[\![\sigma]\!]$ to $([\![\sigma]\!])^\star$ can be mimicked in σ, using variables (only) from $\mathcal{E}$, which yields a substitution $\sigma^\star$, that is well-typed in Γ

and such that $(\llbracket\sigma\rrbracket)^* = \llbracket\sigma^*\rrbracket$. By Lemma 6.9, σ^* is a solution for P. By Theorem 6.10, the length of the substitution terms in $(\llbracket\sigma\rrbracket)^*$ is less than $bound(\llbracket P\rrbracket)$. So the length of the substitution terms in σ^* is less than $\mathsf{F}(bound(\llbracket P\rrbracket))$. Hence σ^* occurs in the enumeration in step 2 of the algorithm. □

Acknowledgements. I would like to thank Erik Barendsen, Gilles Dowek, Jaco van de Pol, and Alex Sellink for their help in writing this paper. The anonymous referees are thanked for their useful suggestions. This paper was prepared using Paul Taylor's packages for diagrams and proof trees.

References

[Bar92] H.P. Barendregt. Lambda calculi with types. In: S. Abramsky, D. M. Gabbay, and T.S.E. Maibaum (eds.), *Handbook of Logic in Computer Science*, volume 2, Oxford University Press, Oxford, 1992, pp. 117–309.

[CH88] T. Coquand and G. Huet. The calculus of constructions. *Information and Control*, 76:95–120, 1988.

[Dow91b] G. Dowek. L'indécidabilité du filtrage du troisième ordre dans les calculs avec types dépendants ou constructeurs de types. *Compte Rendu à l'Académie des Sciences*, 312, Série I:951–956, 1991. With erratum in: ibid., 318, Série I, p. 873, 1994.

[Dow91d] G. Dowek. A second-order pattern matching algorithm for the cube of typed λ-calculi. In: A. Tarlecki (ed.), *Mathematical Foundations of Computer Science '91*, LNCS 520, Springer-Verlag, Berlin, 1991, pp. 151–160.

[Dow93b] G. Dowek. The undecidability of pattern matching in calculi where primitive recursive functions are representable. *Theoretical Computer Science*, 107:349–356, 1993.

[Dow94] G. Dowek. Third order matching is decidable. *Annals of Pure and Applied Logic*, 69:135–155, 1994.

[FH89] A.J. Field and P.G. Harrison. *Functional Programming*. Addison-Wesley, Wokingham, 1989.

[Geu93b] H. Geuvers. *Logics and Type Systems*. PhD thesis, University of Nijmegen, 1993.

[Gir72] J.-Y. Girard. *Interprétation fonctionelle et élimination des coupures dans l'arithmétique d'ordre supérieur*. PhD thesis, Université Paris VII, 1972.

[Gir89] J.-Y. Girard, Y. Lafont, and P. Taylor. *Proofs and Types*. Cambridge Tracts in Theoretical Computer Science, volume 7, Cambridge University Press, 1989.

[Löb76] M.H. Löb. Embedding first order predicate logic in fragments of intuitionistic logic. *Journal of Symbolic Logic*, 41:705–719, 1976.

[Mil92] D. Miller. Unification under a mixed prefix. *Journal of Symbolic Computation*, 14(4):321–359, 1992.

[Pad94a] V. Padovani. Fourth order dual interpolation is decidable. Manuscript, Université Paris VII - C.N.R.S, 1994.

[Pad94b] V. Padovani. On equivalence classes of interpolation equations. In: M. Dezani-Ciancaglini, G. Plotkin (eds.), *Typed Lambda Calculi and Applications*, LNCS 902, Springer-Verlag, Berlin, 1995, pp. 335–349.

[Rey74] J.C. Reynolds. Towards a theory of type structure. In: B. Robinet (ed.), *Proceedings of the Colloque sur la Programmation*, LNCS 19, Springer-Verlag, Berlin, 1974, pp. 408–524.

[Spr95a] J. Springintveld. Third-order matching in the presence of type constructors. Extended Abstract. In: M. Dezani-Ciancaglini, G. Plotkin (eds.), *Typed Lambda Calculi and Applications*, LNCS 902, Springer-Verlag, Berlin, 1995, pp. 428-442.

[Spr95b] J. Springintveld. *Algorithms for Type Theory.* PhD Thesis, Utrecht University, Department of Philosophy, May 1995.

[Wol93] D.A. Wolfram. *The Clausal Theory of Types*, Cambridge Tracts in Theoretical Computer Science, volume 21. Cambridge University Press, Cambridge, 1993.

Higher–Order Algebra with Transfinite Types

L.J. Steggles

Department of Computer Science, University of Newcastle upon Tyne,
Newcastle upon Tyne, NE1 7RU
email: L.J.Steggles@ncl.ac.uk

Abstract. We extend the simple type system of higher–order algebra with transfinite types. We present a general model theory for transfinite higher–order algebra including results on the existence and construction of free and initial models, and a sound and complete equational calculus. We demonstrate the use of transfinite types for modelling polymorphism by specifying a simple polymorphic functional programming language.

1 Introduction

Higher–order algebra provides a natural framework in which to formally develop computing systems and has been shown to be substantially more expressive than first–order algebraic methods (see Kosiuczenko and Meinke [1994] and Meinke [1995]). For an introduction to higher–order algebraic methods see Möller [1987], Möller et al [1988] and Meinke [1992], while for examples of their applications see for example Meinke and Steggles [1994], Meinke [1994] and Steggles [1995].

However one of the limitations of higher–order algebra using only binary product and function types is the difficulty in modelling objects which are parametric in type, such as polymorphic functions, partial functions, generic data structures and infinite families of algorithms and architectures. We propose to overcome this limitation by extending the type system of higher–order algebra with *limit types*. The idea is that given a set of basic types we close up under product and function types until we reach the limit at which point we add a new type, a so called limit type. We then iterate this process to produce a transfinite type hierarchy in which the limit types act as "universal" types in which all the types below them in the transfinite type hierarchy can be embedded. We refer to the resulting theory as *higher–order algebra with transfinite types*. Following the approach of Meinke [1992], we develop a general model theory for higher–order algebra with transfinite types using the framework of many–sorted first–order predicate calculus (usually referred to as finite or simple type theory). Results presented include the existence and construction of free and initial models, and a sound and complete equational calculus. Higher–order algebra with transfinite types provides a natural framework in which to model polymorphism, partiality and error handling. We demonstrate this by considering the specification of a simple polymorphic functional programming language based on the language $\mathcal{PR}$ of primitive recursive functions presented in Thompson [1987]. This specification case study generalises the definitions and results of Meinke [1994].

The structure of this paper is as follows. In Section 2 we introduce a notation for transfinite types and extend the basic definitions of higher–order algebra to take account of transfinite types. In Section 3 we introduce the first–order condition of extensionality, present a representation theorem for extensional algebras and consider those algebraic constructions which preserve or induce extensionality. Section 4 considers the existence of free transfinite higher–order algebras and gives an existence theorem for free algebras in a class of transfinite higher–order algebras. We also consider the existence of initial extensional models and the conditions necessary for a class of extensional algebras to contain an initial extensional model. In Section 5 we present a sound and complete transfinite higher–order equational calculus and give a concrete construction of the initial extensional model for an equational theory. In Section 6 we demonstrate a possible application of transfinite types by specifying a simple polymorphic functional programming language. This specification case study demonstrates how transfinite types can be used to model partial functions and both parametric and overloading polymorphism. Finally, in Section 7 we make some concluding remarks about the results we have presented.

We have attempted to make this paper self contained. For further background material on universal algebra we recommend Cohn [1965], Wechler [1992] and Meinke and Tucker [1993].

2 Transfinite Higher–Order Algebras

In this section we present the syntax and semantics of higher–order algebras with transfinite types. We extend the $\times$ and $\rightarrow$ type system of higher–order algebra (see for example Meinke [1992]) with *limit types* and reformulate the definitions of a type structure S, an S–typed signature Σ and an S–typed Σ algebra. We refer to the resulting theory as ***higher order algebra with transfinite types*** (or simply ***transfinite higher–order algebra***). A simple example of a transfinite higher–order algebra is presented at the end of the section. We begin by recalling some basic definitions of many–sorted first–order universal algebra (see for example Meinke and Tucker [1993]).

By a set S of sorts we mean any non-empty set. Let S^* denote the set of all words in the free monoid generated by S, let λ denote the empty word and define $S^+ = S^* - \{\lambda\}$. An S-sorted signature Σ is an $S^* \times S$-indexed family of sets $\Sigma = \langle \Sigma_{w,s} \mid w \in S^*,\ s \in S \rangle$. For $\lambda \in S^*$ and each sort $s \in S$, each element $c \in \Sigma_{\lambda,s}$ is a constant symbol of sort s. For each non-empty word $w = s(1)\ldots s(n) \in S^+$ and each sort $s \in S$, each element $f \in \Sigma_{w,s}$ is a function symbol of domain type w, codomain type s and arity n.

Let Σ be an S-sorted signature. An S-sorted Σ algebra is an ordered pair (A, Σ^A), consisting of an S-indexed family $A = \langle A_s \mid s \in S \rangle$ of carrier sets A_s and an $S^* \times S$-indexed family $\Sigma^A = \langle \Sigma^A_{w,s} \mid w \in S^*,\ s \in S \rangle$ of sets of constants and algebraic operations. For each sort $s \in S$, $\Sigma^A_{\lambda,s} = \langle c_A \mid c \in \Sigma_{\lambda,s} \rangle$, where $c_A \in A_s$ is a constant that interprets c in A_s. For each $w = s(1)\ldots s(n) \in S^+$ and $s \in S$, $\Sigma^A_{w,s} = \langle f_A \mid f \in \Sigma_{w,s} \rangle$, where $f_A : A^w \rightarrow A_s$ is an operation with domain $A^w = A_{s(1)} \times \cdots \times A_{s(n)}$ and codomain A_s which interprets f over A. As usual, we let A denote both a Σ algebra and its S-indexed family of carrier sets.

Let $A = \langle A_s \mid s \in S \rangle$ and $B = \langle B_s \mid s \in S \rangle$ be S-indexed families of sets then the basic set theoretic operations can be defined pointwise on A and B. We let $A \subseteq B$ denote pointwise inclusion, $A_s \subseteq B_s$, for each sort $s \in S$ and $A \subset B$ denotes the fact that $A \subseteq B$ but $A \neq B$. When no ambiguity arises we let $\emptyset$ denote the unique S-indexed family of empty sets. With A we associate an S-indexed family of cardinals, $|A| = \langle |A|_s = |A_s| \mid s \in S \rangle$ and $|A| \leq |B|$ denotes the fact that $|A|_s \leq |B|_s$ for each $s \in S$. We let $\phi : A \to B$ denote the S-indexed family of mappings $\phi = \langle \phi_s : A_s \to B_s \mid s \in S \rangle$. We say that ϕ is injective (respectively surjective, bijective) if, and only if, each ϕ_s is injective (respectively surjective, bijective). If $C = \langle C_s \mid s \in S \rangle$ is an S-indexed family of sets and $\phi : A \to B$ and $\psi : B \to C$ are S-indexed families of mappings then $\psi \circ \phi : A \to C$ denotes the pointwise composition

$$\psi \circ \phi = \langle (\psi \circ \phi)_s = \psi_s \circ \phi_s : A_s \to C_s \mid s \in S \rangle.$$

Next we recall some basic definitions for ordinals (see for example Devlin [1979] or Henle [1986]). Let *Ord* denote the (proper) class of von Neumann ordinals. Let $\emptyset \in Ord$ denote the *initial ordinal* and for any ordinal $\alpha \in Ord$ let $\alpha+1$ denote the *successor ordinal* $\alpha \cup \{\alpha\} \in Ord$. All other ordinals $\beta \in Ord$ are called *limit ordinals*, the first being $\omega = \{\emptyset, \{\emptyset\}, \{\emptyset, \{\emptyset\}\}, \ldots\}$, the next $\omega+\omega$ and so on. Recall that *Ord* is well–ordered by $\subseteq$. Let $\alpha \in Ord$ be a limit ordinal, if $\alpha \neq \omega$ then define $\alpha^- \in Ord$ to be the greatest limit ordinal β such that $\beta < \alpha$, otherwise if $\alpha = \omega$ then define $\alpha^- = \emptyset$.

We define the rules of type formation for transfinite type theories as follows.

2.1 Definition. By a *type basis* $\mathcal{B}$ we mean a non-empty set. Define the *transfinite type hierarchy* $H(\mathcal{B})$ generated by a type basis $\mathcal{B}$ to be the class

$$H(\mathcal{B}) = \bigcup_{\alpha \in Ord} H_\alpha(\mathcal{B})$$

where the sets $H_\alpha(\mathcal{B})$ are defined by transfinite induction as follows. For $\emptyset \in Ord$ the initial ordinal, define $H_\emptyset(\mathcal{B}) = \mathcal{B}$. For $\alpha+1 \in Ord$ a successor ordinal, define

$$H_{\alpha+1}(\mathcal{B}) = H_\alpha(\mathcal{B}) \cup \{ (\sigma \times \tau), (\sigma \to \tau) \mid \sigma, \tau \in H_\alpha(\mathcal{B}) \}.$$

For $\beta \in Ord$ a limit ordinal, define

$$H_\beta(\mathcal{B}) = \bigcup_{\alpha < \beta} H_\alpha(\mathcal{B}) \cup \{\beta\}.$$

Each limit ordinal $\beta \in H(\mathcal{B})$ is termed a *limit type*. □

We can assign an order $O(\tau) \in Ord$ to each type $\tau \in H(\mathcal{B})$ as follows.

2.2 Definition. Given any type basis $\mathcal{B}$ we define the *order* $O(\tau) \in Ord$ of each type $\tau \in H(\mathcal{B})$ by induction on the complexity of types. Let $\tau \in \mathcal{B}$ be a basic type, then define $O(\tau) = \emptyset$. Let $\sigma, \tau \in H(\mathcal{B})$, then define

$$O(\sigma \times \tau) = sup\{O(\sigma), O(\tau)\}, \quad O(\sigma \to \tau) = sup\{O(\sigma), O(\tau)\} + 1.$$

Let $\beta \in H(\mathcal{B})$ be a limit type, then define $O(\beta) = \beta$. □

In the sequel, we write $\sigma < \tau$ to denote that $O(\sigma) < O(\tau)$, for any types $\sigma, \tau \in H(\mathcal{B})$.

2.3 Definition. A (*transfinite*) *type structure* S over a type basis $\mathcal{B}$ is a subset of $H(\mathcal{B})$ which is closed under subtypes in the sense that: for any $\sigma, \tau \in H(\mathcal{B})$, if $(\sigma \times \tau) \in S$ or $(\sigma \to \tau) \in S$ then both $\sigma \in S$ and $\tau \in S$; and for each limit type $\beta \in S$ we have $\alpha \in S$, for each limit type $\alpha \in H(\mathcal{B})$ such that $\alpha < \beta$. □

For any ordinal $\alpha \in Ord$ and any type structure $S \subseteq H(\mathcal{B})$ over a type basis $\mathcal{B}$, S is said to be a *α–order type structure* if, and only if, the order of each type $\tau \in S$ is strictly less than α.

2.4 Definition. Let $S \subseteq H(\mathcal{B})$ be a type structure over a type basis $\mathcal{B}$. A (*transfinite*) *S-typed signature* Σ is an S-sorted signature such that: for each product type $(\sigma \times \tau) \in S$ we have two unary *projection operation symbols* $proj^{(\sigma\times\tau),1} \in \Sigma_{(\sigma\times\tau),\sigma}$ and $proj^{(\sigma\times\tau),2} \in \Sigma_{(\sigma\times\tau),\tau}$; for each function type $(\sigma \to \tau) \in S$ we have a binary *evaluation operation symbol* $eval^{(\sigma\to\tau)} \in \Sigma_{(\sigma\to\tau)\ \sigma,\tau}$; and for each limit type $\beta \in S$ and for each type $\tau \in S$, such that $\beta^- \leq \tau < \beta$, we have a unary *injection operation symbol* $inj^\tau \in \Sigma_{\tau,\beta}$. □

When the types σ and τ are clear we let $proj^1$ and $proj^2$ denote the projection operation symbols $proj^{(\sigma\times\tau),1}$ and $proj^{(\sigma\times\tau),2}$. An S-typed signature Σ is also termed an *α–order signature* when S is an α–order type structure. Next we introduce the intended interpretations of a transfinite higher–order signature.

2.5 Definition. Let $S \subseteq H(\mathcal{B})$ be a type structure over a type basis $\mathcal{B}$, let Σ be an S-typed signature and let A be an S-sorted Σ algebra. We say that A is *cumulative* if, and only if, for each product type $(\sigma\times\tau) \in S$ we have $A_{(\sigma\times\tau)} \subseteq A_\sigma \times A_\tau$, for each function type $(\sigma \to \tau) \in S$ we have $A_{(\sigma\to\tau)} \subseteq [A_\sigma \to A_\tau]$, and for each limit type $\beta \in S$ we have

$$\coprod_{\langle \tau \in S \mid \beta^- \leq \tau < \beta \rangle} A_\tau \subseteq A_\beta.$$

We say that A is a (*transfinite*) *S–typed Σ algebra* if, and only if, A is cumulative and
(i) for each product type $(\sigma \times \tau) \in S$ the operations

$$proj_A^{(\sigma\times\tau),1} : A_{(\sigma\times\tau)} \to A_\sigma, \quad proj_A^{(\sigma\times\tau),2} : A_{(\sigma\times\tau)} \to A_\tau,$$

are the *first* and *second projection operations*;
(ii) for each function type $(\sigma \to \tau) \in S$ the operation $eval_A^{(\sigma\to\tau)} : A_{(\sigma\to\tau)} \times A_\sigma \to A_\tau$ is the *evaluation operation*; and
(iii) for each limit type $\beta \in S$ and each type $\tau \in S$, such that $\beta^- \leq \tau < \beta$, the operation $inj_A^\tau : A_\tau \to A_\beta$, is the *canonical injection* defined on each $a \in A_\tau$ by $inj_A^\tau(a) = (\tau, a)$.

We let $Alg_{typ}(\Sigma)$ denote the class of all S-typed Σ algebras. □

Given an S–typed Σ algebra A and any function type $(\sigma \to \tau) \in S$ we may write $a(b)$ as an abbreviation for $eval_A^{(\sigma\to\tau)}(a, b)$, for any $a \in A_{(\sigma\to\tau)}$ and $b \in A_\sigma$. We conclude this section with an example of a transfinite higher–order algebra constructed from the term algebra $T(\Sigma, X)$.

2.6 Example. Let $S \subseteq H(\mathcal{B})$ be a type structure over a type basis $\mathcal{B}$ and let Σ be an S-typed signature. Let $X = \langle X_\tau \mid \tau \in S \rangle$ be an S-indexed family of sets of variables. Clearly $T(\Sigma, X)$ is not an S-typed Σ algebra (e.g. $T(\Sigma, X)_{(\sigma\to\tau)} \not\subseteq [T(\Sigma, X)_\sigma \to T(\Sigma, X)_\tau]$). However, we can use $T(\Sigma, X)$ to construct an S-typed Σ algebra $\overline{T(\Sigma, X)}$, which we call the *collapse of* $T(\Sigma, X)$. Define the S-indexed family of carrier sets $\overline{T(\Sigma, X)} = \langle \overline{T(\Sigma, X)}_\tau \mid \tau \in S \rangle$ by

$$\overline{T(\Sigma, X)}_\tau = \langle\, \bar{t} \mid t \in T(\Sigma, X)_\tau \,\rangle,$$

where $\tau \in S$ and each $\bar{t}$ is defined inductively on the complexity of types as follows. For each basic type $\tau \in S$ and each term $t \in T(\Sigma, X)_\tau$, define $\bar{t} = t$. For each product type $(\sigma \times \tau) \in S$ and each term $t \in T(\Sigma, X)_{(\sigma\times\tau)}$, define

$$\bar{t} = (\overline{proj^1(t)}, \overline{proj^2(t)}).$$

For each function type $(\sigma \to \tau) \in S$ and each term $t \in T(\Sigma, X)_{(\sigma\to\tau)}$, define $\bar{t} : \overline{T(\Sigma, X)}_\sigma \to \overline{T(\Sigma, X)}_\tau$ on each $\overline{t_0} \in \overline{T(\Sigma, X)}_\sigma$ by

$$\bar{t}(\overline{t_0}) = \overline{eval^{(\sigma\to\tau)}(t, t_0)}.$$

For each limit type $\beta \in S$ and each term $t \in T(\Sigma, X)_\beta$ define

$$\bar{t} = \begin{cases} (\tau, \overline{t_0}), & \text{if there exists } \tau \in S \text{ with } \beta^- \leq \tau < \beta \text{ and } t_0 \in T(\Sigma, X)_\tau \\ & \text{such that } inj^\tau(t_0) = t; \\ (\beta, t), & \text{otherwise.} \end{cases}$$

The constants and algebraic operations of $\overline{T(\Sigma, X)}$ are defined as follows. For each type $\tau \in S$ and each constant symbol $c \in \Sigma_{\lambda,\tau}$, define $c_{\overline{T(\Sigma,X)}} = \bar{c}$. For each $w = \tau(1)\ldots\tau(n) \in S^+$, $\tau \in S$, each function symbol $f \in \Sigma_{w,\tau}$ and any $(t_1, \ldots, t_n) \in T(\Sigma, X)^w$, define

$$f_{\overline{T(\Sigma,X)}}(\overline{t_1}, \ldots, \overline{t_n}) = \overline{f(t_1, \ldots, t_n)}.$$

Using the fact that $\overline{t_1} = \overline{t_2}$ if, and only if, $t_1 = t_2$ it is easily verified that $\overline{T(\Sigma, X)}$ is a well defined S-typed Σ algebra. In the next section we shall see that $\overline{T(\Sigma, X)}$

arises as a special case of the *collapsing construction* which is used in the proof of Collapsing Theorem 3.6. □

3 Algebraic Constructions

From both the viewpoint of algebra and specification theory we are mainly concerned with the structure of transfinite higher–order algebras up to isomorphism. As we will show this structure can be characterised by a set of first–order sentences which we refer to as the *axioms of extensionality*. We investigate which basic algebraic constructions preserve or induce extensionality.

Let S be a type structure over a type basis $\mathcal{B}$, let Σ be an S-typed signature and let $X = \langle\, X_\tau \mid \tau \in S \,\rangle$ be an S-indexed family of disjoint infinite sets of variables. To avoid the problems associated with empty carrier sets we assume that Σ is non-void, i.e. a ground term exists for each type $\tau \in S$.

3.1 Definition. The set $\mathbf{Ext} = \mathbf{Ext}_\Sigma$ of *extensionality sentences* over Σ is the set of all Σ sentences of the form:
(i) for each product type $(\sigma \times \tau) \in S$ and variables $x, y \in X_{(\sigma\times\tau)}$

$$\forall x\ \forall y\ (proj^1(x) = proj^1(y)\ \wedge\ proj^2(x) = proj^2(y) \Rightarrow x = y);$$

(ii) for each function type $(\sigma \to \tau) \in S$ and variables $x, y \in X_{(\sigma\to\tau)}$, $z \in X_\sigma$

$$\forall x\ \forall y\ (\forall z\ (eval^{(\sigma\to\tau)}(x, z) = eval^{(\sigma\to\tau)}(y, z)) \Rightarrow x = y);$$

(iii) for each limit type $\beta \in S$ and each type $\tau \in S$ such that $\beta^- \le \tau < \beta$, and $x, y \in X_\tau$

$$\forall x\ \forall y\ (inj^\tau(x) = inj^\tau(y)\ \Rightarrow\ x = y);$$

(iv) for each limit type $\beta \in S$ and variables $x \in X_\tau, y \in X_{\tau'}$ for types $\tau, \tau' \in S$ such that $\beta^- \le \tau, \tau' < \beta$ and $\tau \ne \tau'$

$$\forall x\ \forall y\ (inj^\tau(x) \ne inj^{\tau'}(y)).$$

We say that a Σ algebra A is *extensional* if, and only if, $A \models \mathbf{Ext}$. We let $Alg_{Ext}(\Sigma)$ denote the class $Alg(\Sigma, \mathbf{Ext})$ of all extensional Σ algebras, and $Alg_{Ext}(\Sigma, \Phi)$ denote the class $Alg(\Sigma, \mathbf{Ext} \cup \Phi)$, for any set $\Phi \subseteq \mathcal{L}(\Sigma, X)$ of Σ formulas. □

Clearly, every S-typed Σ algebra is extensional. However, there exist extensional Σ algebras which are not S–typed Σ algebras, for example consider the term algebra $T(\Sigma, X)$. Nevertheless, given an extensional Σ algebra A we can always use the following *collapsing construction* to obtain an S–typed Σ algebra $\overline{A}$ which is isomorphic to A.

3.2 Definition. Let A be an extensional S-sorted Σ algebra. We define the *collapse* $\overline{A}$ of A to be the S-typed Σ algebra with S-indexed family of carrier sets $\overline{A} = \langle\ \overline{A}_\tau \mid \tau \in S\ \rangle$, where $\overline{A}_\tau = \{\overline{a} \mid a \in A_\tau\ \}$. We define each element $\overline{a} \in \overline{A}_\tau$ by induction on the complexity of types as follows. For each basic type $\tau \in S$ and each $a \in A_\tau$ define $\overline{a} = a$. For each product type $(\sigma \times \tau) \in S$ and each $a \in A_{(\sigma\times\tau)}$ define $\overline{a} = (\overline{proj^1_A(a)},\ \overline{proj^2_A(a)})$. For each function type $(\sigma \to \tau) \in S$ and each $a \in A_{(\sigma\to\tau)}$ define $\overline{a} : \overline{A}_\sigma \to \overline{A}_\tau$ on each $\overline{b} \in \overline{A}_\sigma$ by

$$\overline{a}(\overline{b}) = \overline{eval^{(\sigma\to\tau)}_A(a, b)}.$$

Finally, for each limit type $\beta \in S$ and each $a \in A_\beta$ define

$$\overline{a} = \begin{cases} (\tau, \overline{b}), & \text{if exists } \tau \in S,\ \beta^- \le \tau < \beta \text{ and } b \in A_\tau \text{ such that } inj^\tau_A(b) = a; \\ (\beta, a), & \text{otherwise.} \end{cases}$$

The constants and algebraic operations of $\overline{A}$ are defined as follows. For each type $\tau \in S$ and each constant symbol $c \in \Sigma_{\lambda,\tau}$, define $c_{\overline{A}} = \overline{c_A}$. For each $\tau \in S$, each $w = \tau(1)\ldots\tau(n) \in S^+$, each function symbol $f \in \Sigma_{w,\tau}$ and any $(a_1, \ldots, a_n) \in A^w$, define $f_{\overline{A}}(\overline{a_1}, \ldots, \overline{a_n}) = \overline{f_A(a_1, \ldots, a_n)}$. □

In order to ensure that the operations of $\overline{A}$ are well defined we need the following result.

3.3 Proposition. *Let A be an extensional S-sorted Σ algebra. For each $\tau \in S$ and any $a, a' \in A_\tau$,*

$$\overline{a} = \overline{a'} \ \Leftrightarrow\ a = a'.$$

Proof. ⇐ Trivial. ⇒ By induction on the complexity of types. □

3.4 Corollary. *Let A be an extensional S-sorted Σ algebra. Then $\overline{A}$, the collapse of A, is a well defined S-typed Σ algebra.*

Proof. It can easily be shown that $\overline{A}$ is cumulative. Clearly, by Proposition 3.3 we know the operations of $\overline{A}$ are well defined. Thus, it only remains to show that the projection, evaluation and injection operation symbols are interpreted correctly in $\overline{A}$. This is straightforward and is left as an exercise for the reader. □

3.5 Lemma. *Let A be an extensional S-sorted Σ algebra, then $A \cong \overline{A}$.*

Proof. Using Proposition 3.3 it is easily verified that the map $\phi : A \to \overline{A}$ defined by $\phi_\tau(a) = \overline{a}$, for each $\tau \in S$ and $a \in A_\tau$, is a well defined isomorphism. □

An immediate consequence of the above Lemma is the following representation theorem for extensional Σ algebras which is a further generalisation of the Collapsing Theorem of Shepherdson and Mostowski.

3.6 Collapsing Theorem. (Mostowski, Shepherdson) *Let A be an S-sorted Σ algebra. Then A is isomorphic to an S-typed Σ algebra if, and only if, A is extensional.*

Proof. $\Rightarrow$ Follows since every S-typed Σ algebra is extensional.
$\Leftarrow$ Follows directly from Lemma 3.5. □

We now consider which basic algebraic constructions preserve or induce extensionality. It can be easily shown that the class of all extensional Σ algebras need not be closed under the formation of subalgebras and this leads us to define a stronger notion of subalgebra.

3.7 Definition. Let A and B be Σ algebras. We say that A is an *extensional subalgebra* of B, written $A \leq_{Ext} B$ if, and only if, $A \leq B$ and $A \models$ **Ext**. Let K be any class of Σ algebras then K is *closed under extensional subalgebras* if, and only if, for any Σ algebras A and B, if $B \in K$ and $A \leq_{Ext} B$ then $A \in K$. □

Clearly, $Alg_{Ext}(\Sigma)$ is closed under the formation of extensional Σ subalgebras. Since the extensionality sentences for any higher–order signature are all Horn sentences which are preserved under direct products it follows that the class of all extensional Σ algebras $Alg_{Ext}(\Sigma)$ is closed under non-empty direct products. Note that empty direct products are excluded since they correspond to unit algebras which are not in general extensional algebras. However, the class of all extensional Σ algebras need not be closed under homomorphic images and this leads us to define a more restrictive notion of homomorphic image.

3.8 Definition. Let A and B be Σ algebras. We say that B is an *extensional homomorphic image* of A if, and only if, B is a homomorphic image of A and B is extensional. For any class K of Σ algebras we say K is *closed under extensional homomorphic images* if, and only if, for any Σ algebras A and B, if $A \in K$ and B is an extensional homomorphic image of A then $B \in K$. □

Clearly, the class of all extensional Σ algebras is closed under extensional homomorphic images. Next we introduce the notion of extensional Σ congruence.

3.9 Definition. Let $\equiv \; = \langle \equiv_\tau \subseteq A_\tau \times A_\tau \mid \tau \in S \rangle$ be a Σ congruence over an S-sorted Σ algebra A. Then $\equiv$ is said to be an *extensional Σ congruence* if, and only if, it satisfies the following four conditions.
(i) For each product type $(\sigma \times \tau) \in S$ and $a, a' \in A_{(\sigma\times\tau)}$,

$$proj^1_A(a) \equiv_\sigma proj^1_A(a') \text{ and } proj^2_A(a) \equiv_\tau proj^2_A(a') \implies a \equiv_{(\sigma\times\tau)} a'.$$

(ii) For each function type $(\sigma \to \tau) \in S$ and $a, a' \in A_{(\sigma\to\tau)}$,

$$\forall b \in A_\sigma \; (eval^{(\sigma\to\tau)}_A(a, b) \equiv_\tau eval^{(\sigma\to\tau)}_A(a', b)) \implies a \equiv_{(\sigma\to\tau)} a'.$$

(iii) For each limit type $\beta \in S$, each type $\tau \in S$, $\beta^- \leq \tau < \beta$ and any $a, a' \in A_\tau$,

$$inj^\tau_A(a) \equiv_\beta inj^\tau_A(a') \implies a \equiv_\tau a'.$$

(iv) For each limit type $\beta \in S$, any types $\tau, \tau' \in S$, $\beta^- \leq \tau, \tau' < \beta$ and $\tau \neq \tau'$,

$$inj^{\tau}_{A}(a) \not\equiv_{\beta} inj^{\tau'}_{A}(a'),$$

for all $a \in A_\tau$, $a' \in A_{\tau'}$. □

Let A be an S-sorted Σ algebra then $Con_{Ext}(A)$ denotes the set of all extensional Σ congruences on A. Note that $Con_{Ext}(A)$ may be empty, for example if S contains a limit type with at least two types below it and A is a unit algebra. A class K of Σ algebras is said to be *closed under extensional quotients* if, and only if, for any Σ algebra A, if $A \in K$ and $\equiv$ is an extensional Σ congruence then $A/\equiv \in K$.

We can characterise those congruences on a Σ algebra A which yield extensional quotient algebras as follows.

3.10 Theorem. *Let A be an S-sorted Σ algebra and $\equiv$ be a Σ congruence on A. Then $A/\equiv$ is extensional if, and only if, $\equiv$ is an extensional Σ congruence on A.*

Proof. $\Rightarrow$ Suppose that $A/\equiv$ is extensional. Since the natural mapping $nat : A \to A/\equiv$ is an epimorphism it is straightforward to show that its kernel $\equiv$ is an extensional Σ congruence on A.

$\Leftarrow$ Suppose that $\equiv$ is extensional. Then we must prove that $A/\equiv$ is extensional. This is straightforward and is left as an exercise for the reader. □

4 Free Algebras

In this section we consider the existence of free and initial extensional algebras and the closure conditions necessary for a class of extensional Σ algebras to contain free and initial algebras.

Let Σ be an S-sorted signature and let K be a class of Σ algebras. Let F be a Σ algebra (not necessarily in K) and let $X \subseteq F$. Recall that F is said to be free for K on X if, and only if, for each $A \in K$ and each assignment $\alpha : X \to A$, there exists a unique homomorphic extension $\overline{\alpha} : F \to A$ of α. The S-indexed family X is termed the family of sets of generators for F and F is said to be freely generated by X. If in addition $F \in K$ then we say that F is free in K on X. If F is free for (respectively in) K on $\emptyset$ then there exists a unique homomorphism from F to each algebra $A \in K$ and we say that F is initial for (respectively in) K.

We now consider the existence and construction of free extensional algebras. Recall that for any S-typed signature Σ and any S-indexed family $X = \langle X_\tau \mid \tau \in S\rangle$ of sets of variables, $T(\Sigma, X)$ is free in $Alg(\Sigma)$ on X and is thus free for each class $K \subseteq Alg(\Sigma)$ on X. It is straigtforward to show that $T(\Sigma, X) \models$ **Ext**. Thus it follows that $T(\Sigma, X)$ is free in $Alg_{Ext}(\Sigma)$ on X and in particular, $T(\Sigma)$ is initial in $Alg_{Ext}(\Sigma)$. Recall the concrete construction of a free algebra for a non-empty class K of Σ algebras.

4.1 Definition. Let Σ be an S-typed signature and let K be any non-empty class of Σ algebras. Define the Σ congruence $\equiv^K = \langle \equiv^K_\tau \subseteq T(\Sigma, X)^2_\tau \mid \tau \in S \rangle$ for each $\tau \in S$ and any terms $t, t' \in T(\Sigma, X)_\tau$ by $t \equiv^K_\tau t'$ if, and only if, $\overline{\alpha}_\tau(t) = \overline{\alpha}_\tau(t')$ for each $A \in K$ and each assignment $\alpha : X \to A$, i.e. $K \models t = t'$. It can easily be shown that if K is a non–empty extensional class of Σ algebras then $\equiv^K$ is an extensional Σ congruence. □

We denote the quotient algebra $T(\Sigma, X)/\equiv^K$ by $T_K(\Sigma, X)$. By a basic result of universal algebra $T_K(\Sigma, X)$ is free for K on X. Furthermore, by a result of Birkhoff [1935] a sufficient condition for $T_K(\Sigma, X)$ to be in K is that K is closed under the formation of isomorphic images, direct products and subalgebras. However, as we have already seen in Section 3, the class $Alg_{Ext}(\Sigma)$ of all extensional Σ algebras need not be closed under subalgebras. So which extensional classes contain free algebras? To answer this question we generalise a result of Meinke [1992] which extended the result of Birkhoff to higher–order algebras with product and function types.

4.2 Theorem. *There exists an S-indexed family κ of cardinals such that for every S-indexed family X of sets of variables and every non-empty class K of extensional Σ algebras which is closed under isomorphic images, non-empty direct products and extensional subalgebras, if $|X| \geq \kappa$ then*

$$T_K(\Sigma, X) \in K.$$

Proof. Define the Σ congruence $\approx^K = \langle \approx^K_\tau \subseteq T(\Sigma, X)^2_\tau \mid \tau \in S \rangle$ by $\approx^K = \bigcap C$, where

$$C = \{\equiv \in Con_{Ext}(T(\Sigma, X)) \mid T(\Sigma, X)/\equiv \text{ embeds in some } A \in K\}.$$

Then we can show there exists an S-indexed family κ of cardinals such that $\equiv^K$ and $\approx^K$ are identical extensional Σ congruences on $T(\Sigma, X)$ for any S-indexed family X of sets of variables with $|X| \geq \kappa$ (see Steggles [1995]). Let I be any indexing set for C such that $\approx^K = \bigcap_{i \in I} \equiv^i$ (note that I is non-empty since $\equiv^K = \approx^K$ is extensional and so cannot be the unit congruence). By a basic result of subdirect products, there exists a subdirect embedding

$$\epsilon : T(\Sigma, X)/\approx^K \to \Pi_{i \in I} T(\Sigma, X)/\equiv^i .$$

For each $i \in I$ we have $T(\Sigma, X)/\equiv^i$ embeds in some $A \in K$ so we can define a Σ embedding $\phi : T(\Sigma, X)/\equiv^i \to A$. By Theorem 3.10, $T(\Sigma, X)/\equiv^i$ is extensional and so $\phi(T(\Sigma, X)/\equiv^i) \leq_{Ext} A$. Furthermore $T(\Sigma, X)/\equiv^i \cong \phi(T(\Sigma, X)/\equiv^i)$. Therefore, since $A \in K$ and K is closed under isomorphic images and extensional subalgebras it follows that $T(\Sigma, X)/\equiv^i \in K$. Since K is closed under the formation of non-empty direct products it follows that $\Pi_{i \in I} T(\Sigma, X)/\equiv^i \in K$. As $\approx^K$ is an extensional Σ congruence it follows by Theorem 3.10 that $T(\Sigma, X)/\approx^K$ is extensional. Since ϵ embeds $T(\Sigma, X)/\approx^K$ in $\Pi_{i \in I} T(\Sigma, X)/\equiv^i$ and K is closed under the formation of isomorphic images and extensional subalgebras it follows that $T(\Sigma, X)/\approx^K \in K$. Thus, $T_K(\Sigma, X) \in K$. □

Unfortunately, Theorem 4.2 provides no information about the existence of initial algebras. We know the class $Alg_{Ext}(\Sigma)$ of all extensional Σ algebras admits an initial algebra, namely $T(\Sigma)$, but this turns out to be unusual. Generally, most subclasses of $Alg_{Ext}(\Sigma)$, for example those satisfying some set of equations E, do not contain initial algebras. Nevertheless, a subclass $K \subseteq Alg_{Ext}(\Sigma)$ satisfying the closure conditions of Theorem 4.2, which admits a *minimal* algebra, does admit a minimal algebra which approximates the properties of an initial model. We refer to this minimal algebra as the (*transfinite*) *initial extensional model.* Given any class K of Σ algebras we let $Min(K)$ denote the class of all minimal Σ algebras in K.

4.3 Theorem. *Let K be any non-empty class of extensional Σ algebras which is closed under the formation of isomorphic images, non-empty direct products and extensional subalgebras. If K admits a minimal algebra then $T_{Min(K)}(\Sigma)$ is initial in $Min(K)$.*

Proof. Define the Σ congruence $\approx^{Min(K)} = \langle \approx_\tau^{Min(K)} \subseteq T(\Sigma)_\tau^2 \mid \tau \in S \rangle$ by

$$\approx^{Min(K)} = \bigcap \{ \equiv \in Con_{Ext}(T(\Sigma)) \mid T(\Sigma)/\equiv \text{ embeds in some } A \in Min(K) \}.$$

Suppose K admits a minimal algebra. Then we can show that $\equiv^{Min(K)}$ and $\approx^{Min(K)}$ are identical extensional Σ congruences on $T(\Sigma)$ (see Steggles [1995]). The remainder of the proof is similar to the proof of Theorem 4.2. □

We term $T_{Min(K)}(\Sigma)$ the (*transfinite*) *initial extensional model* to emphasize the fact that it is initial in a weaker, but still non-trivial, sense. For any equational theory $E \subseteq Eqn(\Sigma, X)$ we denote the initial extensional model in the class $Min_{Ext}(\Sigma, E)$ by $I_{Ext}(\Sigma, E)$.

5 Transfinite Higher–Order Equational Logic

We extend the higher–order equational logic presented in Meinke [1992] to a *higher–order equational logic with transfinite types* (referred to as simply *transfinite higher–order equational logic*). We show that transfinite higher–order equational logic is both sound and complete with respect to extensional Σ, E algebras. We then extend this transfinite higher–order equational logic with an infinitary inference rule for function types and show that this new calculus provides a concrete construction of the *initial extensional model* $I_{Ext}(\Sigma, E)$, which is initial in the class of all minimal extensional Σ, E algebras.

In the sequel let $S \subseteq H(\mathcal{B})$ be a type structure over some type basis $\mathcal{B}$ and let Σ be an S-typed signature. Let $X = \langle X_\tau \mid \tau \in S \rangle$ be an S-indexed family of non-empty sets of variable symbols. To avoid the problems associated with empty carrier sets in the many-sorted case we assume that Σ is non-void.

5.1 Definition. The (ordinary) inference rules of *transfinite higher–order equational logic* are the following.
(i) For any type $\tau \in S$ and any term $t \in T(\Sigma, X)_\tau$,

$$\frac{}{t = t}$$

is a *reflexivity* rule.
(ii) For any type $\tau \in S$ and any terms $t_0, t_1 \in T(\Sigma, X)_\tau$,

$$\frac{t_0 = t_1}{t_1 = t_0}$$

is a *symmetry* rule.
(iii) For any type $\tau \in S$ and any terms $t_0, t_1, t_2 \in T(\Sigma, X)_\tau$,

$$\frac{t_0 = t_1, \quad t_1 = t_2}{t_0 = t_2}$$

is a *transitivity* rule.
(iv) For each type $\sigma \in S$, any terms $t, t' \in T(\Sigma, X)_\sigma$, any type $\tau \in S$, any variable symbol $x \in X_\tau$ and any terms $t_0, t_1 \in T(\Sigma, X)_\tau$,

$$\frac{t = t', \quad t_0 = t_1}{t[x/t_0] = t'[x/t_1]}$$

is a *substitution* rule.
(v) For each product type $(\sigma \times \tau) \in S$ and any terms $t_0, t_1 \in T(\Sigma, X)_{(\sigma\times\tau)}$,

$$\frac{proj^{(\sigma\times\tau),1}(t_0) = proj^{(\sigma\times\tau),1}(t_1), \quad proj^{(\sigma\times\tau),2}(t_0) = proj^{(\sigma\times\tau),2}(t_1)}{t_0 = t_1}$$

is a *projection* rule.
(vi) For each function type $(\sigma \to \tau) \in S$, any terms $t_0, t_1 \in T(\Sigma, X)_{(\sigma\to\tau)}$ and any variable symbol $x \in X_\sigma$ not occurring in t_0 or t_1,

$$\frac{eval^{(\sigma\to\tau)}(t_0, x) = eval^{(\sigma\to\tau)}(t_1, x)}{t_0 = t_1}$$

is an *evaluation* rule.
(vii) For each limit type $\beta \in S$, each type $\tau \in S$ such that $\beta^- \leq \tau < \beta$ and any terms $t_0, t_1 \in T(\Sigma, X)_\tau$,

$$\frac{inj^\tau(t_0) = inj^\tau(t_1)}{t_0 = t_1}$$

is an *injection* rule. □

Let $\vdash$ denote the inference relation between equational theories $E \subseteq Eqn(\Sigma, X)$ and equations $e \in Eqn(\Sigma, X)$, defined by $E \vdash e$ if, and only if, there exists a (finite) proof of e from E using the rules of transfinite higher–order equational logic. We can prove that $\vdash$ is *sound* with respect to extensional Σ, E algebras.

5.2 Soundness Theorem. *Let $E \subseteq Eqn(\Sigma, X)$ be an equational theory and let $e \in Eqn(\Sigma, X)$ be any equation. Then*

$$E \vdash e \Rightarrow Alg_{Ext}(\Sigma, E) \models e.$$

Proof. By induction on the length of proofs. □

Note that an equational theory $E \subseteq Eqn(\Sigma, X)$ may have no extensional models, for example if it contains an equation which contradicts extensionality axiom 3.1.(iv). We say an equational theory $E \subseteq Eqn(\Sigma, X)$ is *consistent* if, and only if, $Alg_{Ext}(\Sigma, E) \neq \emptyset$.

Given an S-typed signature Σ and an equational theory $E \subseteq Eqn(\Sigma, X)$ we define the congruence $\equiv^E$ over the term algebra $T(\Sigma, X)$ by $t \equiv^E_\tau t' \Leftrightarrow E \vdash t = t'$, for $\tau \in S$ and any terms $t, t' \in T(\Sigma, X)_\tau$. Let $E \subseteq Eqn(\Sigma, X)$ be a consistent equational theory. Then it is straightforward to show that $\equiv^E$ is an extensional Σ congruence on $T(\Sigma, X)$. Thus, by Theorem 3.10 we know that $T(\Sigma, X)/\equiv^E$ is extensional.

Before being able to prove that $\vdash$ is complete with respect to extensional Σ, E algebras we need to show that $T(\Sigma, X)/\equiv^E$ is equationally generic.

5.3 Proposition. *Let $E \subseteq Eqn(\Sigma, X)$ be an equational theory and let $e \in Eqn(\Sigma, X)$ be any equation. Then*

$$E \vdash e \Leftrightarrow T(\Sigma, X)/\equiv^E \models e.$$

Proof. The proof is omitted for brevity (see Steggles [1995]). □

We now prove that $\vdash$ is complete with respect to extensional Σ, E algebras.

5.4 Completeness Theorem. *Let $E \subseteq Eqn(\Sigma, X)$ be a consistent equational theory and let $e \in Eqn(\Sigma, X)$ be any equation. Then*

$$E \vdash e \Leftrightarrow Alg_{Ext}(\Sigma, E) \models e.$$

Proof. $\Rightarrow$ By Soundness Theorem 5.2.
$\Leftarrow$ Suppose that $E \not\vdash e$ then by Proposition 5.3 we know $T(\Sigma, X)/\equiv^E \not\models e$. By consistency $Alg_{Ext}(\Sigma, E) \neq \emptyset$ and so it follows that $T(\Sigma, X)/\equiv^E$ is extensional. Also by Proposition 5.3 we have $T(\Sigma, X)/\equiv^E \models E$ but $T(\Sigma, X)/\equiv^E \not\models e$. Thus, $Alg_{Ext}(\Sigma, E) \not\models e$. □

In general the quotient algebra $T(\Sigma)/\equiv^E$ is not extensional and thus cannot be taken as the initial extensional model $I_{Ext}(\Sigma, E)$. To construct a quotient algebra of $T(\Sigma)$ which is both a model of E and extensional we need to add the following (infinitary) ω–evaluation rule to transfinite higher–order equational logic.

5.5 Definition. Let S be a type structure over a type basis $\mathcal{B}$. Let Σ be an S-typed signature. For each function type $(\sigma \to \tau) \in S$ and any terms $t_0, t_1 \in T(\Sigma, X)_{(\sigma\to\tau)}$,

$$\frac{\langle eval^{(\sigma\to\tau)}(t_0, t) = eval^{(\sigma\to\tau)}(t_1, t) \mid t \in T(\Sigma)_\sigma \rangle}{t_0 = t_1}$$

is an (infinitary) *ω–evaluation rule.* □

Let $\vdash_\omega$ denote the inference relation between equational theories $E \subseteq Eqn(\Sigma, X)$ and equations $e \in Eqn(\Sigma, X)$, defined by $E \vdash_\omega e$ if, and only if, there exists a proof of e from E using the inference rules of transfinite higher–order equational logic and the ω-evaluation rule. Clearly, if $E \vdash_\omega e$ then $A \models e$ for every minimal extensional Σ, E algebra A, i.e. the ω-evaluation rule is *sound* with respect to minimal models. Given an equational theory $E \subseteq Eqn(\Sigma, X)$ we can use $\vdash_\omega$ to define a Σ congruence $\equiv^{E,\omega}$ on the ground term algebra $T(\Sigma)$ in the usual way. Let $Min_{Ext}(\Sigma, E)$ denote the class of all minimal, extensional Σ, E algebras. If $Min_{Ext}(\Sigma, E) \neq \emptyset$ then it is straightforward to show that $\equiv^{E,\omega}$ is an extensional Σ congruence on $T(\Sigma)$.

5.6 Proposition. *Let $E \subseteq Eqn(\Sigma, X)$ be an equational theory and let $e \in Eqn(\Sigma, X)$ be any equation. Then*

$$E \vdash_\omega e \Rightarrow T(\Sigma)/\equiv^{E,\omega} \models e.$$

Proof. The proof is omitted for brevity (see Steggles [1995]). □

From the above proposition we know that $T(\Sigma)/\equiv^{E,\omega} \models E$. Thus if there exists a minimal extensional Σ, E algebra then by Theorem 3.10 and Proposition 5.6, $T(\Sigma)/\equiv^{E,\omega} \in Min_{Ext}(\Sigma, E)$.

Finally, we prove that $T(\Sigma)/\equiv^{E,\omega}$ is isomorphic to $T_{Min(K)}(\Sigma)$ and thus, that $T(\Sigma)/\equiv^{E,\omega}$ is initial in $Min_{Ext}(\Sigma, E)$. Therefore it may be taken as a concrete construction of the *initial extensional model* $I_{Ext}(\Sigma, E)$.

5.7 Theorem. *Let $E \subseteq Eqn(\Sigma, X)$ be any equational theory and let $K = Alg_{Ext}(\Sigma, E)$. If $Min_{Ext}(\Sigma, E) \neq \emptyset$ then*

$$T(\Sigma)/\equiv^{E,\omega} \cong T_{Min(K)}(\Sigma),$$

and hence $T(\Sigma)/\equiv^{E,\omega}$ is initial in $Min(K)$.

Proof. Since K is an equational class we know by Theorem 4.3 that $T_{Min(K)}(\Sigma)$ is initial in $Min(K)$. By Theorem 3.10 and Proposition 5.6 we have $T(\Sigma)/\equiv^{E,\omega} \in Min_{Ext}(\Sigma, E)$. Therefore, there exists a homomorphism $\phi : T_{Min(K)}(\Sigma) \to T(\Sigma)/\equiv^{E,\omega}$. Since $T_{Min(K)}(\Sigma)$ and $T(\Sigma)/\equiv^{E,\omega}$ are both minimal Σ, E algebras we need only show that there exists a homomorphism $\psi : T(\Sigma)/\equiv^{E,\omega} \to T_{Min(K)}(\Sigma)$. Define ψ by $\psi_\tau([t]_{E,\omega}) = [t]_{Min(K)}$, for each $\tau \in S$ and each term

$t \in T(\Sigma)_\tau$. Then ψ is well defined since the rules of transfinite higher–order equational logic and the ω-evaluation rule are sound for every minimal extensional Σ, E algebra. It only remains to prove that ψ is a homomorphism which is straightforward to do. □

6 A Simple Specification Example

In this section we demonstrate the use of transfinite types for modelling polymorphism and partiality by considering the specification of a simple polymorphic functional programming language $\mathcal{PR}^\omega$ which is based on the functional programming language $\mathcal{PR}$ of primitive recursive functions (see Thompson [1987]). In this specification example the first limit type ω is used to allow infinite schemes of functions to be replaced by polymorphic functions. These polymorphic functions are partial and return a value of type *undef* whenever they are undefined for a particular type. We present an $\omega+1$–order equational specification of $\mathcal{PR}^\omega$ which we show is correct under higher–order initial algebra semantics. Our definitions and results generalise those for the algebra of primitive recursive functions over the natural numbers presented in Meinke [1994].

For the remainder of this section let S be a sort set such that $nat \in S$ and let Σ be an S–sorted signature such that $0 \in \Sigma_{\lambda,nat}$ and $succ \in \Sigma_{nat,nat}$. Let A be any Σ algebra such that $A_{nat} = \mathbf{N}$, $0_A = 0 \in \mathbf{N}$ and $succ_A(n) = n+1$, for any $n \in \mathbf{N}$, and let $Spec = (\Sigma, E)$ be a correct first–order equational specification of A under first–order initial algebra semantics, i.e. $I(Spec) \cong A$.

We use the first–order signature Σ to define an $\omega+1$–order signature $PR^\omega(\Sigma)$ which names all the basic functions and function building operations in our polymorphic functional programming language. We then use the Σ algebra A to construct a standard $PR^\omega(\Sigma)$ model $PR^\omega(A)$. Using the equational specification $Spec = (\Sigma, E)$, which by assumption is a correct specification of A under first–order initial algebra semantics, we construct a higher–order equational specification $PR^\omega(Spec) = (PR^\omega(\Sigma), PR^\omega(E))$ which we show correctly specifies $PR^\omega(A)$ under higher–order initial algebra semantics.

The following technical definition is needed in the sequel.

6.1 Definition. Let $\mathcal{B}$ be any type basis. For any types $\tau(1), \ldots, \tau(n) \in H(\mathcal{B})$, $n \geq 1$ we let $(\tau(1)\times\cdots\times\tau(n))$ denote the product type $(\tau(1)\times(\tau(2)\times\cdots\times\tau(n)))$.

For each non–empty word $w = \tau(1)\ldots\tau(n) \in \mathcal{B}^+$ define the type $\overline{w} \in H(\mathcal{B})$ by

$$\overline{w} = (\tau(1) \times \cdots \times \tau(n)).$$

□

We begin by constructing a signature $PR^\omega(\Sigma)$ for our polymorphic functional programming language using the signature Σ as follows.

6.2 Definition. Let $\mathcal{B}(S) = S \cup \{undef\}$ be a type basis and let $S_{PR^\omega} \subseteq H(\mathcal{B}(S))$ be a type structure over the type basis $\mathcal{B}(S)$ defined by

$$S_{PR^\omega} = PT \cup \{\omega, (\omega \to \omega)\},$$

where $PT = \bigcup_{i \in \mathbf{N}} PT_i$, and $PT_0 = \mathcal{B}(S)$ and $PT_{n+1} = PT_n \cup \{(\sigma \times \tau) \mid \sigma, \tau \in PT_n\}$. Define the S_{PR^ω}-typed signature $PR^\omega(\Sigma)$ to be the smallest S_{PR^ω}-indexed family of sets satisfying the following conditions.

For each $w \in S^*$ and $\tau \in S$ we have $\Sigma_{w,\tau} \subseteq PR^\omega(\Sigma)_{w,\tau}$ and $error \in PR^\omega(\Sigma)_{\lambda,undef}$. We have $eval^{(\omega \to \omega)} \in PR^\omega(\Sigma)_{(\omega\to\omega)\ \omega,\omega}$, for each product type $(\sigma \times \tau) \in S_{PR^\omega}$,

$$proj^{(\sigma\times\tau),1} \in PR^\omega(\Sigma)_{(\sigma\times\tau),\sigma},\quad proj^{(\sigma\times\tau),2} \in PR^\omega(\Sigma)_{(\sigma\times\tau),\tau},$$

$$\langle . \rangle^{(\sigma\times\tau)} \in PR^\omega(\Sigma)_{\sigma\ \tau,(\sigma\times\tau)},$$

and for each $\tau \in S_{PR^\omega}$ such that $\tau < \omega$, $inj^\tau \in PR^\omega(\Sigma)_{\tau,\omega}$. For each $\tau \in S$ and each constant symbol $c \in \Sigma_{\lambda,\tau}$, $c^\omega \in PR^\omega(\Sigma)_{\lambda,(\omega\to\omega)}$. For each $w \in S^+$, $\tau \in S$ and each function symbol $f \in \Sigma_{w,\tau}$, $f^\omega \in PR^\omega(\Sigma)_{\lambda,(\omega\to\omega)}$. We also have

$$id^\omega,\ und\_fn,\ proj^{\omega,1},\ proj^{\omega,2} \in PR^\omega(\Sigma)_{\lambda,(\omega\to\omega)},$$

$$\langle . \rangle^\omega, und? \in PR^\omega(\Sigma)_{\omega\ \omega,\omega},\quad |, \circ, pr, switch^\perp \in PR^\omega(\Sigma)_{(\omega\to\omega)\ (\omega\to\omega),(\omega\to\omega)},$$

and for each type $\tau \in S_{PR^\omega}$ such that $\tau < \omega$,

$$case^\tau \in PR^\omega(\Sigma)_{(\omega\to\omega)\ (\omega\to\omega),(\omega\to\omega)}.$$

□

Given any S_{PR^ω}-typed $PR^\omega(\Sigma)$ algebra B, any type $(\tau(1) \times \cdots \times \tau(n)) \in S_{PR^\omega}$, $n \geq 1$ and any $x_i \in B_{\tau(i)}$, $1 \leq i \leq n$ we define $\langle x_1, \ldots, x_n \rangle_B$ by $\langle x_1 \rangle_B = x_1$ and for $n > 1$, $\langle x_1, \ldots, x_n \rangle_B = \langle x_1, \langle x_2, \ldots, x_n \rangle_B \rangle_B^{(\tau(1)\times\cdots\times\tau(n))}$.

We can construct a standard $PR^\omega(\Sigma)$ algebra $PR^\omega(A)$ using the Σ algebra A as follows.

6.3 Definition. We construct the S_{PR^ω}-typed $PR^\omega(\Sigma)$ algebra $PR^\omega(A)$ as the minimal subalgebra of a complete $\omega + 1$-order S_{PR^ω}-typed $PR^\omega(\Sigma)$ algebra P which is defined as follows.

Define the carrier sets of P by $P_{undef} = \{\perp\}$, for each type $\tau \in S$, $P_\tau = A_\tau$, for each product type $(\sigma \times \tau) \in S_{PR^\omega}$, $P_{(\sigma\times\tau)} = (P_\sigma \times P_\tau)$, and

$$P_\omega = \coprod_{\langle \tau \in S_{PR^\omega} \mid \tau < \omega \rangle} P_\tau,\quad P_{(\omega\to\omega)} = [P_\omega \to P_\omega].$$

Define the constants and operations of P as follows. For each $w \in S^*$, $\tau \in S$ and each symbol $f \in \Sigma_{w,\tau}$ define $f_P = f_A$ and define $error_P = \perp$. For each product type $(\sigma \times \tau) \in S_{PR^\omega}$ define $proj_P^{(\sigma\times\tau),1} : P_{(\sigma\times\tau)} \to P_\sigma$, $proj_P^{(\sigma\times\tau),2} : P_{(\sigma\times\tau)} \to P_\tau$ to be the left and right projection functions and define the pairing function $\langle . \rangle_P : P_\sigma \times P_\tau \to P_{(\sigma\times\tau)}$ on each $a \in P_\sigma$ and $b \in P_\tau$ by $\langle a, b\rangle_P = (a, b)$. Define $eval_P^{(\omega\to\omega)} : P_{(\omega\to\omega)} \times P_\omega \to P_\omega$ to be the evaluation function and for each type $\tau \in S_{PR^\omega}$ such that $\tau < \omega$ define $inj_P^\tau : P_\tau \to P_\omega$ to be the canonical injection function. For each $\tau \in S$ and each constant symbol $c \in \Sigma_{\lambda,\tau}$ define the *constant function* $c_P^\omega : P_\omega \to P_\omega$ on each $a \in P_\omega$ by $c_P^\omega(a) = (\tau, c_A)$. For each $w \in S^+$, $\tau \in S$ and each function symbol $f \in \Sigma_{w,\tau}$ define $f_P^\omega : P_\omega \to P_\omega$ on each $a \in P_\omega$ by

$$f_P^\omega(a) = \begin{cases} (\tau, f_A(a_1,\ldots,a_n)), & \text{if } a = (\overline{w},(a_1,\ldots,a_n)), \text{ for } (a_1,\ldots,a_n) \in P_{\overline{w}}; \\ (undef, \bot), & \text{otherwise.} \end{cases}$$

For each $a \in P_\omega$ define the *identity function* $id_P^\omega : P_\omega \to P_\omega$ by $id_P^\omega(a) = a$, the *undefined function* $und\_fn_P : P_\omega \to P_\omega$ by $und\_fn_P(a) = (undef, \bot)$, and the *$\omega$-projection functions* $proj_P^{\omega,1}, proj_P^{\omega,2} : P_\omega \to P_\omega$ by

$$proj_P^{\omega,1}(a) = \begin{cases} (\sigma, a_1), & \text{if } a = ((\sigma\times\tau),(a_1,a_2)), \text{ for some product type} \\ & (\sigma\times\tau) \in S_{PR^\omega} \text{ and } (a_1,a_2) \in P_{(\sigma\times\tau)}; \\ (undef,\bot), & \text{otherwise;} \end{cases}$$

$$proj_P^{\omega,2}(a) = \begin{cases} (\tau, a_2), & \text{if } a = ((\sigma\times\tau),(a_1,a_2)), \text{ for some product type} \\ & (\sigma\times\tau) \in S_{PR^\omega} \text{ and } (a_1,a_2) \in P_{(\sigma\times\tau)}; \\ (undef,\bot), & \text{otherwise.} \end{cases}$$

For any $(\sigma, a), (\tau, b) \in P_\omega$ define the *ω-pairing function* $\langle . \rangle_P^\omega : P_\omega \times P_\omega \to P_\omega$ by

$$\langle(\sigma,a),(\tau,b)\rangle_P^\omega = ((\sigma\times\tau),(a,b)).$$

For any $a, b \in P_\omega$ define $und?_P : P_\omega \times P_\omega \to P_\omega$ by

$$und?_P(a,b) = \begin{cases} b, & \text{if } a = (undef,\bot); \\ a, & \text{otherwise.} \end{cases}$$

For any $f, g \in P_{(\omega\to\omega)}$ and any $a \in P_\omega$ define the *vectorisation function*

$$|_P : P_{(\omega\to\omega)} \times P_{(\omega\to\omega)} \to P_{(\omega\to\omega)}, \quad (f|_P g)(a) = \langle f(a), g(a)\rangle_P^\omega,$$

and the *composition function*

$$\circ_P : P_{(\omega\to\omega)} \times P_{(\omega\to\omega)} \to P_{(\omega\to\omega)}, \quad (f \circ_P g)(a) = f(g(a)).$$

We define the *primitive recursion operator* $pr_P : P_{(\omega\to\omega)} \times P_{(\omega\to\omega)} \to P_{(\omega\to\omega)}$ on any $f, g \in P_{(\omega\to\omega)}$ as follows. For each product type of the form $(nat\times\tau) \in S_{PR^\omega}$ and any $b \in P_\tau$ define

$$pr_P(f,g)((nat\times\tau),(0,b)) = f(\tau,b),$$

and for any $n \in P_{nat}$ define

$$pr_P(f,g)((nat\times\tau),(n+1,b)) =$$
$$g(\langle(nat,n),\langle(\tau,b),pr_P(f,g)((nat\times\tau),(n,b))\rangle_P^\omega\rangle_P^\omega).$$

For each $(\tau, b) \in P_\omega$ such that $\tau \neq (nat \times \sigma)$, for any $\sigma \in S_{PR^\omega}$ define

$$pr_P(f,g)(\tau,b) = (undef, \bot).$$

Define $switch^{\bot}_P : P_{(\omega\to\omega)} \times P_{(\omega\to\omega)} \to P_{(\omega\to\omega)}$ on each $a \in P_\omega$ by

$$switch^{\bot}_P(f,g)(a) = \begin{cases} g(a), & \text{if } f(a) = (undef, \bot); \\ f(a), & \text{otherwise.} \end{cases}$$

For each type $\tau \in S_{PR^\omega}$, $\tau < \omega$ define $case^{\tau}_P : P_{(\omega\to\omega)} \times P_{(\omega\to\omega)} \to P_{(\omega\to\omega)}$ on each $a \in P_\omega$ by

$$case^{\tau}_P(f,g)(a) = \begin{cases} f(a), & \text{if } a = (\tau, b) \text{, for some } b \in P_\tau; \\ g(a), & \text{otherwise.} \end{cases}$$

Define $PR^\omega(A)$ to be the minimal subalgebra of P. Note that $PR^\omega(A)$ is a countable algebra and that the Σ algebra A is preserved in $PR^\omega(A)$, i.e. $A \cong PR^\omega(A)|_\Sigma$. □

We now present some simple examples of polymorphic functions that can be constructed using the functional programming language $\mathcal{PR}^\omega$.

6.4 Examples.
(i) Consider the projection function $projn : PR^\omega(A)_\omega \to PR^\omega(A)_\omega$ defined on each $b \in PR^\omega(A)_\omega$ by

$$projn(b) = \begin{cases} (\tau_{i+1}, a_{i+1}), & \text{if } b = ((nat \times \tau(1) \times \cdots \times \tau(n)), (i, a_1, \ldots, a_n)), \\ & \text{for some } n \geq 1 \text{ and } i < n; \\ (undef, \bot), & \text{otherwise.} \end{cases}$$

Define the function term

$$PROJ \equiv (switch^{\bot}(proj^{\omega,1}, id^\omega) \circ (pr(id^\omega, (proj^{\omega,2} \circ (proj^{\omega,2} \circ proj^{\omega,2}))))).$$

Then we can show that $PROJ_{PR^\omega(A)} = projn$.
(ii) Let S_{NB} be a sort set such that $nat, bool \in S_{NB}$ and let Σ^{NB} be an S_{NB}–sorted signature such that

$$0 \in \Sigma^{NB}_{\lambda,nat},\ succ \in \Sigma^{NB}_{nat,nat},\ True, False \in \Sigma^{NB}_{\lambda,bool},\ EOR \in \Sigma^{NB}_{bool\ bool,bool}.$$

Let B be an S_{NB}–sorted Σ^{NB} algebra with carrier sets $B_{nat} = \mathbf{N}$, $B_{bool} = \mathbf{B}$, and constants and functions defined by $0_B = 0$, $True_B = tt$, $False_B = ff$, $succ_B(n) = n + 1$, and

$$EOR_B(b_1, b_2) = \begin{cases} tt, & \text{if either } b_1 = tt \text{ or } b_2 = tt \text{ but not both;} \\ ff, & \text{otherwise;} \end{cases}$$

for any $n \in \mathbf{N}$ and $b_1, b_2 \in \mathbf{B}$.

Consider the addition function $add : PR^{\omega}(B)_{\omega} \to PR^{\omega}(B)_{\omega}$ defined on each $a \in PR^{\omega}(B)_{\omega}$ by

$$add(a) = \begin{cases} (nat, n+m), & \text{if } a = ((nat \times nat), (n, m)); \\ (undef, \bot), & \text{otherwise.} \end{cases}$$

Define the function term

$$ADD \equiv pr(case^{nat}(id^{\omega}, und\_fn), succ^{\omega} \circ proj^{\omega,2} \circ proj^{\omega,2}).$$

Then it is straightforward to show $ADD_{PR^{\omega}(B)} = add$. We can use $case^{\tau}_{PR^{\omega}(B)}$ and $switch^{\bot}_{PR^{\omega}(B)}$ to combine function terms to produce overloaded polymorphic functions. For example consider the function $add\_eor : PR^{\omega}(B)_{\omega} \to PR^{\omega}(B)_{\omega}$ defined on each $a \in PR^{\omega}(B)_{\omega}$ by

$$add\_eor(a) = \begin{cases} (nat, n+m), & \text{if } a = ((nat \times nat), (n, m)); \\ (bool, EOR_B(b_1, b_2)), & \text{if } a = ((bool \times bool), (b_1, b_2)); \\ (undef, \bot), & \text{otherwise.} \end{cases}$$

Then we can show that

$$case^{(nat \times nat)}(ADD, EOR^{\omega})_{PR^{\omega}(B)} = add\_eor,$$

$$switch^{\bot}(ADD, EOR^{\omega})_{PR^{\omega}(B)} = add\_eor.$$

□

We construct an $\omega+1$–order equational specification of $PR^{\omega}(A)$ based on the first–order equational specification $Spec = (\Sigma, E)$ for the Σ algebra A as follows.

6.5 Definition. Let X be an $S_{PR^{\omega}}$–indexed family of sets of variables. Define the $\omega+1$–order equational specification $PR^{\omega}(Spec)$ by

$$PR^{\omega}(Spec) = (PR^{\omega}(\Sigma), PR^{\omega}(E)),$$

where $PR^{\omega}(E) \subseteq Eqn(PR^{\omega}(\Sigma), X)$ is an $\omega+1$–order equational theory consisting of the first–order equations E of $Spec$ and the following equations.

Let $Y, Z \in X_{(\omega\to\omega)}$, $x, y \in X_{\omega}$, $t \in X_{nat}$ and for each type $\tau \in S_{PR^{\omega}}$, $x^{\tau}, y^{\tau} \in X_{\tau}$. For each product type $(\sigma \times \tau) \in S_{PR^{\omega}}$ we have the equations

$$proj^1(\langle x^{\sigma}, y^{\tau}\rangle) = x^{\sigma}, \; proj^2(\langle x^{\sigma}, y^{\tau}\rangle) = y^{\tau}. \qquad (1a, b)$$

For each $\tau \in S$ and each constant symbol $c \in \Sigma_{\lambda,\tau}$ we have the equation

$$c^{\omega}(x) = inj^{\tau}(c). \qquad (2)$$

For each $w = \tau(1)\ldots\tau(n) \in S^+$, $\tau \in S$, each function symbol $f \in \Sigma_{w,\tau}$ and variables $x_1 \in X_{\tau(1)}, \ldots, x_n \in X_{\tau(n)}$ we have the equation

$$f^{\omega}(inj^{\overline{w}}(\langle x_1, \ldots, x_n\rangle)) = inj^{\tau}(f(x_1, \ldots, x_n)), \qquad (3a)$$

and for each $\sigma \in S_{PR^\omega}$ such that $\sigma < \omega$ and $\sigma \neq \overline{w}$,

$$f^\omega(inj^\sigma(x^\sigma)) = inj^{undef}(error). \tag{3b}$$

We have the equations

$$id^\omega(x) = x, \tag{4}$$

$$und\_fn(x) = inj^{undef}(error). \tag{5}$$

For each product type $(\sigma \times \tau) \in S_{PR^\omega}$ we have the equations

$$proj^{\omega,1}(inj^{(\sigma \times \tau)}(\langle x^\sigma, y^\tau \rangle)) = inj^\sigma(x^\sigma), \tag{6a}$$

$$proj^{\omega,2}(inj^{(\sigma \times \tau)}(\langle x^\sigma, y^\tau \rangle)) = inj^\tau(y^\tau). \tag{6b}$$

For each basic type $\tau \in \mathcal{B}(S)$ we have the equations

$$proj^{\omega,1}(inj^\tau(x^\tau)) = inj^{undef}(error), \tag{6c}$$

$$proj^{\omega,2}(inj^\tau(x^\tau)) = inj^{undef}(error). \tag{6d}$$

For types $\sigma, \tau \in S_{PR^\omega}$ such that $\sigma, \tau < \omega$ we have the equation

$$\langle inj^\sigma(x^\sigma), inj^\tau(y^\tau) \rangle^\omega = inj^{(\sigma \times \tau)}(\langle x^\sigma, y^\tau \rangle). \tag{7}$$

For each type $\tau \in S_{PR^\omega}$ such that $\tau < \omega$ and $\tau \neq undef$ we have the equations

$$und?(inj^{undef}(error), x) = x, \quad und?(inj^\tau(y^\tau), x) = inj^\tau(y^\tau). \tag{8a, b}$$

We have the equations

$$(Y|Z)(x) = \langle Y(x), Z(x) \rangle^\omega, \tag{9}$$

$$(Y \circ Z)(x) = Y(Z(x)). \tag{10}$$

For each product type $(nat \times \tau) \in S_{PR^\omega}$ we have the equations

$$pr(Y, Z)(inj^{(nat \times \tau)}(\langle 0, x^\tau \rangle)) = Y(inj^\tau(x^\tau)), \tag{11a}$$

$$pr(Y, Z)(inj^{(nat \times \tau)}(\langle succ(t), x^\tau \rangle)) =$$
$$Z(\langle inj^{nat}(t), \langle inj^\tau(x^\tau), pr(Y, Z)(inj^{(nat \times \tau)}(\langle t, x^\tau \rangle)) \rangle^\omega \rangle^\omega). \tag{11b}$$

For each type $\tau \in S_{PR^\omega}$ such that $\tau < \omega$ and $\tau \neq (nat \times \sigma)$, for any $\sigma \in S_{PR^\omega}$ we have the equation

$$pr(Y, Z)(inj^\tau(x^\tau)) = inj^{undef}(error). \tag{11c}$$

We have the equation

$$switch^\perp(Y, Z)(x) = und?(Y(x), Z(x)). \tag{12}$$

For each type $\tau \in S_{PR^\omega}$ such that $\tau < \omega$ we have the equations

$$case^\tau(Y, Z)(inj^\tau(x^\tau)) = Y(inj^\tau(x^\tau)), \tag{13a}$$

and for each type $\sigma \in S_{PR^\omega}$ such that $\sigma < \omega$ and $\sigma \neq \tau$,

$$case^\tau(Y, Z)(inj^\sigma(x^\sigma)) = Z(inj^\sigma(x^\sigma)). \tag{13b}$$

□

We need to show that $PR^\omega(Spec)$ is a correct $\omega + 1$–order equational specification of $PR^\omega(A)$ under higher–order initial algebra semantics. Following the approach of Meinke [1994] we do this using a normalisation result for $PR^\omega(\Sigma)$ terms. We begin by showing that $PR^\omega(E)$ is a consistent equational theory.

6.6 Proposition. $PR^\omega(A) \models PR^\omega(E)$.

Proof. Clearly $PR^\omega(A) \models E$ since $A \cong PR^\omega(A)|_\Sigma$ and $A \models E$. Thus we need only show that $PR^\omega(A)$ is a model of the remaining equations of $PR^\omega(E)$. This is straightforward and is left as an exercise to the reader. □

Since *Spec* is a correct equational specification of A under first–order initial algebra semantics we know there must exist an isomorphism $\psi : A \to T(\Sigma)/\equiv^E$. We can decompose ψ such that $nat^E \circ \theta = \psi$, where $\theta : A \to T(\Sigma)$ may be assumed to be a homomorphism (since ψ is a homomorphism) and $nat^E : T(\Sigma) \to T(\Sigma)/\equiv^E$ is the natural mapping associated with the term congruence $\equiv^E$. We use the homomorphism θ to define a term representation for the first–order and ω–order elements of $PR^\omega(A)$.

6.7 Definition. Let $\theta : A \to T(\Sigma)$ be the homomorphism given above. For each type $\tau \in S_{PR^\omega}$ such that $\tau \neq (\omega \to \omega)$ define the mapping $\alpha(\theta)_\tau : PR^\omega(A)_\tau \to T(PR^\omega(\Sigma))_\tau$ by

$$\alpha(\theta)_{undef}(\bot) = error,$$

for any type $\tau \in S$ and $a \in PR^\omega(A)_\tau$

$$\alpha(\theta)_\tau(a) = \theta_\tau(a),$$

for any product type $(\sigma \times \tau) \in S_{PR^\omega}$ and any $(a, b) \in PR^\omega(A)_{(\sigma\times\tau)}$

$$\alpha(\theta)_{(\sigma\times\tau)}(a, b) = \langle \alpha(\theta)_\sigma(a), \alpha(\theta)_\tau(b) \rangle,$$

and for any type $\sigma \in S_{PR^\omega}$ such that $\sigma < \omega$ and $a \in PR^\omega(A)_\sigma$

$$\alpha(\theta)_\omega(\sigma, a) = inj^\sigma(\alpha(\theta)_\sigma(a)).$$

□

We now prove the following normalisation result.

6.8 Normalisation Lemma. *For any type $\tau \in S_{PR^\omega}$ such that $\tau \neq (\omega \to \omega)$ and any term $t \in T(PR^\omega(\Sigma))_\tau$,*

$$PR^\omega(E) \vdash t = \alpha(\theta)_\tau(t_{PR^\omega(A)}).$$

Proof. By induction on the complexity of terms.

Basis. Suppose $t \equiv error$ or $t \equiv c$, for some $\tau \in S$ and constant symbol $c \in \Sigma_{\lambda,\tau}$. Then by definition of $PR^{\omega}(A)$, $\alpha(\theta)$ and since θ is a homomorphism we have using reflexivity

$$PR^{\omega}(E) \vdash t = \alpha(\theta)_{\tau}(t_{PR^{\omega}(A)}).$$

Induction step. We need to show that for each $w = \tau(1)\ldots\tau(n) \in S^{+}_{PR^{\omega}}$, $\tau \in S_{PR^{\omega}}$ such that $\tau \neq (\omega \to \omega)$, $f \in PR^{\omega}(\Sigma)_{w,\tau}$ and any terms $t_1 \in T(PR^{\omega}(\Sigma))_{\tau(1)}, \ldots, t_n \in T(PR^{\omega}(\Sigma))_{\tau(n)}$ that

$$PR^{\omega}(E) \vdash f(t_1, \ldots, t_n) = \alpha(\theta)_{\tau}(f(t_1, \ldots, t_n)_{PR^{\omega}(A)}).$$

We prove this only for $\langle . \rangle \in PR^{\omega}(\Sigma)_{\sigma\ \tau,(\sigma\times\tau)}$, for each product type $(\sigma \times \tau) \in S_{PR^{\omega}}$, $und? \in PR^{\omega}(\Sigma)_{\omega\ \omega,\omega}$ and $eval^{(\omega\to\omega)} \in PR^{\omega}(\Sigma)_{(\omega\to\omega)\ \omega,\omega}$. We leave the proofs for the remaining four cases (which follow along similar lines) as an exercise for the reader.

Case (1) Suppose $t \equiv \langle t^1, t^2\rangle$, for $(\sigma \times \tau) \in S_{PR^{\omega}}$ and $t^1 \in T(PR^{\omega}(\Sigma))_{\sigma}$, $t^2 \in T(PR^{\omega}(\Sigma))_{\tau}$. Then by the induction hypothesis we have

$$PR^{\omega}(E) \vdash \langle t^1, t^2\rangle = \langle \alpha(\theta)_{\sigma}(t^1_{PR^{\omega}(A)}), \alpha(\theta)_{\tau}(t^2_{PR^{\omega}(A)})\rangle. \tag{1}$$

Also by the definition of $\alpha(\theta)$ and $\langle . \rangle_{PR^{\omega}(A)}$

$$\alpha(\theta)_{(\sigma\times\tau)}(\langle t^1, t^2\rangle_{PR^{\omega}(A)}) = \langle \alpha(\theta)_{\sigma}(t^1_{PR^{\omega}(A)}), \alpha(\theta)_{\tau}(t^2_{PR^{\omega}(A)})\rangle. \tag{2}$$

So using (1) and (2) above we have

$$PR^{\omega}(E) \vdash \langle t^1, t^2\rangle = \alpha(\theta)_{(\sigma\times\tau)}(\langle t^1, t^2\rangle_{PR^{\omega}(A)}).$$

Case (2) Suppose $t \equiv und?(t^1, t^2)$, for some terms $t^1, t^2 \in T(PR^{\omega}(\Sigma))_{\omega}$. Then we have two possible cases to consider.

Subcase (1) Suppose $t^1_{PR^{\omega}(A)} = (undef, \bot)$. Then by the induction hypothesis and definition of $\alpha(\theta)$ we have

$$PR^{\omega}(E) \vdash und?(t^1, t^2) = und?(inj^{undef}(error), \alpha(\theta)_{\omega}(t^2_{PR^{\omega}(A)})). \tag{3}$$

By equation 6.5.(8a) and definition of $und?_{PR^{\omega}(A)}$ we have

$$\begin{aligned} PR^{\omega}(E) \vdash und?(inj^{undef}(error), \alpha(\theta)_{\omega}(t^2_{PR^{\omega}(A)})) = \\ \alpha(\theta)_{\omega}(und?(t^1, t^2)_{PR^{\omega}(A)}). \end{aligned} \tag{4}$$

Thus using transitivity, (3) and (4) above

$$PR^{\omega}(E) \vdash und?(t^1, t^2) = \alpha(\theta)_{\omega}(und?(t^1, t^2)_{PR^{\omega}(A)}).$$

Subcase (2) Suppose $t^1_{PR^\omega(A)} = (\tau, a)$, for some type $\tau \in S_{PR^\omega}$ such that $\tau < \omega$ and $\tau \neq \mathit{undef}$. Then the proof is similar to subcase (1) above except equation 6.5.(8b) is used instead of 6.5.(8a).

Case (3) Finally, suppose $t \equiv \mathit{eval}^{(\omega\to\omega)}(t^1, t^2)$, for terms $t^1 \in T(PR^\omega(\Sigma))_{(\omega\to\omega)}$ and $t^2 \in T(PR^\omega(\Sigma))_\omega$. By the induction hypothesis

$$PR^\omega(E) \vdash t^1(t^2) = t^1(\alpha(\theta)_\omega(t^2_{PR^\omega(A)})). \tag{5}$$

By induction on the complexity of t^1 we can prove (see Steggles [1995]) that

$$PR^\omega(E) \vdash t^1(\alpha(\theta)_\omega(t^2_{PR^\omega(A)})) = \alpha(\theta)_\omega(t^1(t^2)_{PR^\omega(A)}),$$

from which using transitivity and (5) above the result follows. □

We may now prove that $PR^\omega(\mathit{Spec})$ is a correct $\omega + 1$–order equational specification of $PR^\omega(A)$ under higher–order initial algebra semantics.

6.9 Correctness Theorem. $I_{\mathit{Ext}}(PR^\omega(\mathit{Spec})) \cong PR^\omega(A)$.

Proof. Since $PR^\omega(A)$ and $I_{\mathit{Ext}}(PR^\omega(\mathit{Spec}))$ are minimal, extensional $PR^\omega(\Sigma)$, $PR^\omega(E)$ algebras, and $I_{\mathit{Ext}}(PR^\omega(\mathit{Spec}))$ is initial in $\mathit{Min}_{\mathit{Ext}}(PR^\omega(\mathit{Spec}))$ it suffices to construct a homomorphism

$$\phi : PR^\omega(A) \to I_{\mathit{Ext}}(PR^\omega(\mathit{Spec})).$$

For each type $\tau \in S_{PR^\omega}$ and each term $t \in T(PR^\omega(\Sigma))_\tau$, define $\phi_\tau(t_{PR^\omega(A)}) = \mathit{nat}^{PR^\omega(E),\omega}(t)$, where $\mathit{nat}^{PR^\omega(E),\omega} : T(PR^\omega(\Sigma)) \to T(PR^\omega(\Sigma))/\equiv^{PR^\omega(E),\omega}$ is the natural mapping associated with the term congruence $\equiv^{PR^\omega(E),\omega}$. We need to show that ϕ is well–defined on each element $a \in PR^\omega(A)_\tau$.

Consider any type $\tau \in S_{PR^\omega}$ such that $\tau \neq (\omega \to \omega)$ and any terms $t, t' \in T(PR^\omega(\Sigma))_\tau$. Suppose that $t_{PR^\omega(A)} = t'_{PR^\omega(A)}$. Then using the Normalisation Lemma 6.8 we can show $PR^\omega(E) \vdash t = t'$, and so by the definition of ϕ and $\equiv^{PR^\omega(E),\omega}$ we have $\phi_\tau(t_{PR^\omega(A)}) = \phi_\tau(t'_{PR^\omega(A)})$.

Consider the function type $(\omega \to \omega) \in S_{PR^\omega}$ and any $t, t' \in T(PR^\omega(\Sigma))_{(\omega\to\omega)}$. Suppose that $t_{PR^\omega(A)} = t'_{PR^\omega(A)}$. Then for any term $t'' \in T(PR^\omega(\Sigma))_\omega$ we know that $t(t'')_{PR^\omega(A)} = t'(t'')_{PR^\omega(A)}$ and so by the Normalisation Lemma 6.8,

$$PR^\omega(E) \vdash t(t'') = t'(t'').$$

Since t'' was arbitrarily chosen we have that $PR^\omega(E) \vdash_\omega t = t'$ by the ω–evaluation rule and so by the definition of ϕ and $\equiv^{PR^\omega(E),\omega}$ it follows that $\phi_{(\omega\to\omega)}(t_{PR^\omega(A)}) = \phi_{(\omega\to\omega)}(t'_{PR^\omega(A)})$.

Clearly ϕ is homomorphism since the valuation mapping $(.)_{PR^\omega(A)}$ and the natural mapping $\mathit{nat}^{PR^\omega(E),\omega}$ are homomorphisms. □

7 Conclusions

We have presented a theory of ***higher–order algebra with transfinite types*** which extends the simple binary product and function type system of higher–order algebra with ***limit types***. The aim of this extension was to provide a higher–order algebraic framework in which the problematic phenomena of polymorphism and partiality could be naturally modelled. We demonstrated the use of transfinite types for modelling partial functions and both parametric and overloading polymorphism by considering the specification of a simple polymorphic functional programming language $\mathcal{PR}^{\omega}$. In this specification case study we used the first limit type ω to replace an infinite scheme of functions by a single polymorphic function. The problem of coping with partial functions was addressed by having a basic type *undef* for undefined values and defining partial functions to be polymorphic functions which return an element of type *undef* when they are undefined. This approach allows the undefined values to be partitioned from the "normal" data values and provides a very natural axiomatisation of partial functions (compare this approach to the so called *error algebra* approach, see for example Goguen [1978] and Goguen et al [1978]).

Transfinite types can be used to model Π–types (see Steggles [1995]) and in fact this provided one of the original motivations for considering a theory of transfinite types. We note that transfinite types provide a decidable typing discipline whereas Π–types are undecidable, since the equality of two typing (recursive) functions is undecidable. Thus there are strong practical reasons for considering the theory of transfinite types.

A range of algebraic methods exist in the literature for coping with polymorphism, partiality and error handling. For example *order–sorted algebra* (see Poigné [1991], Goguen and Meseguer [1992], Goguen and Diaconescu [1994] and for higher–order order–sorted algebra see Qian [1993]), *unified algebras* (see Mosses [1989, 1994]) and *equational type logic* (see Manca et al [1989, 1990]). However, our approach of using transfinite types is new. For a survey of the different approaches to algebraically modelling polymorphism and partiality we recommend Mosses [1993] and Poigné [1994]. For work on classical transfinite type theory we refer the interested reader to L'Abbé [1953], Andrews [1965] and Pottinger [1993].

The theory of higher–order algebra with transfinite types we have presented is still in its infancy and needs to be further developed and investigated. One area in particular that needs to be addressed is the provision of case studies demonstrating the use of the above algebraic techniques in practice. We note that a selection of specification case studies demonstrating some of the possible applications of transfinite types can be found in Steggles [1995]. Further work is also needed to investigate the possible uses of the higher limit types.

We are very grateful for the helpful advice and comments of K. Meinke and J. R. Hindley during the preparation of this paper. We also acknowledge the financial support of the Engineering and Physical Sciences Research Council and the HCM MeDiCiS project.

8 References

P. B. Andrews. *A Transfinite Type Theory with Type Variables.* Studies in Logic and the foundations of Mathematics, North Holland, 1965.

G. Birkhoff. On the Structure of Abstract Algebras. In: *Proceedings of the Cambridge Philosophical Society*, Vol. 31, 1935.

P.M. Cohn. *Universal Algebra.* Harper and Row, New York, 1965.

K. J. Devlin. *Fundamentals of Contemporary Set Theory.* Springer–Verlag, 1979.

J. A. Goguen. Abstract Errors for Abstract Data Types. In: E. J. Neuhold (ed), *Formal Descriptions of Programming Concepts*, p. 491–522, North-Holland, 1978.

J. A. Goguen, J. Thatcher and E. G. Wagner. An initial algebra approach to the specification, correctness and implementations of abstract data types. In: R. T. Yeh (ed), *Current Trends in Programming Methodology IV*, pages 80–149, Prentice–Hall, 1978.

J.A. Goguen and J. Meseguer. Order–sorted algebra I: Equational Deduction for Multiple Inheritance, Overloading, Exceptions and Partial Operations. *Theoretical Computer Science*, 105:217–273, 1992.

J. A. Goguen and R. Diaconescu. An Oxford survey of order-sorted algebra. In: *Mathematical Structures in Computer Science*, Vol. 4, pages 363-392, Cambridge University Press, 1994.

J. M. Henle. *An Outline of Set Theory.* Springer–Verlag, 1986.

P. Kosiuczenko and K. Meinke. On the Power of Higher–Order Algebraic Specification Methods. Report CSR 13–94, Department of Computer Science, University College of Swansea, 1994.

M. L'Abbé. Systems of Transfinite Types Involving λ–Conversion. *Journal of Symbolic Logic.* 18:209–224, 1953.

V. Manca, A. Salibra and G. Scollo. On the nature of TELLUS: a Typed Equational Logic Look over Uniform Specifications. *Proceedings of 1989 Symposium on Mathematical Foundations of Computer Science*, Lecture Notes in Computer Science 379, pages 338–349, Springer–Verlag, 1989.

V. Manca, A. Salibra and G. Scollo. Equational Type Logic. *Theoretical Computer Science*, 77:131–159, 1990.

K. Meinke. Universal algebra in higher types. *Theoretical Computer Science*, 100:385–417, 1992.

K. Meinke. A recursive second order initial algebra specification of primitive recursion. *Acta Informatica*, 31:329–340, 1994.

K. Meinke. A completeness theorem for the expressive power of higher–order algebraic specifications. Technical Report CSR–13–95, Department of Computer Science, University of Wales, Swansea, 1995.

K. Meinke and L. J. Steggles. Specification and Verification in Higher Order Algebra: A Case Study of Convolution. In: J. Heering, K. Meinke, B. Möller and T. Nipkow (eds), *Proc. of HOA '93: An International Workshop on Higher Order Algebra, Logic and Term Rewriting*, Lecture Notes in Computer Science 816, pages 189–222, Springer–Verlag, 1994.

K. Meinke and J.V. Tucker. Universal algebra. In: S. Abramsky, D. Gabbay and

T.S.E. Maibaum, (eds) *Handbook of Logic in Computer Science*, Volume I, pages 189–412. Oxford University Press, Oxford, 1993.

B. Möller. Algebraic specifications with higher–order operators. In: L.G.L.T. Meertens (ed), *Program specification and transformation*, North–Holland, 1987.

B. Möller, A. Tarlecki and M. Wirsing. Algebraic specifications of reachable higher–order algebras. In: D. Sannella and A. Tarlecki (eds), *Recent Trends in Data Type Specification*, Lecture Notes in Computer Science 332, pages 154–169, Springer-Verlag, 1988.

P. D. Mosses. Unified Algebras and Action Semantics. In: B. Monien and R. Cori (eds), *Proc. of STACS 89 : Annual Symposium on Theoretical Aspects of Computer Science*, Lecture Notes in Computer Science 349. Springer-Verlag, 1989.

P. D. Mosses. The use of sorts in algebraic specifications. In: *Proceedings of 8th Workshop on Abstract Data Types and 3rd COMPASS Workshop*, Lecture Notes in Computer Science 655, pages 66–91, Springer–Verlag, 1993.

P. D. Mosses. Unified Algebras and Abstract Syntax. In: H. Ehrig and R. Orejas (eds), *Recent Trends in Data Type Specification*, Lecture Notes in Computer Science 785, pages 280–294, Springer–Verlag, 1994.

A. Poigné. Once more on Order–Sorted Algebra. In: *Mathematical Foundations of Computer Science 1991*, Lecture Notes in Computer Science 520, Springer–Verlag, 1991.

A. Poigné. Identity and Existence, and Types in Algebra. In: H. Ehrig and R. Orejas (eds), *Recent Trends in Data Type Specification*, Lecture Notes in Computer Science 785, pages 53–78, Springer–Verlag, 1994.

G. Pottinger. A Classical Type Theory with Transfinite Types. In: L. J. M. Claesen and M. J. C. Gordon (eds), *Higher Order Logic Theorem Proving and its Applications*, Elsevier Science Publishers (North–Holland), 1993.

Z. Qian. Higher–Order Order–Sorted Algebra. *Acta Informatica*, 30:569–607, 1993.

L. J. Steggles. *Extensions of Higher-Order Algebra: Fundamental Theory and Case Studies.* Ph. D. Thesis, Computer Science Department, University of Wales, Swansea, submitted 1995.

B. C. Thompson. *Mathematical theory of synchronous concurrent algorithms.* Ph. D. Thesis, School of Computer Studies, University of Leeds, 1987.

W. Wechler. *Universal Algebra for Computer Scientists.* EATCS Monographs on Theoretical Computer Science 25, Springer-Verlag, 1992.

Abstraction of Hardware Construction

Li-Guo Wang[1]* and Michael Mendler[2]**.

[1] Department of Computer Science, University of Edinburgh, The King's Buildings, Edinburgh EH9 3JZ, Scotland, U.K.

[2] Fakultät für Mathematik und Informatik, Universität Passau, Innstraße 33, D-94032 Passau, Deutschland

Abstract. We concentrate on the foundation of formal hardware construction and present a new hardware-centered concept and methodology. It is a development from, and also contrast to, the traditional logic-centered approach. It introduces a higher-order variable construction model of hardware and the notion of generic construction schemes. We demonstrate how the model and the schemes formalize the construction (and verification as well) for a class of computers, as an example to illustrate our concepts and method.

1 Introduction

1.1 From Application to Theory

The formal method for hardware design presented in this paper was stimulated primarily by application, concretely by the formal derivation of a class of microcomputers [39]. The search for a formal approach that is adequate at a non-academic scale on realistic computers, naturally leads us to reconsider and depart from the rather low level of formalization currently used in the area. While often "formal method" is reduced to the application of a specific calculus of formal logic or specific theorem prover, in our work it is taken in its straight meaning: as the formal manipulation of a class of objects (here computers at different levels of abstraction,) where 'formal' refers to the fact that these objects are represented by abstract syntax; the objects themselves, or rather their behaviours, are given by a well-defined semantics outside of the formal calculus. In this paper we will summarize the central aspects of this work.

1.2 Synthesis versus Verification

The classical application of formal methods to the design of microcomputers is concerned with *formal verification*, *i.e.* the process of proving the consistency between a given implementation and a given specification. The most well-known

* Li-Guo Wang was supported by a scholarship from Siemens AG Munich and under SERC GR/F 35890.

** Michael Mendler was supported by a Human Capital and Mobility fellowship in the *EuroForm* network examples are the verification of the FM8501 by Hunt [21], of the VIPER by Cohn [9, 10], and of the TAMARACK-3 by Joyce [24]. A number of verification examples followed these pioneering works, such as the SECD [3], Mini Cayuga [31], DLX [36], just to mention a few. The approaches mainly differ in the number of abstraction levels they consider, and what logic or theorem prover they use. However, they all share the essential feature of ***post-hoc*** formal verification: specification, implementation and all intermediate level descriptions are given ***a priori***, while the nontrivial task consists in finding a proof of correctness for each step.

When initially the research focused on the correctness of specific well-defined hardware designs, the importance of ***genericity*** was soon realized, and the inherent potential of the formal approach to verify whole classes of related hardware was stressed [37, 25, 20]. This trend towards proof genericity is natural not least because of the great efforts required for and the resource-demanding nature of formal verification. A side-effect of the strive for generality is to identify and formalize common hardware structure and abstraction levels along which the verification task can be organized and systematized. Examples of such work for the context of microcomputers is the notion of generic interpreter [41, 42] or the algebraic models of [19].

There are many ways to make formal proofs more generic. However, the ultimate level of genericity is not obtained by formal verification but by ***formal synthesis***. In contrast to verification, formal synthesis subjects the very description of a computer to the process of construction and not only the correctness proof. This, of course, offers maximal genericity. It also pays pragmatic tribute to the fact that in real life a specification is never completely determined. One usually starts off merely with a first approximation, which is later refined and modified as more details of the implementation become known.

The idea of applying formal methods to the design process itself has been introduced rather early, for instance with the theorem provers LAMBDA [11, 12] or VERITAS [17]. Though quite some work has been done on the formal synthesis of actual hardware, *e.g.* [35, 1, 2, 6], the formal synthesis of realistic (or at least nearly realistic) computers remains an open challenge. A first attempt in this direction is the formal derivation of the DDD-FM9001 [5].

As we see it, the current way of realizing formal synthesis must be seen essentially as a refinement of the tradition of formal verification. Thus, it suffers from being limited into the corset of a particular logic or theorem prover. For instance, [35] is based on Martin-Löf type theory implemented in ISABELLE [30], [1, 2] is based on Martin-Löf type theory implemented in NUPRL, [6] is based on higher-order logic implemented in the LAMBDA prover, and finally [5] uses (a fragment of) first-order logic plus induction implemented in the Boyer-Moore theorem prover. We believe that in order for formal synthesis to get off the ground this is the wrong level of formalization. After all, the main concern of synthesis should be the construction of correct computers, not in fact the construction of proofs in some logic on some theorem prover. A large stock of practical experience in formal verification has shown that for a well-structured design, once this structure has been identified, the actual verification may be quite a trivial task. We believe that synthesis in the first place should deal with the formal construction and refinement of hardware structure, and only in second instance with the construction of proofs.

1.3 Hardware-Centered versus Logic-Centered

The previous substantial efforts in applying formal methods to the hardware design process are *logic centered*, typically so are those based on HOL [15], LAMBDA [11] and VERITAS [17]. These tools are all more or less directly motivated by their application to hardware, and yet they are full-fledged general-purpose theorem provers. For example, regarding HOL we have had several international conferences which showed various interesting applications of HOL in very different areas. A similar statement is true for LAMBDA. The first author successfully used LAMBDA to the mechanical synthesis of nondeterministic logic programs [38]. Thus, the traditional applications of formal methods to hardware are mainly centered around very powerful logic systems whose expressiveness is beyond the relatively narrow needs of typical hardware designs. Therefore, we may call these approaches logic-centered.

However, it appears that the strength of the logic-centered approach is at the same time its main weakness: it is far too general to deal with realistically complex hardware designs in an efficient way. This problem is hinted at in [18]:

> *"[...] it is still, in the authors' opinion, a very open question as to whether these theoretical gains can be translated into practical ones. Thus far, our own computational trials have been limited to very simple circuits and it is not clear how both the computational load and the amount of human guidance required will scale with increasing size of circuit."*

Concretely, the weakness of the encode-it-in-a-general-purpose-theorem-prover approach lies in the low level of syntactic description and the expensive granularity of elementary design steps. Synthesizing a computer in terms of logical connectives and modus ponens is similar to programming an operating system in an assembler language. More subtly, the generic structure of computers is too rich to be handled by simple formula schemes. For instance, consider a specification of the machine instruction cycle in terms of a global next-state function, such as

$$\begin{aligned} \text{state}\,(t+1) = \ & \text{opcode}\,t = c_1 \ \Rightarrow \ \text{instruction}_1\ (\text{state}\,t) \ \downarrow \\ & \text{opcode}\,t = c_2 \ \Rightarrow \ \text{instruction}_2\ (\text{state}\,t) \ \downarrow \\ & \qquad \ldots \\ & \text{opcode}\,t = c_n \ \Rightarrow \ \text{instruction}_n\ (\text{state}\,t). \end{aligned}$$

where $P \Rightarrow Q \downarrow R$ stands for "if P then Q else R." A simple refinement of this specification might consist in pushing the top-level equation inwards, say in order to separate control and data:

$$\begin{aligned}\text{opcode}\,t = c_1 &\Rightarrow \text{state}\,(t+1) = \text{instruction}_1\,(\text{state}\,t)\ \downarrow\\ \text{opcode}\,t = c_2 &\Rightarrow \text{state}\,(t+1) = \text{instruction}_2\,(\text{state}\,t)\ \downarrow\\ &\cdots\\ \text{opcode}\,t = c_n &\Rightarrow \text{state}\,(t+1) = \text{instruction}_n\,(\text{state}\,t).\end{aligned}$$

Now suppose we wanted to implement this step on a theorem prover, as a refinement that is generic in the number n of machine instructions. The most direct solution, namely to exploit the elementary genericity of syntactic substitution, does not work. The syntax of standard logic is not discriminative enough to define the class of all well-formed specifications as the substitution instances of a formula scheme (and the refinement as instances of a rule scheme).

In principle, there are two opposite directions we can take to achieve genericity. The solution usually adopted is to implement the refinement as a tactic. This, however, pushes the formalization to the meta-level at the price of a significant "*correctness leak*": who verifies the tactic? Although we can be sure that tactics are correct relative to the low level of formulas, there is no guarantee that they are actually sensible as refinements of hardware. A tactic must neither be too timid, so that it succeeds in all interesting cases, nor must it be too general, so that it does not mess up the proof state. For example, the synthesis tactics may introduce short-circuits or other inconsistencies. To specify and verify a tactic is a nontrivial problem which is rarely, if ever, attempted.

The second possibility to achieve genericity is by encoding the generic structure in terms of inductive data-types, say natural numbers or lists, and recursive functions and induction rules. Such an approach is suggested for example in [20, 18]. This way of going about avoids the correctness leak of tactics. However, by pushing the important structure into the object level it introduces a "*formalization leak*": Explicit structure becomes implicit and the manipulation of this structure (*e.g.* by recursion equations) must be implemented *within* rather than *by* the calculus. In this way the formalization becomes clumsy and in the extreme case may get close to a Gödel numbering of hardware.

When it comes to applying formal methods to hardware design then choosing the right level of formalization is a rather crucial part of the exercise. We believe that to deal with real world designs the current logic-centered approach is too low level and that, therefore, a more refined approach is needed. This new approach should be *hardware-centered* and restricted to the hardware construction area of interest, rather than being general purpose. The purpose of this paper is to present such an approach.

1.4 The New Concepts

The main concepts advanced in this paper are the notions of specification scheme, construction scheme, and construction model. They form the basis for a formal construction framework that has a status very similar to a formal logic, but with a higher level of description, larger degree of genericity, and focus on a particular class of hardware.

Specification schemes are abstract syntax schemes to cover the generic structure of a class of concrete computers, at different levels of abstraction. They are the objects of our constructions just as formulas are the objects manipulated by a theorem prover. The semantics of a specification scheme is induced by a syntactic translation into higher-order logic.

A *construction scheme* is a generic translation of one specification scheme to another. It might formalize a refinement from one level of description to another one, or a semantics-preserving transformation. The application of construction schemes drives the formal synthesis just as rules drive the derivation of formulas in a theorem-prover.

Finally, the *construction model*, in our framework, is the most generic form of a specification scheme. It is meant to capture the behaviour of a hardware component as a combination of a functional aspect and a control aspect. This model applies to the whole construction process of a class of computers as shown in [39, 40]. In contrast to the traditional verification-oriented hardware model, it admits finer-grained design descriptions in which aspects of hardware and software may coexist, and it abstracts from the concrete mode of operation. The choice, for example, of a synchronous, asynchronous, or pipelined realization may be postponed until later design stages.

1.5 The Higher-Order Aspect

The hardware-centered concept and methodology presented in this paper can be implemented in several logic systems such as HOL [15], LAMBDA [11], or ISABELLE [30]. As its purpose is to handle generic structure, a minimum of higher-order features is required for an adequate implementation of (inductive) specification and construction schemes. Also, we use higher-order variables for the logic (entry predicates) and space (signal/port) parameters in the construction model.

2 Specification Schemes

Construction is inherently formal and generic. It is *formal* in the sense that it manipulates syntactic structure. The track of the construction process is a series of syntactic descriptions, called specifications, in our case of hardware, from a behavioural to a structural level. The construction is *generic* in two dimensions. The specification is a scheme that captures the common structure of a class of objects rather than that of a particular instance. This we call the *horizontal* genericity of construction. Moreover, a specification scheme captures the structure only at a particular level of description. It is built up from "primitive" components that are uninterpreted and whose structure is irrelevant at the given level. They may be filled in by later refinements. This we call the *vertical* genericity.

To express specification schemes we use a formal language approach. A specification scheme is a context-free formal grammar presented in Backus-Naur form (BNF). Although BNF does not give us maximal control over syntactic structure, it does extend the possibilities of usual logic formulas considerably. Making use of the technology of abstract syntax BNF works well to keep all constructions and specification schemes generic and exact. To get an instance specification from the scheme we instantiate the uninterpreted primitive components (vertical instantiation) and the generic structure of the scheme (horizontal instantiation). So, the specification scheme is our main interface between formal construction and the real world. It should be generic enough to reflect the real world and exact enough, both in the sense of rigorous and informative, such that the formal construction can work on it.

Let us take a class of computers as an example. Taking a bird's eye view a computer is nothing but a big state transition function working on a number of state signals. We can fix this structure by the following simple specification scheme:

$comp$	$::= os = iss$	$comp$	: computer
os	$::= (sg\ y, \ldots, sg\ y)$	os	: output state
	$\mid sg\ y$	iss	: input-state structure
		sg	: signal
		y	: distinguished time variable.

The scheme is exact. It says that the output state os of an instruction cycle is computed according to an input-state structure iss. The output state is formed by applying a number of state signals $sg_1, \ldots, sg_n$ to a distinguished time variable y, which represents the completion time of the cycle. Thus, $os = (sg_1\, y, \ldots, sg_n\, y)$. The scheme is generic. It does not restrict how many state signals a computer has and what information is maintained in them. It does not say what the next-state structure is except that it can be evaluated to give a possible state of the computer at a particular point in time.

We may refine the specification scheme by providing specific structure for iss. Let us adopt a programmer's model according to which a computer is a set of machine instructions. These instructions are executed under different conditions, *e.g.* depending on the value of the operation code (*cf.* Windley's notion of generic interpreters [42].) Let us suppose further that instructions may have sub-instructions in common. For such a class of computers, we may set up the following refinement:

iss	$::= is$	is	: input state
	$\mid$ let A in iss	A	: sharing
	$\mid P \Rightarrow iss \downarrow iss$	P	: predicate
A	$::= v = comm$	$comm$	: common part
		v	: variable.

Though we have supplied more detail we are still generic: We do not restrict how many instructions a computer has, and in which language they are formulated; we have not said how the instructions are selected, *i.e.* what kind of predicates can be used. These are generic components of the scheme, left to be filled in in different ways by different instances from the real world. Different refinements and design methods may require different kind of structure for instructions and predicates. However, what is described exactly by the scheme is the relation between these generic components. It is an inductive and recursive structure. Its base case is an instruction and the step case tells us that through let-in and condition structure we can get any computer, no matter how complex it is.

Although this is still quite an abstract scheme, certain construction steps may find enough structure to work on, already at this level of description, without the need to supply further details. Some constructions may push inwards the global equation $os = iss$ to localize the state transitions and separate control structure from data-structure (called ***setting state transitions*** in [39, 40].) Other constructions may rearrange, massage, or normalize the let-in and condition structure in *iss*. Yet another type of constructions can try to break up the atomic calculation of *iss* into a collection of sequential and parallel computations.

At some point in the synthesis it will be important to provide the details of the input state *is*, predicates P, and common part *comm*. There are many possibilities, depending on how powerful a language we wish to have and on what primitives we wish to build. For our purpose a simple first-order language of terms suffices:

is	$::= tm$	*is*	: input state
	$\mid (tm, \ldots, tm)$	*tm*	: term
P	$::= tm$	P	: predicate
comm	$::= tm$	*comm*	: common part
tm	$::= c$	c	: constant
	$\mid v$	v	: variable
	$\mid sg\ x$	*sg*	: signal
		x	: distinguished time variable
	$\mid f(tm, \ldots, tm)$	f	: function.

According to this scheme an input state is a tuple of terms, and predicate and common parts are terms. A term is built up from constants, variables, the value of signals at a distinguished point in time (x stands for the start time of the instruction cycle,) and functions applied to term.

Every construction works on a characteristic syntactic structure, which in an implementation might be established automatically by unification. However, the crucial point is that the specification schemes allow us to specify the effect and purpose of a particular construction syntactically in a precise way. It allows us to control genericity, which in traditional systems either goes astray in the meta-level of tactics or is obscured in the internals of a clumsy object-level encoding of the structure. The constructions considered in this paper require the structure that is made explicit by the above schemes for *comp*, *iss*, and *is*. We will call this level of description the ***machine instruction level*** (MIL).

A final remark is in order. Not all constraints we might wish to impose are (naturally) context-free. This is a limitation which sometimes forces us to add extra side-conditions to capture the well-formed specification instances completely. For example, we might want to constrain the variable v of a sharing 'let $v = tm$ in iss' to occur at least twice in iss (otherwise the v is unnecessary,) and further require that no two sharings must use the same variable. Further, we might restrict v not to appear in any *comm*. Though this might be too restrictive in other circumstances, for a computer's specification it is useful to eliminate nested or recursive sharing. Another side-condition requires that the output state os and all input states is have the same number of components. Such restrictions, which will often exist for a syntax scheme, could be handled formally, too, but are ignored in this paper to keep matters simple.

In order to be able to define (and verify) the correctness of construction steps we need some form of semantics. We adopt the simplest possibility: the meaning of a scheme is induced by a translation into higher-order predicate logic. This logic will not be formally introduced as it can be assumed, at an intuitive level at least, to be understood throughout. The translation is almost immediate in the sense that instances of our schemes can be viewed directly as formulas in higher-order logic containing some syntactic sugar. This would not be strictly necessary, but it stresses our point that the construction framework has the status of a formal logic.

The primitive elements of the specification scheme above, *e.g.* signals sg, the time variables x, y, and terms tm, are assumed to be part of the higher-order logic. Also, the meaning of composite structures like tuples and equations are taken directly from the logic. The let_in and conditional structures are translated as follows:

$$(\text{let } v = A \text{ in } iss)^* = \exists v.\ v = A \ \wedge\ iss^*$$
$$(P \Rightarrow iss_1 \downarrow iss_2)^* = (P^* \ \wedge\ iss_1^*) \ \vee\ (\neg P^* \ \wedge\ iss_2^*).$$

In this way the semantics of a computer *comp* is a predicate *comp** over a number of free signal variables $sg_1, \ldots, sg_n$ and the two distinguished time variables x, y referring to the start and termination time of a machine cycle. Let us assume that time is represented by natural numbers $\mathbf{N}$, and that $\mathbf{N} \rightarrow \alpha_i$, for $i = 1, \ldots, n$, is the type of sg_i. Then *comp** is a relation over $\Sigma \times \mathbf{N} \times \mathbf{N}$ where $\Sigma = (\mathbf{N} \rightarrow \alpha_1) \times \cdots \times (\mathbf{N} \rightarrow \alpha_n)$ is the set of possible state trajectories of the computer. Given $s \in \Sigma$ and $t_1, t_2 \in \mathbf{N}$, $(s, t_1, t_2) \in comp^*$ intuitively means that t_1 is the start time and t_2 the termination time of a machine cycle in trajectory s.

3 The Construction Model

In the field of formal hardware verification a variety of different ways of modelling behaviour have been introduced. The two salient variants are the functional and the relational paradigm. The former is well known from the work of W. Hunt [21] and usually adopted in verification research based on the Boyer-Moore theorem prover. The latter, on the other hand, probably is the most widely accepted and successfully used hardware model in the HOL community [16, 14, 12], and the basis for formal hardware design methodologies such as [23]. Though one method of specifying hardware often can be encoded in terms of the other, the relational paradigm is usually considered to be the more general one, covering the functional variant as a special case. For instance, in terms of relations we can directly capture phenomena which are not elementary in a purely functional setting such as bidirectionality, partiality, and nondeterminism.

This work presents an extension of the components-as-relations, or to be more precise, the components-as-predicates paradigm. This paradigm, which we simply call the (first-order) *verification model*, can be characterized by three features: (*i*) Components are represented by predicates with the free variables as their ports, (*ii*) Internal port connection is realized by existential quantification, and (*iii*) Composition of components is achieved by logical conjunction. For example, the delay and inverter circuits would be defined as

$$\mathrm{Del}\,(i,z) \equiv \forall t.\; z\,(t+1) = i\;t$$
$$\mathrm{Inv}\,(z,o) \equiv \forall t.\; o\,(t+1) = \neg(z\;t)$$

and the structure of the delay-inverter

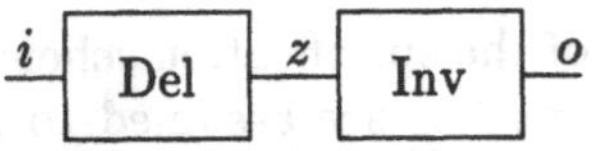

be described as $\exists z.\,\mathrm{Del}\,(i,z) \wedge \mathrm{Inv}\,(z,o)$.

The verification model originates from the obvious static-topological understanding of physical hardware focusing on the port connection structure of hardware and spatial structural abstraction. This is expressed for instance in [28]:

> *"The type of abstraction most fundamental to hardware verification is* structural *abstraction – the suppression of information about a device's internal structure."*

Despite its generality a weakness of the verification model is that it is not flexible enough to support a process of formal synthesis and construction. It exhibits the spatial structure of hardware but fails to capture important other structure arising in a top-down construction of hardware. Examples of such non-spatial structure are the interface between software and hardware, microprograms, or behavioural constraints. Therefore the traditional verification model, which flattens away this extra conceptual structure, is intrinsically limited to *post-hoc* verification and transformation of *complete* hardware components [9, 10, 27, 13, 26, 28]. To capture the top-down construction of *incomplete* hardware, in particular for the construction of microprogrammed computers, more sophisticated structure is needed. We have introduced such a new and more general hardware model, the *construction model* [40]. From our point of view the main aspect in the formal design process is *logic* rather than *spatial* refinement. In order to support this logical decomposition our construction model allows for extra logical communication of a component with its environment.

With the model the inverter is written as

$$\mathrm{Inv}\,(z,o)\;[E_i,\;E_o] \equiv \forall x.\; \exists y.\; E_i\,x \Rightarrow o\,y = \neg(z\,x) \wedge E_o\,y.$$

This formulation states that whenever a certain condition E_i is true at the start time point x then the inversion of the input will be effected by some time y, and after the execution the predicate E_o will be true. The predicates E_i and E_o are the logical interface to the environment by which the entry and exit of the component is controlled. We call E_i the *entry* and E_o the *leaving* predicate of the model. The difference to the verification model is the fact that it focuses on *what* makes up the behaviour of an inverter, *viz.* the equation $o\,y = \neg(z\,x)$, rather than *how* this is achieved. The original absolute reference to a fixed propagation delay has been abstracted away and replaced by a generic time interval $\langle x, y\rangle$ demarcated only by the conditions $E_i\,x$ and $E_o\,y$ but otherwise undetermined. Here the predicates E_i, E_o are considered parameters, *i.e.* free variables, of the specification rather than concrete predicates. This results in a generic description, which can be instantiated in many ways to obtain a variety of different realizations of an inverter. For instance, we get back to the original formulation by taking $E_i\,x \;\equiv\; x = t$ and $E_o\,y \;\equiv\; y = t+1$, where t is a free variable referring to a global notion of time. The substitution gives

$$\forall x.\; \exists y.\; x = t \;\Rightarrow\; o y = \neg(z\,x) \wedge y = t + 1,$$

which is logically equivalent to the old $\mathrm{Inv}\,(z,o)$, but it makes explicit the essential mode of control implicit in the original formulation, *viz.* unconstrained operation with a fixed propagation delay relative to a global and absolute concept of time. Other modes of operation can be obtained by other choices of E_i, E_o. For instance, viewing the inverter function as part of a microprogram might lead us to choose $E_i\,x \;\equiv\; \mathrm{addr}\,x = 3$ and $E_o\,y \;\equiv\; \mathrm{addr}\,y = 4$, where addr refers to the microprogram address counter and x, y to the sequence number of the executed microinstruction. Another possibility is given by viewing the inverter as part of a synchronous circuit: we instantiate $E_i\,x \;\equiv\; \mathrm{stab}(z,x,\Delta) \wedge \mathrm{clk}(x,n)$ and $E_o\,y \;\equiv\; \mathrm{clk}(y,n)$ such that $\mathrm{stab}(z,x,\Delta)$ means input z has been stable for at least a period of length Δ prior to time x, and $\mathrm{clk}(x,n)$ means x is the time of the n-th clock tick. We leave it to the reader to add other instantiations, say to capture input constraints, multiple phase synchronous realizations, or asynchronous handshake.

The semantics of the entry and leaving predicates is non-trivial in the sense that not all possible instantiations are permitted. In particular, it is not permitted to let the entry predicate be inconsistent, say $E_i\,x \equiv 0 = 1$. In this case the construction model would trivialize to the full relation which does not constrain the implementation at all. Our constructive framework avoids these fallacies by keeping a rigid discipline in the use of the entry and leaving predicates.

To sum up, the construction model allows us to capture a wide range of possible modes of control and interaction, and — this is the crucial point — to keep the decomposition at an abstract level as long as convenient and defer instantiating to a particular realization until much later in the design process.

Construction Model	Syntactic Class:
$cm ::= \; mp\,[E, lps]$	cm : construction model
$lps ::= \; E$	lps : leaving predicate structure
$\quad\mid P\,(lps, lps)$	
	mp : model predicate
	E : (entry or leaving) predicate
	P : Boolean term

Fig. 1. Abstract Syntax of the Construction Model

To finish off this section let us sum up our definitions formally. Extending what has been introduced above, the refinements will use a construction model in which the leaving predicate is generalized to a *leaving predicate structure*. This allows us to express non-linear control structure in terms of entry and leaving predicates. With this extension the abstract syntax is as given in Fig. 1. In the following we will call any instance of the scheme in Fig. 1 a construction model or simply a model.

As mentioned in Sec. 2, as far as 'meaning' is concerned, we adopt the simplest possible approach, *viz.* a translation into a (higher-order) predicate logic. In this vein, the semantics of the construction model is defined by the following inductive translation:

$$\begin{aligned} (mp\,[E, lps])^* &= \forall x.\exists y.\, E\,x \Rightarrow mp^* \;\wedge\; lps^* \\ (P(lps_1, lps_2))^* &= P \Rightarrow lps_1^* \downarrow lps_2^* \\ E^* &= E\,y, \end{aligned}$$

where mp^* is the translation of mp. To give an example we may write $(pc\,y = pc\,x)\,[E_2, b_1\,x\,(E_3, b_2\,x\,(E_4, E_5))]$ to stand for $\forall x.\exists y.\, E_2\,x \Rightarrow pc\,y = pc\,x \wedge (b_1\,x \Rightarrow E_3\,y \downarrow (b_2\,x \Rightarrow E_4\,y \downarrow E_5\,y))$.

Note, this syntactic translation hardwires the special rôle of the time variables x, y as the generic start and end time points of a construction model: formally speaking, the operator $[E_i, E_o]$ is a binding operator for x and y which are assumed to be distinguished throughout this paper. For this to work, of course, we must make sure x and y are used consistently according to this convention.

The construction model $mp\,[E_i, lps]$ being introduced the refinement process may be viewed as a repeated process of providing particular structure for the body mp of the model which then is broken down along this structure into smaller pieces $mp_1\,[E_{1i}, lps_{1o}] \wedge \cdots \wedge mp_n[E_{ni}, lps_{no}]$ until the mp_i can be considered implementable. In our case the model predicates will be instances of the scheme *comp* and conjunctions thereof. The main means of decomposition — as explored in this paper — is to break up terms syntactically into sub-terms and to introduce new entry predicates and new ports. For convenience we adopt the following assumption:

Assumption [existential variables] *All new entry predicates and new signals (ports) introduced in the construction process are existential variables.*

With this assumption we may drop the existential quantifier $\exists z.$ hiding an internal signal in a composite circuit and simply turn z into an existential variable. The same is done with internal entry and leaving predicates.

Existential variables are supported in some theorem provers, like LAMBDA [11] ('flexible' variables) or ISABELLE [30] ('meta' variables) and have been shown to be of great advantage in practice. In particular the work based on LAMBDA (e.g. [11]) makes heavy use of existential variables to support formal hardware synthesis.

4 Construction Schemes

Our objects, to be formally manipulated, are not just arbitrary formulas but computers. Just as a theorem prover usually does not derive ground formulas but formula schemes, our framework constructs specification schemes describing classes of computers (at different levels of description), rather than concrete instances.

The specification scheme is the high-level analogue to a formula scheme, the notion of a construction scheme discussed in this section is the high-level analogue to a derivation rule. An ordinary rule like

$$\forall x.\, \phi \;\wedge\; \forall x.\, \psi \vdash \forall x.\, (\phi \;\wedge\; \psi)$$

in predicate logic is not derived as an entailment relation between two concrete formulas but between formula schemes. The premise might be the schematic description of an implementation, and the conclusion that of a specification:

$$\mathit{Impl} \vdash \mathit{Spec}. \qquad (1)$$

Then, the rule can be seen as a *construction scheme*, which allows us to construct from every concrete specification S a concrete implementation I, such that $I \vdash S$. Given S, the construction consists in computing a suitable substitution θ such that $S = \mathit{Spec}[\theta]$, from which the implementation can be obtained as $I = \mathit{Impl}[\theta]$. The computation of θ is usually done by some form of unification.

Our plan is to lift the same idea to the case where *Impl* and *Spec* are specification schemes, such as those considered in the previous sections. A rule, then, becomes a high-level construction. In this way we do not only raise the level of description. In addition to the vertical genericity of syntactic substitution, we get the horizontal genericity of the BNF syntax schemes. Of course, we cannot any longer represent a construction as a finite rule, though in trivial cases such as

$$\forall x_1.\, \phi_1 \;\wedge\; \cdots \;\wedge\; \forall x_n.\, \phi_n \vdash \forall x_1.\; \cdots\; \forall x_n.\, (\phi_1 \wedge\; \cdots\; \wedge \phi_n)$$

a pseudo-formal representation using indices and ellipses is possible. We need a different, more general, way of looking at a rule (1) where *Impl* and *Spec* are specification schemes. The most general interpretation is to view it as a function Ψ which translates (or: refines) any instance S of scheme *Spec* into an instance $\Psi(S)$ of scheme *Impl* such that $\Psi(S) \vdash S$ is provable. In other words, we use (1) as an abbreviation for the condition:

$$\Psi : Spec \rightarrow Impl \quad \forall S \in Spec.\, \Psi(S) \vdash S.$$

If Ψ satisfies these conditions, then we write

$$\Psi : Impl \vdash Spec.$$

Notice, the correctness involves two aspects: Ψ must be a *total* function from well-formed instances of *Spec* to well-formed instances of *Impl*, and it must be *sound*, *i.e.* the result of applying Ψ must be a provable refinement. Such a construction Ψ may be defined by recursion over the structure of *Spec* and its correctness verified by structural induction.

It is useful to be a bit more general and consider also non-deterministic constructions Ψ, which for a given input specification determine a *set* of refinements. In this case the correctness conditions have to be modified accordingly:

$$\forall S \in Spec.\, \emptyset \neq \Psi(S) \subseteq Impl \quad \forall S \in Spec.\, \forall I \in \Psi(S).\, I \vdash S.$$

Nondeterministic constructions arise in an ordinary logic calculus naturally as a set of rules with overlapping conclusions, or by rules producing more than one matching for the conclusion. An example of the latter kind is the equality rule $x = e \wedge \psi \vdash \psi\{e/x\}$. In our framework the nondeterminism is essential, and models the potential external selections to steer the synthesis in one direction or another. This can be either by user interaction or by external heuristic algorithms, say for scheduling or allocation.

In this paper we concentrate on the syntactic structure decomposition aspects of construction. However, generally speaking, the construction of hardware includes not only syntactic but also semantical aspects. The main purpose of semantic construction steps, apart from optimization, is to make each function in the specification implementable *w.r.t.* a given implementation level and component library.

As an example of a semantical step consider a non-available 13 bit counter Inc_{13} as in the model

$$(s\,y = \mathrm{Inc}_{13}\,(r\,x))\ [E_i,\, E_o].$$

It needs to be transformed semantically to the composition of three available functions Cut_{16_13}, Inc_{16}, and Pad_{13_16}:

$$(s\,y = \mathrm{Cut}_{16\_13}\,(\mathrm{Inc}_{16}\,(\mathrm{Pad}_{13\_16}\,(r\,x))))\ [E_i,\, E_o].$$

The syntactic aspect of a construction is to divide the specification into sub-specifications by introducing new entry predicates in logic, new signals (port) in space, and new points in time. For example, decomposing the instruction cycle

$$(s\,y = execute(fetch(pc\ x, mem\ x), mem\ x))\ [E, E]$$

into a separate fetch and execution phase

$$(r\ y = fetch(pc\ x, mem\ x)\ \wedge\ mem\ y = mem\ x)\ [E, E_a]\ \wedge$$
$$(s\ y = execute(r\ x, mem\ x))\ [E_a, E]$$

is an example of a syntactic decomposition step. An atomic computation is made non-atomic by breaking it into two pieces. For this to be correct we need to add a *signal protection* constraint: the condition $mem\ y = mem\ x$ makes sure that the *fetch* and the *execute* phase work on an identical *mem* state.

Definition [Input and output signal protection] *For a given syntactic construct S the input signal protection S^i for S is $(r_1\ y = r_1\ x) \wedge \cdots \wedge (r_n\ y = r_n\ x)$, where r_i $(1 \leq i \leq n)$ are all the input signals in S. Similarly, $S^o \triangleq (s_1\ y = s_1\ x) \wedge \cdots \wedge (s_m\ y = s_m\ x)$, where s_i $(1 \leq i \leq m)$ are all the output signals, is the output protection for S. In the special case where S does not contain any input or output signals we take the boolean constant T.*

Let us now look at a number of construction schemes that perform relevant syntactic decompositions, taking the specification scheme of the microinstruction level (see Sec. 2) as our starting point:

$$spec ::= comp\ [E, E].$$

It presents a computer as a dynamic process $comp\ [E, E] = (os = iss)\ [E, E]$ that executes a global state transition $os = iss$ in a single loop $[E, E]$, where E represents a generic invariant. A natural refinement goal is to implement the global transition in terms of a set of elementary local state transitions, which are executed partly in sequence and partly in parallel. To synchronize them the single loop will have to be broken up into richer control structure. In other words, we want a refinement

$$\Psi : \text{simple control} + \text{complex data} \rightarrow \text{complex control} + \text{simple data}.$$

where the domain coincides with the scheme *spec*. But what is the structure of our target specification scheme, capturing the idea of complex control and simple data? The simplest (nontrivial) form of a computer are the instances of the scheme $tt\ ::=\ sg\ y = tm$, *i.e.* those with a single output state signal and the input state structure consisting of a single term. We call the sub-class *tt* of *comp* the *term transitions.* With the term transitions our target scheme now can be specified by

$$impl ::= (tt^+\ [E, lps])^+,$$

where for a given scheme $\mathcal{S}$, we use the notation $\mathcal{S}^+$ to stand for the finite, nonempty conjunctions of (instances of) $\mathcal{S}$. Thus, an implementation is a conjunction of models

$$(s_{11}\ y = tm_{11}\ \wedge\ \cdots\ \wedge\ s_{1k_1}\ y = tm_{1k_1})\ [E_{1i}, lps_{1o}]$$
$$\wedge\ \cdots\ \wedge$$
$$(s_{n1}\ y = tm_{n1}\ \wedge\ \cdots\ \wedge\ s_{nk_n}\ y = tm_{nk_n})\ [E_{ni}, lps_{no}].$$

Intuitively, each model i consists of a set of term transitions $s_{i1}\ y = tm_{i1} \wedge \cdots \wedge\ s_{ik_1}\ y = tm_{ik_1}$ executed in parallel in one control step. These sets are synchronized by the $[E_{ii}, lps_{io}]$, which determine how the control steps are linked together. Recall that *lps* is the leaving predicate structure (see Sec. 3.)

The level of description expressed by the scheme *impl* has been called *term transition level* [39, 40]. It establishes, so we believe, a rather natural and generic level of abstraction between the machine instruction level above and the register transfer level below. The term transition level can be viewed as a generalization of the computational model of UNITY [7] or SYNCHRONIZED TRANSITIONS [33]. A term transition model $tt^+\ [E, lps]$ here corresponds to a basic transition, *i.e.* a multi-assignment, there. A conjunction $(tt^+\ [E, lps])^+$ of term transition models, then, corresponds to a basic UNITY or SYNCHRONIZED TRANSITIONS program. However, whereas in these languages all control is flattened away and effected implicitly through state changes, in our model a rather generic form of control structure is still explicit through the entry and leaving predicates.

To get back to our main theme, we are looking for a (nondeterministic) construction scheme Ψ that satisfies the specification $impl \vdash spec$, or more explicitly,

$$(tt^+\ [E, lps])^+\ \vdash\ comp\ [E, E].$$

As explained above, the correctness of Ψ amounts to the two conditions

$$\forall S \in comp\ [E, E].\ \emptyset \neq \Psi(S) \subseteq (tt^+\ [E, lps])^+$$
$$\forall S \in comp\ [E, E].\ \forall I \in \Psi(S).\ I \vdash S.$$

Usually a construction such as Ψ will not be defined in one step. It will be composed of a number of smaller constructions, which realize certain more or less independent aspects of the refinement. The composition is by way of certain combinators, which allow us to obtain new constructions from constructions already defined previously. In the following let us see how this would look like.

A first construction step might consist in transforming some of the data into explicit control structure. Formally, this is done by pushing the global equation inwards into *iss*. This way we can break up the input state structure *iss* into smaller pieces while at the same time producing a richer control structure, *i.e.* more equations. Let this construction be called Ψ^a. It can be defined as follows:

4.a.1 $\Psi^a(comp)\ \widehat{=}\ comp\ \mid\ \Psi^a_1(comp)$

4.a.2 $\Psi^a_1(os = is)\ \widehat{=}\ os = is$

4.a.3 $\Psi^a_1(os = \text{let } A \text{ in } iss)\ \widehat{=}\ \text{let } A \text{ in } \Psi^a(os = iss)$

4.a.4 $\Psi^a_1(os = (P \Rightarrow iss_1 \downarrow iss_2))\ \widehat{=}\ P \Rightarrow \Psi^a(os = iss_1) \downarrow \Psi^a(os = iss_2)$.

Ψ^a is a nondeterministic construction scheme defined by induction on the structure of *comp*. It serves to push the outermost equation across an arbitrary number of let-ins and conditionals. In the first clause, 4.a.1, the vertical bar denotes the choice either to stop the construction or to break out one more control structure (either let-in or conditional) with the sub-construction Ψ^a_1. The new specification scheme describing the result of this construction is as follows:

$comps ::= comp$	$comps$: computer structure
$\quad\mid$ let A in $comps$	$comp$: computer
$\quad\mid P \Rightarrow comps \downarrow comps$	A : sharing
	P : predicate.

Theorem 1. *The construction Ψ^a satisfies the specification $comps \vdash comp$, i.e. for all $S \in comp$, $\emptyset \neq \Psi^a(S) \subseteq comps$ and for all $I \in \Psi^a(S)$, $I \vdash S$.*

We can apply Ψ^a directly, or under an entry and leaving predicate structure $[E, lps]$. In general, if a construction Θ satisfies $\mathcal{S} \vdash \mathcal{T}$, then the construction $\Theta' = \Theta\ [E, lps]$, defined as $\Theta'(T\ [E, lps]) = (\Theta(T))\ [E, lps]$, where $T \in \mathcal{T}$, satisfies $\mathcal{S}\,[E, lps] \vdash \mathcal{T}\,[E, lps]$. For instance,

$$\Psi^a\,[E, lps] : comps\,[E, lps] \vdash comp\,[E, lps].$$

The operation $\Theta \mapsto \Theta\,[E, lps]$ is a first example of a combinator on constructions.

Once we have singled out some control structure with Ψ^a, or $\Psi^a\,[E, lps]$, we might want to decompose the computer so that the control becomes a separate computation. Before we get to the general case let us first discuss two special cases, *viz.* the decomposition of a condition and the decomposition of a let-in structure.

Case 1. [decomposing condition structure]: A condition structure is formed as

$$(P \Rightarrow R \downarrow S)\ [E_i,\ E_o],$$

where P is a predicate with time variable x and R and S are predicates with time variables x and y. We can implement the conditional model in three stages:

$$\begin{aligned} (b\,y = P \wedge R^i \wedge S^i)\ [E_i,\ E_a] \wedge \qquad & \text{(i)} \\ (R^i \wedge S^i)\ [E_a,\ b\,x\,(E_b, E_c)] \wedge \qquad & \text{(ii)} \\ R\,[E_b,\ E_o] \wedge S\,[E_c,\ E_o]. \qquad & \text{(iii)} \end{aligned}$$

The first component (i) realizes the computation of the predicate P and supplies its value on a new signal b. The second component (ii), then, performs the branching in the leaving predicate structure. The third stage (iii) is the computation of R or S, depending on whether control is passed to entry predicate E_b or E_c. Notice, this decomposition separates the *computation* of the condition from its actual *execution*; the computation is done in the data, the decision is part of the control. The construction introduces predicate variables E_a, E_b and E_c together with their implicit time variables, and port variable b. As mentioned before these variables are existential variables. The decomposition may be illustrated in a flow-graph as follows:

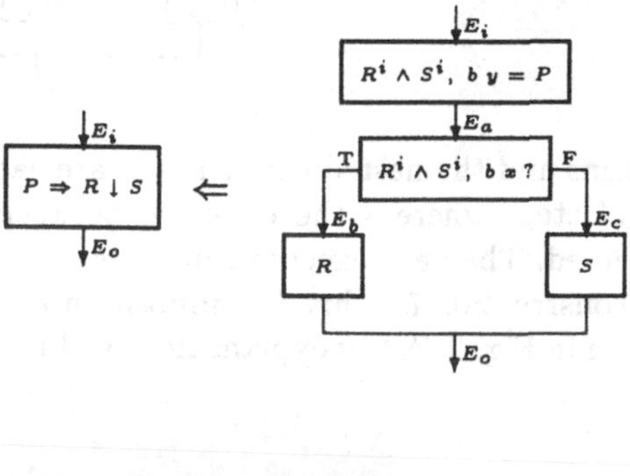

Case 2. [decomposing let-in structure]: A let-in structure is formed as

$$(\text{let } v = tm \text{ in } S)\ [E_i,\ E_o],$$

where tm is a term (see Sec. 2) with time point variable x; v is a variable; S is a predicate with time point variables x and y. By decomposition we get

$$(s_v\, y = tm \wedge S^i)\ [E_i,\ E_a] \wedge \tag{2}$$
$$S\ \{s_v\ x/v\}\ [E_a,\ E_o], \tag{3}$$

where s_v is a fresh signal variable and $S\ \{s_v\ x/v\}$ denotes the syntactic substitution of $s_v\ x$ for v in S. Component (2) first computes the value of tm, which is supplied on signal s_v. In the next control step, S picks up this value from s_v by (3).

One can generalize the constructions to decompose a nested sequence of let-ins and conditionals such as

$$(\text{let } v_1 = tm_1 \text{ in let } v_2 = tm_2 \text{ in } P_1 \Rightarrow P_2 \Rightarrow (cp_1 \downarrow cp_2) \downarrow cp_3)\ [E_i, E_o]$$

into the models

$$\begin{aligned}(s_{v_1}\ y = tm_1 \wedge s_{v_2}\ y = tm_2 \wedge P_1^i \wedge P_2^i \wedge cp_1^i \wedge cp_2^i \wedge cp_3^i)\ [E_i, E_a] \wedge \\ (s_{P_1}\ y = P_1' \wedge s_{P_2}\ y = P_2' \wedge {cp'}_1^i \wedge {cp'}_2^i \wedge {cp'}_3^i)\ [E_a, E_b] \wedge \\ ({cp'}_1^i \wedge {cp'}_2^i \wedge {cp'}_3^i)\ [E_b, (s_{P_1}\ x)((s_{P_2}\ x)(E_c, E_d), E_e)] \wedge \\ {cp'}_1\ [E_c, E_o] \wedge {cp'}_2\ [E_d, E_o] \wedge {cp'}_3\ [E_e, E_o],\end{aligned}$$

where P_1' abbreviates the substitution $P_1\{s_{v_1}/v_1, s_{v_2}/v_2\}$, and similarly for P_2', ${cp'}_1, {cp'}_2, {cp'}_3$. The decomposition can be illustrated as seen in the picture below, where for simplicity all signal protections have been omitted:

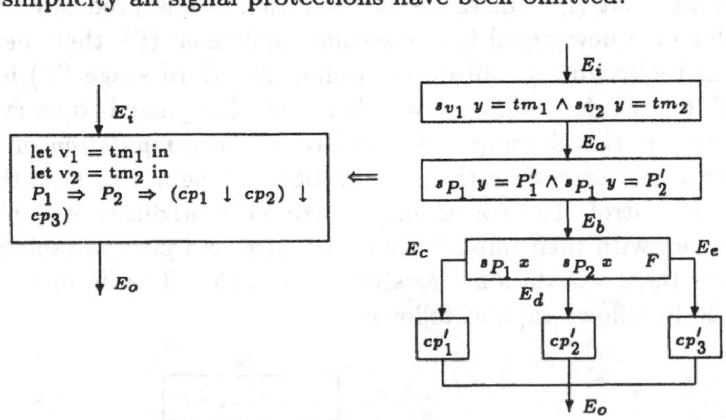

Note, the nested let_ins and the nested conditionals are each executed for themselves in one control step, whereas the ordering between the let_ins and the conditionals is preserved. They are executed in series.

Let the general construction for this decomposition be called Ψ^b. Then Ψ^b is given by the definition in Fig. 2. A few explanations will be helpful to understand

4.b.1	$\Psi^b\ ((os = iss)\ [E_i,\ E_o]) \mathrel{\widehat{=}} (os = iss)\ [E_i,\ E_o]$
4.b.2	$\Psi^b\ ((\text{let } A \text{ in } cs)\ [E_i,\ E_o]) \mathrel{\widehat{=}}$ $(\Psi_{l1}\ (\text{let } A \text{ in } cs))\ [E_i,\ E_m]\ \wedge\ \Psi^b\ ((\Psi_{l2}\ (\text{let } A \text{ in } cs,\ [\,]))\ [E_m,\ E_o])$
4.b.3	$\Psi^b\ ((P \Rightarrow cs_1 \downarrow cs_2)\ [E_i,\ E_o]) \mathrel{\widehat{=}}$ $\text{hds}\ (\Psi_{p1}\ (P \Rightarrow cs_1 \downarrow cs_2)\ \wedge\ iprt)\ [E_i, E_m]\ \wedge$ $iprt\ [E_i,\ \Psi_{p2}\ (P \Rightarrow cs_1 \downarrow cs_2)]\ \wedge\ \Psi_{p3}\ (P \Rightarrow cs_1 \downarrow cs_2,\ E_o)$ where $iprt = (\Psi_{p3}\ (P \Rightarrow cs_1 \downarrow cs_2,\ E_o))^i$.
4.b.4	$\Psi_{l1}\ (\text{let } v = tm \text{ in } cs) \mathrel{\widehat{=}} s_v\ y = tm \wedge \Psi_{l1}(cs)$
4.b.5	$\Psi_{l1}\ (os = iss) \mathrel{\widehat{=}} iss^i$
4.b.6	$\Psi_{l1}\ (P \Rightarrow cs_1 \downarrow cs_2) \mathrel{\widehat{=}} P^i \wedge cs_1^i \wedge cs_2^i$
4.b.7	$\Psi_{l2}\ (\text{let } v = tm \text{ in } cs,\ L) \mathrel{\widehat{=}} \Psi_{l2}\ (cs,\ [s_v\ x/v] :: L)$
4.b.8	$\Psi_{l2}\ (os = iss,\ L) \mathrel{\widehat{=}} os = (iss\ L)$ (substitution L on iss)
4.b.9	$\Psi_{l2}\ (P \Rightarrow cs_1 \downarrow cs_2,\ L) \mathrel{\widehat{=}} (P\ L) \Rightarrow (cs_1\ L) \downarrow (cs_2\ L)$ (substitution)
4.b.10	$\Psi_{p1}\ (P \Rightarrow cs_1 \downarrow cs_2) \mathrel{\widehat{=}} s_P\ y = P\ \wedge\ \Psi_{p1}\ (cs_1)\ \wedge\ \Psi_{p1}\ (cs_2)$
4.b.11	$\Psi_{p1}\ (os = iss) \mathrel{\widehat{=}} T$
4.b.12	$\Psi_{p1}\ (\text{let } A \text{ in } cs) \mathrel{\widehat{=}} T$
4.b.13	$\Psi_{p2}\ (P \Rightarrow cs_1 \downarrow cs_2) \mathrel{\widehat{=}} (s_P\ x)\ (\Psi_{p2}\ (cs_1),\ \Psi_{p2}\ (cs_2))$
4.b.14	$\Psi_{p2}\ (os = iss) \mathrel{\widehat{=}} E_{iss}$
4.b.15	$\Psi_{p2}\ (\text{let } A \text{ in } cs) \mathrel{\widehat{=}} E_{cs}$
4.b.16	$\Psi_{p3}\ (P \Rightarrow cs_1 \downarrow cs_2,\ E_o) \mathrel{\widehat{=}} \Psi_{p3}\ (cs_1,\ E_o) \wedge \Psi_{p3}\ (cs_2,\ E_o)$
4.b.17	$\Psi_{p3}\ (os = iss,\ E_o) \mathrel{\widehat{=}} (os = iss)\ [E_{iss},\ E_o]$
4.b.18	$\Psi_{p3}\ (\text{let } A \text{ in } cs,\ E_o) \mathrel{\widehat{=}} \Psi^b\ ((\text{let } A \text{ in } cs)\ [E_{cs},\ E_o])$

Fig. 2. Construction Scheme Ψ^b

Ψ^b. The input structure is always of the form *comps* $[E_i, E_o]$ where the computer structure *comps* is one of three possibilities, dealt with by Ψ^b in the equations 4.b.1, 4.b.2 and 4.b.3: if *comps* is a computer *comp*, case 4.b.1, then nothing needs to be done. This is the base case of the construction. If it is a let_in structure then 4.b.2 composes the result as the conjunction of two parts determined by the sub-constructions Ψ_{l1} and Ψ_{l2}: first Ψ_{l1} follows a consecutive sequence of nested let_ins and realizes the bindings as a set of parallel transitions, by introducing fresh signals for the bound variables. It also computes the appropriate signal protections; Ψ_{l2} eliminates the let_ins by substitution, the result of which is then continued to be decomposed by Ψ^b. If *comps* is a condition structure then equation 4.b.3 builds on the three sub-constructions Ψ_{p1}, Ψ_{p2}, and Ψ_{p3}. Here Ψ_{p1} follows a consecutive sequence of nested conditions and realizes the evaluation of all predicates in one parallel step. The second conjunct in 4.b.3 is the decision tree in the form of a leaving predicate structure. This is done by Ψ_{p2}. Finally, Ψ_{p3} deals with the decomposition of the continuation processes that are invoked by the decision tree.

The construction is deterministic except for the generation of fresh variables and the choice of their names. In 4.b.4, 4.b.7, 4.b.10, and 4.b.13 new (existential) signal variables s_v, s_P, and in 4.b.2/3, 4.b.14/15, and 4.b.17/18 new (existential) entry and leaving predicate variables, E_m, E_{iss}, E_{cs}, are introduced. The choice of E_m in 4.b.2/3 is arbitrary except that it must be fresh, *i.e.* not occur already anywhere else in the term (Since they are existential they always can be identified later if necessary.) The same holds for variable s_v in 4.b.4 and 4.b.7, and s_P in 4.b.10 and 4.b.13. The choice of s_v and s_P, however, must be functional, *i.e.* be uniquely determined by variable v and predicate P, respectively. Similarly, the choices of E_{iss} and E_{cs} in 4.b.14/15 and 4.17/18 must be functional, *i.e.* (in some way) uniquely depend on the sub-terms *iss* and *cs*, respectively.

Theorem 2. *Assume that the trivial Boolean T is generally omitted in a conjunction. Then,*

$$\Psi^b : ((tt^+ \cup comp)\,[E, lps])^+ \vdash comps\,[E, lps],$$

where $\cup$ *denotes the ordinary set-theoretic union of sets (of languages.)*

We can now combine our two constructions Ψ^a and Ψ^b to obtain a new construction Ψ^c, that first applies $\Psi^a\,[E, lps]$ and then Ψ^b. Formally,

$$\Psi^c = \Psi^b \circ (\Psi^a\,[E, lps]),$$

such that

$$\Psi^c : ((tt^+ \cup comp)\,[E, lps])^+ \vdash comp\,[E, lps].$$

Here $\circ$ is the *functional composition* of constructions. In general, if $\Theta_1 : \mathcal{R} \vdash \mathcal{S}$ and $\Theta_2 : \mathcal{S} \vdash \mathcal{T}$, then $(\Theta_1 \circ \Theta_2) : \mathcal{R} \vdash \mathcal{T}$ is defined by $(\Theta_1 \circ \Theta_2)(T) = \Theta_1(\Theta_2(T))$, for $T \in \mathcal{T}$.

The effect of Ψ^c as a refinement of the specification $comp\,[E, lps]$ is to reach into *comp* up to a certain depth and to break out the appropriate control structure (by $\Psi^a\,[E, lps]$.) This control structure is then implemented in terms of term transitions (by Ψ^b.) In the resulting specification, which is an instance of $((tt^+ \cup comp)\,[E, lps])^+$, there may still be decomposable conjuncts of form $comp\,[E, lps]$. On these we could apply Ψ^c again, until there are no more of them left. We then end up in $(tt^+\,[E, lps])^+$. Let Ψ^e be this construction. Ψ^e can also be expressed using general combinators. The first combinator we need is the *union operation*: if $\Theta : \mathcal{S}_1 \vdash \mathcal{S}_2$ and $\mathcal{T}$ a specification scheme, then $\mathcal{T} \cup \Theta : \mathcal{T} \cup \mathcal{S}_1 \vdash \mathcal{T} \cup \mathcal{S}_2$, where $\mathcal{T} \cup \Theta$ is defined in the obvious way as

$$\begin{aligned} (\mathcal{T} \cup \Theta)(X) &\mathrel{\widehat{=}} \Theta(X) \text{ if } X \in \mathcal{S}_2 \\ (\mathcal{T} \cup \Theta)(X) &\mathrel{\widehat{=}} X \quad\;\; \text{ if } X \in \mathcal{T}. \end{aligned}$$

Notice, if $\mathcal{S}_2$ and $\mathcal{T}$ overlap, then we have a choice. This means that $\mathcal{T} \cup \Theta$ may be nondeterministic even if Θ is not. Another combinator that is relevant is the *conjunctive mapping*: if $\Theta : \mathcal{S} \vdash \mathcal{T}$, then $\Theta^+ : \mathcal{S}^+ \vdash \mathcal{T}^+$, where Θ^+ is defined by

$$\begin{aligned} \Theta^+(X) &\mathrel{\widehat{=}} \Theta(X) & &\text{if } X \in \mathcal{T} \\ \Theta^+(X \wedge R) &\mathrel{\widehat{=}} \Theta(X) \wedge \Theta^+(R) & &\text{otherwise.} \end{aligned}$$

With these two combinators we can now form the construction $\Psi^d = (tt^+ \,[E, lps] \cup \Psi^c)^+$, by which we can apply Ψ^c to every conjunctive element of the form $comp\,[E, lps]$ in an instance of $((tt^+ \cup comp)\,[E, lps])^+$. For Ψ^d we obtain the specification

$$\Psi^d : (tt^+\,[E, lps] \cup ((tt^+ \cup comp)\,[E, lps])^+)^+ \vdash (tt^+\,[E, lps] \cup comp\,[E, lps])^+.$$

It is now easy to verify that for the lhs $(tt^+\,[E, lps] \cup ((tt^+ \cup comp)\,[E, lps])^+)^+ = ((tt^+ \cup comp)\,[E, lps])^+$, and also that for the rhs $(tt^+\,[E, lps] \cup comp\,[E, lps])^+ = ((tt^+ \cup comp)\,[E, lps])^+$. Thus, we get

$$\Psi^d : ((tt^+ \cup comp)\,[E, lps])^+ \vdash ((tt^+ \cup comp)\,[E, lps])^+.$$

All we need to do to get Ψ^e is to iterate Ψ^d until we reach a *fixed-point*, *i.e.* until there is no control structure left to be implemented. Formally, $\Psi^e = \mu\Psi^d$, where

$$\begin{aligned} Y \in (\mu\Psi^d)(X) \;\; \textit{iff} \;\; & \exists X_0, \ldots, X_n.\; X = X_0 \wedge X_n = Y \wedge \\ & (\forall i < n.\; X_{i+1} \in \Psi^d(X_i)) \wedge (\forall Y' \in \Psi^d(X_n).\; Y' = X_n). \end{aligned}$$

For arbitrary constructions $\Theta : \mathcal{S} \vdash \mathcal{S}$, $\mu\Theta$ need not be total, *i.e.* for some X we may have $(\mu\Theta)(X) = \emptyset$. However, for Ψ^d it can be shown that $\mu\Psi^d$ is indeed total. Viewing Ψ^d as a rewriting system, totality of $\mu\Psi^d$ corresponds to the condition of weak normalization. One can further show that Ψ^e satisfies the specification

$$\Psi^e : ((tt^+ \cup os = is)\,[E, lps])^+ \vdash ((tt^+ \cup comp)\,[E, lps])^+.$$

To round off our tool set we might come up with a further construction Ψ^f, using the logical equivalence

$$(sg_1\; y = tm_1) \;\wedge\; (sg_2\; y = tm_2) \equiv (sg_1\; y,\; sg_2\; y) = (tm_1,\; tm_2)$$

in the obvious way, that satisfies

$$\Psi^f : tt^+ \vdash os = is.$$

Now finally we can assemble our desired refinement Ψ from the machine instruction level to the term transition level as follows:

$$\begin{aligned} \Psi &= ((tt^+ \cup \Psi^f)\,[E, lps])^+ \;\circ\; \Psi^e \\ &= ((tt^+ \cup \Psi^f)\,[E, lps])^+ \;\circ\; (\mu\Psi^d) \\ &= ((tt^+ \cup \Psi^f)\,[E, lps])^+ \;\circ\; (\mu(tt^+\,[E, lps] \cup \Psi^c)^+) \\ &= ((tt^+ \cup \Psi^f)\,[E, lps])^+ \;\circ\; \mu(tt^+\,[E, lps] \cup (\Psi^b \;\circ\; (\Psi^a\,[E, lps])))^+. \end{aligned}$$

Here Ψ^a, Ψ^b, and Ψ^f are given directly, though they could be composed systematically from smaller pieces as well. For Ψ we derive the specification

$$\Psi : (tt^+\,[E, lps])^+ \vdash ((tt^+ \cup comp)\,[E, lps])^+,$$

or by restriction, if we observe that *comp* $[E, lps]$ is a sub-specification (subset of languages) of $((tt^+ \cup comp)\,[E, lps])^+$:

$$\Psi : (tt^+\,[E, lps])^+ \vdash comp\,[E, lps]$$

as desired.

We should point out that the constructions presented above are just examples. Many more constructions are necessary to refine a computer from machine instruction level to, say, the register transfer level. The gist of this section is to show that constructions can be treated in a way similar to functions in a functional language; and that specification schemes can be used to specify constructions rigorously, similar to the way types specify functions. However, since BNF specification schemes have a richer structure than ordinary functional types the expressibility is considerably higher.

In this context it is useful to contrast our constructions with the tactics in a typical theorem prover: The only specification usually available for a tactic is that it has the type *theorem* $\rightharpoonup$ *theorem*, where the function arrow $\rightharpoonup$ means "partial function". This is not extremely informative and the reason for the correctness leak mentioned in the introduction.

5 Case Study

In [39, 40] a complete sequence of generic refinement steps has been worked out, from the microinstruction level all the way down to the register-transfer level. The refinements involve 10 major levels of specification schemes together with 9 major construction steps. All of these construction steps are rigorously defined and formally verified, using the framework of specification and constructions schemes described in this paper. The framework has been applied to construct several different versions of Gordon's [13] microprogrammed computer. Here, we describe briefly the main steps.

1. Machine Instruction (MIL): The programmer's view of a computer as a global next-state transition. The specification scheme capturing MIL was introduced in Sec. 2. This level is characterized by complex data (next-state function) but simple control (single-transition loop) structure.

2. State Transition (STL): At this level the control structure has been separated from the data part, but not as yet implemented. Each instruction is decomposed into a set of concurrent state transitions. There is still one global control loop.

3. Term Transition (TTL): This is the mixed behaviour-structural specification at term transition level. The control structure is realized by entry/leaving predicates as an abstract microprogram, while the data functions appear as synchronized term transitions. In contrast to the machine instruction level, the term transition level is characterized by a complex control structure of synchronized local state transformations, each of which is governed by a simple function term.

4. Register Transfer (RTL): A model in the register transfer specification is either a function, which is considered implementable at the register transfer level, a port-connection, bus, or register. The control has been flattened into a linear microprogram, microinstruction and microcode.

6 Conclusion and Further Work

The work reported in this paper is concerned with the formal synthesis of hardware, with the particular goal of achieving the rigour of formal verification while considerably increasing the amount of genericity and the level of formalization. The proposed framework is centered around the notion of a specification scheme, construction scheme, and construction model.

We argue that the traditional logic-centered approach to formal synthesis (based on theorem-proving) in terms of formulas and rules employs too low a level of formalization. Instead of arbitrary formula schemes our calculus constructs specification schemes, each of which captures the generic structure of a class of hardware objects at a specific level of abstraction. This increases the level of formalization but also the amount of genericity: where formula schemes only exhibit vertical genericity a specification scheme also exploits the horizontal genericity of inductive BNF syntax. Construction schemes, then, generalize ordinary rule schemes in a natural way. They are logically correct refinements of one specification scheme to another one.

Yet another source of genericity in our approach lies in the use of the construction model. Besides the obvious signal parameters of the traditional verification model (components-as-relations paradigm) the construction model allows components to communicate via additional entry and leaving predicates. Through this extra dimension of logical communication we can separate functional from control aspects, and refine both independently whenever appropriate. In particular, we can concentrate on functional decomposition, even perform steps of abstract scheduling and allocation, without fixing the particular mode of control. We may instantiate to synchronous or asynchronous, micro-programmed or hard-wired control, at a fairly late stage in the design.

The rather informal character of the presentation in this paper might be misleading. It should be stressed that the proposed construction framework is a rigorous calculus based on formally specified and verified construction schemes. How this can be done is demonstrated in [39], where a suite of concrete specification and construction schemes is presented to derive a class of computers. In this paper we concentrated essentially on motivating the basic ideas from a more general perspective.

Further work will be in two directions. We plan to develop a semi-automatic computer design system which can design a register-transfer level microprocessor from a specification at machine instruction level. It will be based on the ideas discussed in this paper. The user interaction consists in the selections for high-level allocation and scheduling, for which standard algorithms will be integrated. Exploiting the genericity of our approach the tool will be extendible to realize various alternative forms of control, such as multi-phase synchronous, pipelined, or asynchronous handshake. This is a long term goal, however. Other work aims at the theoretical foundations of our framework, in particular concerning the algebraic and logical properties of the construction model.

References

1. D. A. Basin, G. M. Brown, and M. E. Leeser, Formally verified synthesis of combinational CMOS circuits. In L. J. M. Claesen, editor, *Formal VLSI Specification and Synthesis*, pages 197-206, North-Holland, 1990.
2. D. A. Basin, Extracting Circuits from Constructive Proofs. In 1991 International Workshop on *Formal Verification in VLSI Design*. ACM IFIP WG 10.2, Jan. 1991.
3. G. Birtwistle and B. Graham, Verifying the SECD in HOL. In [32].
4. G. Birtwistle and P. Subrahmanyam, eds., *VLSI Specification, Verification and Synthesis*, Kluwer Academic Publishers, Boston, 1988.
5. B. Bose and S. D. Johnson, DDD-FM9001: Derivation of a Verified Microprocessor. In [29].
6. H. Busch, Proof-based transformation of formal hardware models. In [22], pp. 271–296.
7. K. M. Chandy and J. Mishra, *Parallel Program Design. A Foundation.* Addison Wesley, 1988.
8. L. Claesen, editor, IMEC-IFIP International Workshop on *Applied Formal Methods for Correct VLSI Design*, Volume 1+2, Elsevier/North-Holland, 1989.
9. A. Cohn, A Proof of Correctness of the Viper Microprocessor: First Level. In [4], pp. 27-71.
10. A. Cohn, Correctness Properties of the Viper Block Model: The Second Level. In G. Birtwistle and P. Subrahmanyam, eds., *Current Trends in Hardware Verification and Automated Theorem Proving*, Springer-Verlag, 1989, pp.1-91.
11. M. P. Fourman, R. L. Harris, Lambda-Logic and Mathematics Behind Design Automation, 26th ACM/IEEE Design Automation Conference, 1988.
12. M. P. Fourman, Formal System Design. In [32], pp 191-236.
13. M. J. C. Gordon, Proving a Computer Correct with the LCF_LSM Hardware Verification System. Technical Report No. 42, Computer Laboratory, University of Cambridge, 1983.
14. M. J. C. Gordon, Why higher-order logic is a good formalism for specifying and verifying hardware. in: G. Milne and P. Subrahmanyam, eds., *Formal Aspects of VLSI Design*, North-Holland, 1986, pp. 153-177.
15. M. J. C. Gordon and T. F. Melham, *Introduction to HOL.* Cambridge University Press, 1993.
16. F. K. Hanna and N. Daeche, Specification and Verification of Digital Systems using Higher-Order Predicate Logic. IEE Proceedings, Vol. 133, Part E, No. 5, September 1986, pp. 242-254.
17. F. K. Hanna, M. Longley, and N. Daeche, Formal synthesis of digital systems. In [8], pages 532-548.
18. F. K. Hanna and N. Daeche, Strongly-Typed Theory of Structure and Behaviors. In [29], pp. 39-54.
19. N. A. Harman and J. V. Tucker, Algebraic Models and the Correctness of Microprocessors. In [29], pp. 92-108.
20. J. M. J. Herbert, Incremental Design and Formal Verification of Microcoded Microprocessors. In [34], pp. 157–174.

21. W. A. Hunt, FM8501, *A Verified Microprocessor.* Ph.D. Thesis, Report No. 47, Institute for Computing Science, University of Texas, Austin, December 1985.
22. G. Jones and M. Sheeran, *Designing Correct Circuits.* Springer, 1991.
23. G. Jones and M. Sheeran, Circuit Design in RUBY. In [32].
24. J. J. Joyce, *Multi-Level Verification of Microprocessor-Based Systems*, Ph.D. Thesis, Computer Laboratory, Cambridge University, December 1989 (Technical Report No. 195, May 1990).
25. J. J. Joyce, Generic Specification of Digital Hardware. In [22], pp. 68–91.
26. J. J. Joyce, G. Birtwistle and M. Gordon, Proving a Computer Correct in Higher Order Logic. Report No. 100, Computer Laboratory, Cambridge University, 1986.
27. M. Langevin and E. Cerny, Verification of Processor-like Circuits. In *Advanced Work on Correct Hardware Design Methodology*, Turin, 12-14 June 1991.
28. T. F. Melham, Abstraction mechanism for hardware verification. In G. Birtwistle and P.A. Subrahmanyam, eds., *VLSI Specification, Verification, and Synthesis*, pages 267-291. Kluwer Academic Publishers, 1988.
29. G. J. Milne and L. Pierre, eds., *Correct Hardware Design and Verification Methods*, LNCS 683, Springer-Verlag, May 1993.
30. L. C. Paulson. *Isabelle Tutorial and User's Manual*, 1990.
31. M. Srivas and M. Bickford, *Formal Verification of a Pipelined Microprocessor*, In IEEE Software, September 1990, pp. 52–64.
32. J. Staunstrup, editor, IFIP WG 10.5 *Formal Methods for VLSI Design*, North-Holland, 1990.
33. J. Staunstrup, *A Formal Approach to Hardware Design.* Kluwer Academic Publishers, 1994.
34. V. Stavridou, T. F. Melham, and R. T. Boute, eds., *Theorem Provers in Circuit Design: Theory, Practice and Experience.* IFIP TC10/WG 10.2, North Holland, June 1992.
35. Dany Suk, Hardware Synthesis in Constructive Type Theory. In G. Jones and M. Sheeran, eds., *Designing Correct Circuits*, pp 29-49, Oxford, Springer-Verlag, 1990.
36. S. Tahar and R. Kumar, Towards a Methodology for the Formal Verification of RISC Processors. In Proceedings IEEE International Conference on Computer Design (ICCD'93), 1993, pp. 58–62.
37. D. Verkest and L. Claesen and H. De Man, On the use of the Boyer-Moore theorem prover for correctness proofs of parametrized hardware modules. In [8], pp. 405–422.
38. Li-Guo Wang, Synthesis of Nondeterministic Logic Programs, Chinese Journal of Software, Vo. 1, No. 1, ISSN 1000-9825, CN 11 - 2560, January 1990.
39. Li-Guo Wang, *Formal Derivation of A Class of Computers.* PhD Thesis, LFCS, Department of Computer Science, University of Edinburgh, ECS-CST-119-95, September 1995.
40. Li-Guo Wang and M. Mendler, Formal Derivation of A Class of Computers: Its high stage — abstract microprogramming. In H. Eveking and P. Camurati, eds., *Correct Hardware Design and Verification Methods* (CHARME'95), Springer LNCS 987, 1995, pp. 84–102.
41. P. J. Windley, A Hierarchical Methodology for the Verification of Microprogrammed Microprocessors. In IEEE Symposium on *Security and Privacy*, May 1990.
42. P. J. Windley, A Theory of Generic Interpreters. In [29], pp. 122–134.

Springer-Verlag and the Environment

Lecture Notes in Computer Science

For information about Vols. 1–1013

please contact your bookseller or Springer-Verlag

Vol. 1014: A.P. del Pobil, M.A. Serna, Spatial Representation and Motion Planning. XII, 242 pages. 1995.

Vol. 1015: B. Blumenthal, J. Gornostaev, C. Unger (Eds.), Human-Computer Interaction. Proceedings, 1995. VIII, 203 pages. 1995.

VOL. 1016: R. Cipolla, Active Visual Inference of Surface Shape. XII, 194 pages. 1995.

Vol. 1017: M. Nagl (Ed.), Graph-Theoretic Concepts in Computer Science. Proceedings, 1995. XI, 406 pages. 1995.

Vol. 1018: T.D.C. Little, R. Gusella (Eds.), Network and Operating Systems Support for Digital Audio and Video. Proceedings, 1995. XI, 357 pages. 1995.

Vol. 1019: E. Brinksma, W.R. Cleaveland, K.G. Larsen, T. Margaria, B. Steffen (Eds.), Tools and Algorithms for the Construction and Analysis of Systems. Selected Papers, 1995. VII, 291 pages. 1995.

Vol. 1020: I.D. Watson (Ed.), Progress in Case-Based Reasoning. Proceedings, 1995. VIII, 209 pages. 1995. (Subseries LNAI).

Vol. 1021: M.P. Papazoglou (Ed.), OOER '95: Object-Oriented and Entity-Relationship Modeling. Proceedings, 1995. XVII, 451 pages. 1995.

Vol. 1022: P.H. Hartel, R. Plasmeijer (Eds.), Functional Programming Languages in Education. Proceedings, 1995. X, 309 pages. 1995.

Vol. 1023: K. Kanchanasut, J.-J. Lévy (Eds.), Algorithms, Concurrency and Knowlwdge. Proceedings, 1995. X, 410 pages. 1995.

Vol. 1024: R.T. Chin, H.H.S. Ip, A.C. Naiman, T.-C. Pong (Eds.), Image Analysis Applications and Computer Graphics. Proceedings, 1995. XVI, 533 pages. 1995.

Vol. 1025: C. Boyd (Ed.), Cryptography and Coding. Proceedings, 1995. IX, 291 pages. 1995.

Vol. 1026: P.S. Thiagarajan (Ed.), Foundations of Software Technology and Theoretical Computer Science. Proceedings, 1995. XII, 515 pages. 1995.

Vol. 1027: F.J. Brandenburg (Ed.), Graph Drawing. Proceedings, 1995. XII, 526 pages. 1996.

Vol. 1028: N.R. Adam, Y. Yesha (Eds.), Electronic Commerce. X, 155 pages. 1996.

Vol. 1029: E. Dawson, J. Golić (Eds.), Cryptography: Policy and Algorithms. Proceedings, 1995. XI, 327 pages. 1996.

Vol. 1030: F. Pichler, R. Moreno-Díaz, R. Albrecht (Eds.), Computer Aided Systems Theory - EUROCAST '95. Proceedings, 1995. XII, 539 pages. 1996.

Vol.1031: M. Toussaint (Ed.), Ada in Europe. Proceedings, 1995. XI, 455 pages. 1996.

Vol. 1032: P. Godefroid, Partial-Order Methods for the Verification of Concurrent Systems. IV, 143 pages. 1996.

Vol. 1033: C.-H. Huang, P. Sadayappan, U. Banerjee, D. Gelernter, A. Nicolau, D. Padua (Eds.), Languages and Compilers for Parallel Computing. Proceedings, 1995. XIII, 597 pages. 1996.

Vol. 1034: G. Kuper, M. Wallace (Eds.), Constraint Databases and Applications. Proceedings, 1995. VII, 185 pages. 1996.

Vol. 1035: S.Z. Li, D.P. Mital, E.K. Teoh, H. Wang (Eds.), Recent Developments in Computer Vision. Proceedings, 1995. XI, 604 pages. 1996.

Vol. 1036: G. Adorni, M. Zock (Eds.), Trends in Natural Language Generation - An Artificial Intelligence Perspective. Proceedings, 1993. IX, 382 pages. 1996. (Subseries LNAI).

Vol. 1037: M. Wooldridge, J.P. Müller, M. Tambe (Eds.), Intelligent Agents II. Proceedings, 1995. XVI, 437 pages. 1996. (Subseries LNAI).

Vol. 1038: W: Van de Velde, J.W. Perram (Eds.), Agents Breaking Away. Proceedings, 1996. XIV, 232 pages. 1996. (Subseries LNAI).

Vol. 1039: D. Gollmann (Ed.), Fast Software Encryption. Proceedings, 1996. X, 219 pages. 1996.

Vol. 1040: S. Wermter, E. Riloff, G. Scheler (Eds.), Connectionist, Statistical, and Symbolic Approaches to Learning for Natural Language Processing. IX, 468 pages. 1996. (Subseries LNAI).

Vol. 1041: J. Dongarra, K. Madsen, J. Waśniewski (Eds.), Applied Parallel Computing. Proceedings, 1995. XII, 562 pages. 1996.

Vol. 1042: G. Weiß, S. Sen (Eds.), Adaption and Learning in Multi-Agent Systems. Proceedings, 1995. X, 238 pages. 1996. (Subseries LNAI).

Vol. 1043: F. Moller, G. Birtwistle (Eds.), Logics for Concurrency. XI, 266 pages. 1996.

Vol. 1044: B. Plattner (Ed.), Broadband Communications. Proceedings, 1996. XIV, 359 pages. 1996.

Vol. 1045: B. Butscher, E. Moeller, H. Pusch (Eds.), Interactive Distributed Multimedia Systems and Services. Proceedings, 1996. XI, 333 pages. 1996.

Vol. 1046: C. Puech, R. Reischuk (Eds.), STACS 96. Proceedings, 1996. XII, 690 pages. 1996.

Vol. 1047: E. Hajnicz, Time Structures. IX, 244 pages. 1996. (Subseries LNAI).

Vol. 1048: M. Proietti (Ed.), Logic Program Syynthesis and Transformation. Proceedings, 1995. X, 267 pages. 1996.

Vol. 1049: K. Futatsugi, S. Matsuoka (Eds.), Object Technologies for Advanced Software. Proceedings, 1996. X, 309 pages. 1996.

Vol. 1050: R. Dyckhoff, H. Herre, P. Schroeder-Heister (Eds.), Extensions of Logic Programming. Proceedings, 1996. VII, 318 pages. 1996. (Subseries LNAI).

Vol. 1051: M.-C. Gaudel, J. Woodcock (Eds.), FME'96: Industrial Benefit and Advances in Formal Methods. Proceedings, 1996. XII, 704 pages. 1996.

Vol. 1052: D. Hutchison, H. Christiansen, G. Coulson, A. Danthine (Eds.), Teleservices and Multimedia Communications. Proceedings, 1995. XII, 277 pages. 1996.

Vol. 1053: P. Graf, Term Indexing. XVI, 284 pages. 1996. (Subseries LNAI).

Vol. 1054: A. Ferreira, P. Pardalos (Eds.), Solving Combinatorial Optimization Problems in Parallel. VII, 274 pages. 1996.

Vol. 1055: T. Margaria, B. Steffen (Eds.), Tools and Algorithms for the Construction and Analysis of Systems. Proceedings, 1996. XI, 435 pages. 1996.

Vol. 1056: A. Haddadi, Communication and Cooperation in Agent Systems. XIII, 148 pages. 1996. (Subseries LNAI).

Vol. 1057: P. Apers, M. Bouzeghoub, G. Gardarin (Eds.), Advances in Database Technology — EDBT '96. Proceedings, 1996. XII, 636 pages. 1996.

Vol. 1058: H. R. Nielson (Ed.), Programming Languages and Systems – ESOP '96. Proceedings, 1996. X, 405 pages. 1996.

Vol. 1059: H. Kirchner (Ed.), Trees in Algebra and Programming – CAAP '96. Proceedings, 1996. VIII, 331 pages. 1996.

Vol. 1060: T. Gyimóthy (Ed.), Compiler Construction. Proceedings, 1996. X, 355 pages. 1996.

Vol. 1061: P. Ciancarini, C. Hankin (Eds.), Coordination Languages and Models. Proceedings, 1996. XI, 443 pages. 1996.

Vol. 1062: E. Sanchez, M. Tomassini (Eds.), Towards Evolvable Hardware. IX, 265 pages. 1996.

Vol. 1063: J.-M. Alliot, E. Lutton, E. Ronald, M. Schoenauer, D. Snyers (Eds.), Artificial Evolution. Proceedings, 1995. XIII, 396 pages. 1996.

Vol. 1064: B. Buxton, R. Cipolla (Eds.), Computer Vision – ECCV '96. Volume I. Proceedings, 1996. XXI, 725 pages. 1996.

Vol. 1065: B. Buxton, R. Cipolla (Eds.), Computer Vision – ECCV '96. Volume II. Proceedings, 1996. XXI, 723 pages. 1996.

Vol. 1066: R. Alur, T.A. Henzinger, E.D. Sontag (Eds.), Hybrid Systems III. IX, 618 pages. 1996.

Vol. 1067: H. Liddell, A. Colbrook, B. Hertzberger, P. Sloot (Eds.), High-Performance Computing and Networking. Proceedings, 1996. XXV, 1040 pages. 1996.

Vol. 1068: T. Ito, R.H. Halstead, Jr., C. Queinnec (Eds.), Parallel Symbolic Languages and Systems. Proceedings, 1995. X, 363 pages. 1996.

Vol. 1069: J.W. Perram, J.-P. Müller (Eds.), Distributed Software Agents and Applications. Proceedings, 1994. VIII, 219 pages. 1996. (Subseries LNAI).

Vol. 1070: U. Maurer (Ed.), Advances in Cryptology – EUROCRYPT '96. Proceedings, 1996. XII, 417 pages. 1996.

Vol. 1071: P. Miglioli, U. Moscato, D. Mundici, M. Ornaghi (Eds.), Theorem Proving with Analytic Tableaux and Related Methods. Proceedings, 1996. X, 330 pages. 1996. (Subseries LNAI).

Vol. 1072: R. Kasturi, K. Tombre (Eds.), Graphics Recognition. Proceedings, 1995. X, 308 pages. 1996.

Vol. 1073: J. Cuny, H. Ehrig, G. Engels, G. Rozenberg (Eds.), Graph Grammars and Their Application to Computer Science. Proceedings, 1994. X, 565 pages. 1996.

Vol. 1074: G. Dowek, J. Heering, K. Meinke, B. Möller (Eds.), Higher-Order Algebra, Logic, and Term Rewriting. Proceedings, 1995. VII, 287 pages. 1996.

Vol. 1075: D. Hirschberg, G. Myers (Eds.), Combinatorial Pattern Matching. Proceedings, 1996. VIII, 392 pages. 1996.

Vol. 1076: N. Shadbolt, K. O'Hara, G. Schreiber (Eds.), Advances in Knowledge Acquisition. Proceedings, 1996. XII, 371 pages. 1996. (Subseries LNAI).

Vol. 1077: P. Brusilovsky, P. Kommers, N. Streitz (Eds.), Mulimedia, Hypermedia, and Virtual Reality. Proceedings, 1994. IX, 311 pages. 1996.

Vol. 1078: D.A. Lamb (Ed.), Studies of Software Design. Proceedings, 1993. VI, 188 pages. 1996.

Vol. 1079: Z.W. Raś, M. Michalewicz (Eds.), Foundations of Intelligent Systems. Proceedings, 1996. XI, 664 pages. 1996. (Subseries LNAI).

Vol. 1080: P. Constantopoulos, J. Mylopoulos, Y. Vassiliou (Eds.), Advanced Information Systems Engineering. Proceedings, 1996. XI, 582 pages. 1996.

Vol. 1081: G. McCalla (Ed.), Advances in Artificial Intelligence. Proceedings, 1996. XII, 459 pages. 1996. (Subseries LNAI).

Vol. 1083: K. Sparck Jones, J.R. Galliers, Evaluating Natural Language Processing Systems. XV, 228 pages. 1996. (Subseries LNAI).

Vol. 1084: W.H. Cunningham, S.T. McCormick, M. Queyranne (Eds.), Integer Programming and Combinatorial Optimization. Proceedings, 1996. X, 505 pages. 1996.

Vol. 1085: D.M. Gabbay, H.J. Ohlbach (Eds.), Practical Reasoning. Proceedings, 1996. XV, 721 pages. 1996. (Subseries LNAI).

Vol. 1087: C. Zhang, D. Lukose (Eds.), Distributed Artificial Intelliegence. Proceedings, 1995. VIII, 232 pages. 1996. (Subseries LNAI).

Vol. 1088: A. Strohmeier (Ed.), Reliable Software Technologies – Ada-Europe '96. Proceedings, 1996. XI, 513 pages. 1996.